CRUSTACEAN BIOGEOGRAPHY

CRUSTACEAN ISSUES

4

General editor:

FREDERICK R.SCHRAM

San Diego Natural History Museum, California

A.A.BALKEMA / ROTTERDAM / BOSTON / 1986

CRUSTACEAN BIOGEOGRAPHY

Edited by
ROBERT H. GORE & KENNETH L. HECK
Academy of Natural Sciences of Philadelphia, Pennsylvania

A.A. BALKEMA / ROTTERDAM / BOSTON / 1986

CIP-DATA KONINKLIJKE BIBLIOTHEEK, DEN HAAG

Crustacean

Crustacean biogeography / ed. by Robert H. Gore & Kenneth L. Heck. – Rotterdam [etc.]:
Balkema. – Ill. – (Crustacean issues, ISSN 0168-6356; 4)
ISBN 90-6191-593-7 bound
SISO 597.6 UDC 574.9:595.3
Subject heading: crustaceans; biogeography.

ISSN 0168-6356

ISBN 90 6191 593 7

Distributed in USA & Canada by: A.A.Balkema Publishers, P.O.Box 230, Accord, MA 02018

Printed in the Netherlands

Table of contents

Preface

It is a pleasure to acknowledge the following persons or their institutions for helping make the presented papers session in Biogeography of the Crustacea a success. Foremost acknowledgment must go to the Academy of Natural Sciences of Philadelphia for providing funds, personnel and aid before, during and after the Symposium. Particular thanks are jointly extended by the editors to the Division of Ecology and Limnology, and the Department of Malacology. Among the individuals who contributed time, space, or aid are Dr George M.Davis, Chairman, Department of Malacology, and his staff, all of whom made the first editor's tenure at the Academy such a rewarding experience. Mss Sandy Devan and Helen Anderson provided clerical aid above and beyond the call of duty. Special thanks also go to Mary Wiley, business manager of the American Society of Zoologists for her untiring help in formulating the arrangements for both the Symposium papers session, and the contributed papers session on the following day. Dr F.Lynn Carpenter, Chairwoman, Division of Ecology, ASZ, generously provided some funding allowing partial defrayment of travel expenses for some participants. Dr Patsy A.McLaughlin, program officer for the Crustacean Society, first approached one of us (R.H.G.) with the idea for the symposium and helped initiate the requisite action to see it through. The support of the presiding officers of the Crustacean Society, and many of the members, is also appreciated; the Symposium would certainly have been less successful without such help. The first editor (R.H.G.) also wishes to express a very special thanks to Kim A.Wilson, for her tolerance and forbearance for so long. Finally, of course, we gratefully acknowledge all of the participants at the Symposium papers session, and particularly those authors whose papers appear herein. In the final analysis it will be they who are responsible for the success of this volume.

Introduction

This symposium had its origin in a simple and easily observed fact. One need only scan the recent pages of several well-known journals or textbooks that treat biogeography to see the often heated debate, acrimonious rhetoric, and fervent exhortations as to which process, or functioning geological or biological event, would more adequately explain the distribution of this animal or the dispersal of that plant. One is reminded of Victorian missionaries attempting to proselytize a large, and unruly group of all too recalcitrant natives who often are more interested in how to carve up the proselytizer for their colleagues' philosophical dinner table than in paying attention to the way, the truth and light of vicariantist or dispersalist dogma. Yet, in these controversies crustaceans have played but a minor role, if they played any role at all. A perusal of several recent and classical texts in biogeography shows that if crustaceans are referenced at all in the literature cited, they are usually noted as species-listings or other taxonomic-oriented publications, or in an ecological or ethological framework. We believe that the time has come to change this, and this volume is intended to be a positive step in this direction.

A wag, commenting on biogeography in general and crustacean zoogeography in particular, once said 'The more we investigate the distribution of crustaceans, the more it seems they occur'. There is a nugget of truth in this facetious statement. Crustaceans do indeed occur; from the hadal oceanic trenches to the highest mountains; in stygian terrestrial caves on continents as well as on mid-oceanic islands; in anchialine pools, ephemeral desert ponds, hot springs, deep wells, glaciated lakes, swift montane creeks, on sawgrass prairies, brackish sloughs, hypersaline lagoons and major continental rivers. They are found under rocks, in burrows, tubes, and crevices; in shells, on, in, and with other crustaceans as well as members of other invertebrate and vertebrate phyla; they live in sponges, on stony corals, under seastars, between the spines of echinoids, clinging to holothurians, sharing worm tubes, attached to seagrasses, hitching rides on turtles, fouling almost every conceivable type of hard (and many soft) substrates; they curl up in the dirt of household potted plants, engage in explosive population growth in swimming pools, destroy pilings, climb trees, and inhabit epiphytes on other trees. They are as close as one's backyard or as remote as Ascension or Kerguelen Island, in areas as easily collectable as the seashore, or as difficult to sample as hydrothermal vents. In short, Crustacea are ubiquitous. If they do not now occur in a specific habitat, the chances are that they did at one time in their history, or will in their future if given half a chance.

In the papers that comprise this volume we will be given fascinating and intriguing glimpses of some of this ubiquity. Papers treating palaeontological distributions will be balanced by those examining historical, and even man-made dispersals. Distributional explanations or hypotheses will range from groundwater transport to planktonic dispersal, from overland migrations to piggyback radiation via parasitized hosts. Biotopes from the darkest subterranean caves to the hadal depths of the ocean will be treated herein, using fresh, brackish and marine water forms as well as terrestrial groups. Traditional dispersalist views will be defended, but vicariance explanations will also be proposed, to account for the observed distributions. Even a melding of these two often antithetically opposed hypotheses will be given consideration.

In assembling the participants for this volume we leaned towards neither dispersalist nor vicariantist viewpoints, accepting both for the sake of arguments but adhering toward neither. Spatial limitations prevented treating every crustacean group, but we hope that those groups which are addressed might provide some indicative concepts toward the phylum Crustacea in general. After editing the 12 papers that comprise this volume it has become clear to us that the data supporting one viewpoint in one group may outweigh those in another, but that the proposed explanations may also only be valid for the particular group being considered. This is, of course, not unexpected, given that the Crustacea are a large, phylogenetically complex, geologically old, geographically widespread taxon. During the course of their evolutionary history they have evolved numerous, often unique, methods to move around on this planet, and once in place, to ensure that they stay there. Today, if distributions seem disjunct, if ranges seemed skewed, if populations appear to be curiously relict, these may all be the joint results of the animal's efforts to remain in place, and nature's attempts to undo these efforts.

Our task, as biogeographers, is to attempt to make some sense out of these distributions, and in so doing, to ask pertinent questions as to how such distributions might have taken place. A major problem, and a recurring one, will be the generation of acceptable and testable hypotheses. Because the fossil record is often poor, patchy or incomplete, many previous ranges can only be suspected, many distributions only implied. For all of our records both fossil and recent, we are seeing only the remnant, the collectable specimens, and not always the complete previous ranges.

Whether we believe one group to have dispersed from a center of origin or radiation, or another to have undergone a vicariant sundering, we should not lose track among the general tracks of the primary goal in biogeography — the attempt to explain the distribution of organisms as we see such today. It may be, for example, that present day ranges are a combination of both dispersal and vicariance, as some have suggested in the papers that follow. If so, can we say that one occurred before the other? Can not a dispersal event take place after a vicariant event, as well as before? Are not all vicariant distributions a consequence of a previous, and much older dispersion? Is not every dispersion a potential vicariance event — be it a behavioral, physiological, ecological, or reproductive sundering? Does not isolation occur intrapopulationally as well as extrapopulationally? What are the upper and lower limits of any biogeographic event, whether it is a dispersal or a vicariant determination? These are but a few of the philosophical questions that might be applied to any biogeographical study, regardless of the (often almost religiously held) convictions of the researcher.

We may also ask in what way does it matter, if it matters at all. Organisms are notoriously disrespectful of man-imposed boundaries, and man-delimited ranges. If biogeography is to be a dynamic discipline rather than a static recording of ranges we must take every factor of the organism's biology into consideration. Certainly, crustaceans have employed many factors, either during larval stages, or in adulthood, or perhaps during both, to arrive where we find them today. To ignore such factors is to produce incomplete assessments not only of the organism's distributions, but of its ecological responses, biological relationships within the distribution, and indeed even the ability to distribute.

We are indeed fortunate to have assembled such a varied slate of authors to address these and other topics within *Crustacean Biogeography.* When the participants were solicited for the oral presentations during the 1984 meeting of the American Society of Zoologists at Philadelphia, and later for the chapters that appear herein, one important stipulation was made. All contributors were requested to let their imaginations run a little more freely than perhaps they were accustomed to doing. We wanted ideas, perhaps unpopular — speculations, perhaps bold or even tenuous — and conclusions, perhaps controversial. Our goal was not only to inform, but more importantly, to make the student of biogeography think, or even rethink, previously held (and perhaps all too sacred) concepts. Herein we have radiation, centers of origin, island biogeography; but also herein are body size factors in relation to environmental constraints, Lessepsian migrations and tracks, global conclusions, and false cosmopolitanism. In editing these papers, and we hope in the reader's assessment of each, it seems apparent that the field of *Crustacean Biogeography* is ripe for forment, for controversy, and most of all for change. We believe that the Crustacea can and will play a major role, of increasing importance, in the field of biogeography as a whole. We trust that the contributions in this volume will foster continued interest in this regard.

Robert H.Gore
Kenneth L.Heck

1 Realized distributions

Prologue

While not intending to categorize or assign any of the authors who have contributed to this volume by lumping them under one title or another, we believe that the contributions can be separated into three main categories. In Part 1 five authors examine zoogeographical aspects of Decapoda, Cirrepedia, Syncarida, Peracarida and Coepoda primarily from the position of what may best be called 'realized distributions'. Thus, Rodney Feldmann's paper provides new data and insight into possible palaeozoogeographic distributions of some rather poorly known decapod taxa, including a novel hypothesis which employs both plate tectonics and a generalized track of dispersal to explain the occurrences of marine raninid, and freshwater aeglid crabs within the southern hemisphere. His data support similar distributions seen in the Mollusca within the same region.

William Newman's paper takes the position that the Hawaiian biota is probably best explained as the result of long range dispersal from the Indo-Pacific. He examines this concept in relation to several factors, including extinction and island integration, using the Cirrepedia. The hypothesis of high and low island stepping stones, which occurred during eustatic lowering and raising of sea levels across the Pleistocene, is promulgated. The question as to whether island raising or lowering is truly indicative of a dispersal, rather than a vicariant, event should spark some controversy in future studies.

Gilberto Rodriguez discusses the freshwater terrestrial regions of Central America where he investigates the present day occurrences of pseudothelphusid and trichodactylid freshwater crabs. Using a suite of cladistically derived characters, Rodriguez suggests that shared synapomorphies in the respiratory structures of these crabs strongly imply a monophyletic origin for the families. However, both a vicariant and a dispersalist viewpoint are advanced, in which a Gondwanan distribution was severed by one or more vicariant events during the mid-Cretaceous, followed by further dispersal and subsequent vicariant events in the Pliocene. The latter phenomenon is invoked to explain a puzzling disjunct distribution of some pseudothelphusids between the Greater Antilles and the distributional center of northern South America. Rodriguez' paper does much to reconcile the polarity that too often occurs between dispersal and vicariance zoogeographers.

However, migrations certainly have their place in understanding the biogeography of the Crustacea, as is amply shown by Dov Por's concise summary of dispersal and vicariance events in the Levantine landbridge. The formation of this barrier was important in that it severed a once-continuous Tethys Sea, isolating peracaridan and syncaridan groups (among others) from their taxonomic relatives. This more or less ongoing series of events in the

Triassic, coupled with the eventual salinity crisis and a nearly complete drying out of the Mediterranean sea was followed by later transgressions, all of which left indelible marks on many previously united crustacean groups. The establishment of the Suez Canal has again affected distributions, allowing some groups to migrate back into the Mediterranean. Por examines the several aspects of these events using subterranean, interstitial and epigaeic forms, and follows their distributions from the Red Sea-Levantine area into the Mediterranean.

On another continent on the other side of the world John Holsinger also uses subterranean distributions, in this case of amphipods, to show that three major zoogeographic patterns can be discerned: an ancient freshwater distribution of forms that extends possibly from the early Mesozoic; a marine relict distribution which is postulated to be the result of a stranding of ancestral forms previously marine during late Cretaceous-Tertiary times; and an island-associated group of taxa living fully marine to marginally freshwater in or on West Indian islands, which are also hypothesized to have become isolated during island emergence in the Cenozoic. Holsinger invokes both early Laurasian-Pangaean vicariant events, and subsequent destruction of groundwater habitats by glaciation (as another vicariant sundering of distributions) to explain present day distributions of some groups. Other groups are postulated to have radiated from deep groundwater refugia beneath glacial ice coverings. Holsinger's paper is a compilation of fascinating vicariant and dispersal events within an extremely interesting group of familial taxa, most of which are stygobionts that apparently colonize much of their ranges either by remaining subterranean, or by becoming stranded and eventually adapting to freshwater during events of island emergence.

None of these authors can be considered strict dispersalists or vicariance biogeographers. Instead, they have incorporated both concepts to suggest some plausible explanations for distributions seen today. As such, these papers form an ample springboard to jump as deeply as one might wish into the complexities involved with the novel vicariant concepts and reconciliations with traditional dispersals from centers of origins.

Paleobiogeography of two decapod crustacean taxa in the Southern Hemisphere: Global conclusions with sparse data

RODNEY M.FELDMANN
Department of Geology, Kent State University, Kent, Ohio, USA

ABSTRACT

Examination of the fossil record of two decapod crustacean groups, the Aeglidae and the raninid genus *Lyreidus*, shows that they originated in the southern circum-Pacific region in Late Cretaceous and Eocene time, respectively, and that dispersal into lower latitudes probably occurred in the early Oligocene. Fragmentation of the southern Pacific landmass, dominated by Antarctica and Australia, and concomitant development of the circum-Antarctic current system severely reduced the probability of further dispersals along this route. These interpretations, based on extremely limited fossil data, parallel conclusions drawn regarding the paleobiogeographic distribution of mollusks in the early Tertiary Wedellian Province.

1 INTRODUCTION

Ortmann (1902a) introduced an article on the zoogeography of freshwater decapods by emphasizing that the final goal of zoogeography was, 'the retracing of the present animal distribution to its beginning in the past . . . a corollary of this is the recontruction of the ancient physical features of the earth's surface, since these in the first place have guided the development'. He went on to say that, 'In the latter respect, the distribution of land and water in past times is all-important and the easiest to be traced'. His observations are equally true today. We continue to talk about the same kinds of points relative to paleozoogeography and the thrust of his statement forms the basis for this essay.

The distribution of any organism in time and space can be portrayed as an irregularly shaped polygon (Fig.1) when its distribution is plotted against the axes of time, geography and ecology. A taxon, as defined by a human observer, is said to evolve at a single point in time and space and become extinct at some other point in time and space. Between the two events, however, one is faced with the problem of attempting to recontruct the shape of the polygon in such a way as to correctly portray the history of growth, dispersal, radiation and ultimately, diminution and extinction. The question I would like to address is whether a reasonable pattern of the paleozoogeographic history of taxa can be constructed when only limited fossil material is available. Under such circumstances, the data points represent checkpoints to at least begin to define the trends of distribution

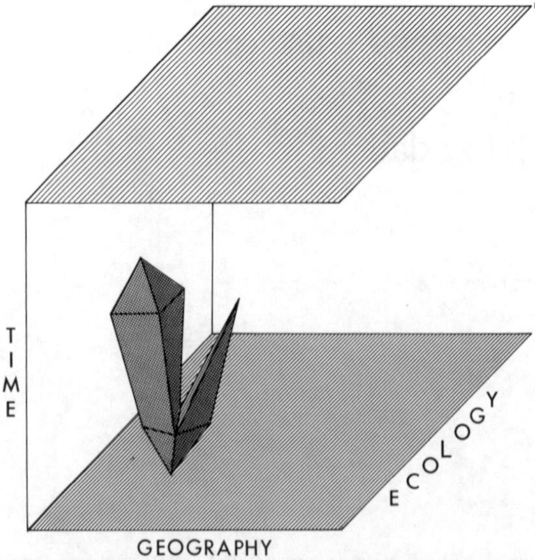

Figure 1. Time-space diagram showing the total distribution in time, space, and ecological setting during the evolution, radiation, and extinction of a hypothetical taxon.

within the continuum. The only test for the adequacy of the system is to compare distributional patterns derived from limited data with established patterns of dispersal based on more frequently collected organisms. I contend that, at least in the case of some of the Cenozoic and late Mesozoic decapods of the southern hemisphere, adequate information is available to recognize some important new dispersal routes, even in the absence of large collections. The conclusion is significant because decapods are almost invariably rare elements in fossil assemblages. If it were always the case that large numbers of individuals, from numerous localities, were essential to analyses of paleobiogeography, little or nothing could be learned about the distributional patterns of decapods.

Description of the paleobiogeography of any group of animals presuppoes that one has an adequate understanding of the geographic setting during the time intervals under consideration; that enough data exist to demonstrate time equivalence from one area to another; that the group of organisms in question is adequately identified and described so that proper biological relationships can be defined; and, finally, that an adequate sample exists on which biogeographic distribution can be defined. All of these factors must then be cast in the framework of modern biogeographic principles so that a coherent description of ancient distributions can be made. The biogeographic distribution of two groups of fossil and recent decapod crustaceans from the southern hemisphere, the anomuran family Aeglidae and the raninid genus *Lyreidus*, will be examined. Both demonstrate that even though we have relatively small samples, enough information is, in fact, available to draw meaningful conclusions about their biogeographic distribution.

2 CONTEXT OF PALEOBIOGEOGRAPHY

2.1 *Paleobiogeographic models*

Until relatively recently, models of paleobiogeographic distribution of organisms were cast in a rigid framework owing to the accepted attitude that continents maintained a fixed position throughout most, or all, of geologic time. Thus, significant biological expeditions, such as that sponsored by Princeton University at about the turn of the century (Ortmann 1902b) and devoted to the study of the biology of Patagonia, provided a tremendous data base documenting affiliations of the fauna of southern South America with those of other southern hemisphere continents. They were, however, forced to rely on what today appears to be a rather fanciful interpretation of land bridges uniting distantly separated continental areas. An analysis of refinements in our understanding of changes in continental positions through time, based on plate tectonic theory, now allows extremely detailed paleogeographic maps to be drawn (Fig.2) which give a clear picture of the relative geographic position of continental land masses throughout most of Phanerozoic time. Several sets of reconstructions of paleo-continental positions are frequently referred to with regard to the Cenozoic and Mesozoic eras (Firstbrook et al. 1979, Scotese 1979, Smith & Briden 1977). Although some details of continental placements differ, the general patterns are in agreement and provide excellent bases for interpretation of paleocurrents. The former two map bases are used herein, essentially interchangeably, depending upon the projection necessary to convey distributional data.

Paleocurrent reconstructions shown in Figure 2, as well as those constructed in all other contemporary works (Schopf 1980) are based on first cause deductions. Their limitations reflect lack of knowledge of paleobathymetry and are subject to the ever-present problem of generalization. On the other hand, global patterns, those necessary to define dispersal routes, would seem to fall well within the limits of usefulness.

Zinsmeister (1982) summarized data on the distribution of circum-Pacific molluscan faunas varying in age from Late Cretaceous to Tertiary and incorporated the seminal work on southern hemispheric biogeography by Flemming (e.g. 1962, 1963, 1967) to propose a pattern of distribution to explain the fossil occurrences. Employing the paleobiogeographic reconstructions and the style of paleocurrent analyses referred to above, Zinsmeister postulated the development of a highly endemic molluscan fauna in the southern circum-Pacific in what he termed the Wedellian Province. This fauna persisted intact until such time as break-up of the southern continents, northward migration of Australia and New Zealand, and development of circum-Antarctic current patterns subjected the organisms to a series of thermal crises that either induced dispersal into favorable habitats elsewhere, or led to extinction. Because Zinsmeister's data base was large and encompassed molluscan species from Australia, New Zealand, Antarctica, and South America, it forms an excellent framework against which decapod distributions can be compared. The timing and significance of the dispersal event has been more strongly defined by the demonstration that 11 genera in five classes of the Mollusca, Echinodermata, and Arthropoda underwent their early development in the Wedellian Province and dispersed northward from the Antarctic in the Neogene (Zinsmeister & Feldmann 1984). At least in the case of the arthropods, the northward dispersal was probably triggered by the final fragmentation of Gondwana near the onset of the Oligocene.

2.2 *Geologic age dating and paleotemperature analysis*

Until relatively recently, it has been difficult to date accurately geologic occurrences from widely separated areas of the globe and from different paleoecological settings. The development of relatively routine techniques for determining radiometric ages, the addition

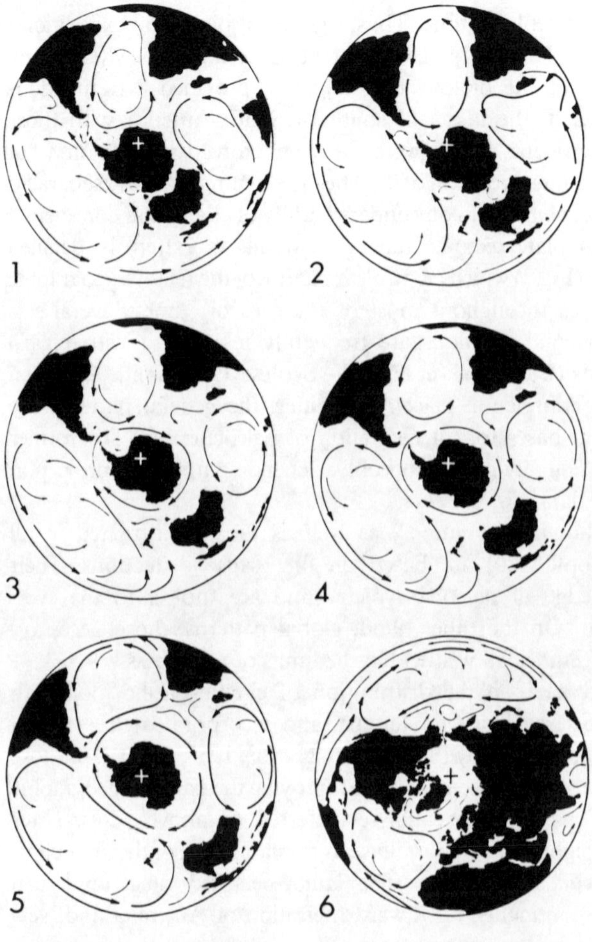

Figure 2. Polar views of paleogeographic maps (from Firstbrook et al. 1979) showing generalized major oceanic current circulation patterns. (1) and (2) 50 MABP and 40 MABP, respectively, showing the circulation of the southern oceans prior to development of the circum-Antarctic current. (3), (4) and (5) 30 MABP, 20 MABP and Recent maps, respectively, showing changes in relative positions of the southern continents subsequent to the development of the circum-Antarctic current system. (6) Recent north polar view showing current circulation pattern analogous to that of the southern polar region prior to northward movement of Australia and New Zealand.

of relatively precise techniques of magnetic stratigraphy and dating, based on the polarity reversal model, and a much improved biostratigraphic correlation, based on the voluminous information provided by the Deep Sea Drilling Project, makes it possible to correlate from one region to another with far greater precision than has previously been possible (see, for example, Hickey et al. 1983). A worthwhile summary of the application of magnetic polarity stratigraphy to the solution of a paleontological problem is given by Opdyke et al. (1979).

These refinements in techniques of correlation have permitted not only the demonstration of age equivalence of faunas in different regions but also have permitted the faunal evidence to be superimposed on data related to paleogeographic changes and other changes in the physical environment. For example, a wealth of information has been accumulating, as a result of the Deep Sea Drilling Project, upon which detailed paleotemperature analyses can be made. Kennett and others (Shackleton & Kennett 1975, Kennett 1978) have conducted a large number of studies using $^{16}O/^{18}O$ ratios on foraminiferans from circum-Antarctic deep sea ocean cores and have provided the detailed documentation of changes in sea surface temperature with time (Fig.3) which can be used to correlate biotic, sedimentational, and paleogeographic events.

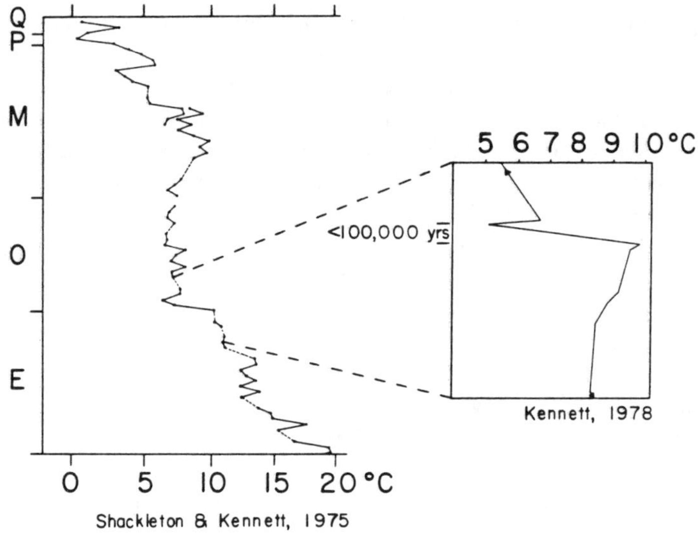

Figure 3. Changes in sea-surface temperature of the circum-Antarctic region as determined from measurement of $^{16}O/^{18}O$ ratios on planktonic foraminiferans collected from the Tasman Sea (left) and the Palmer Peninsula region, Antarctica (right). E — Eocene; O — Oligocene; M — Miocene; P — Pliocene; Q — Pleistocene and Recent.

2.3 *Taxonomic data base*

Factors related to quality of identification of organisms has always been an inherent weak spot in effecting global studies in paleobiogeography. Within the last 25 years, however, there has been an upsurge in interest in fossil decapods. One of the results has been the publication of a large number of substantial taxonomic revisions. With particular reference to the biota of the southern hemisphere, fossil decapods from Australia and New Zealand have been described by Glaessner (1960, 1980) and additional works on New Zealand decapods are forthcoming (Feldmann 1984, and in preparation). Faunas of Cretaceous age from Madagascar have been described in detail by Secretan (1964). Studies by Ball (1960), Taylor (1979) and Feldmann & Zinsmeister (1983, 1984a, b) have drawn together the description and interpretation of decapods from Antarctica. Specimens presently in hand document several fossiliferous localities from South America but, thus far, modern work on this decapod fauna is sparse. This collection of works on the fossil decapods of the southern hemisphere vastly exceeds the sum total of all previous work in the area and has provided excellent information on the organisms from well-studied sites.

The examination of fossil decapod faunas is reinforced by numerous outstanding monographic treatments of recent decapod groups. Two that are central to the present study are those of Schmitt (1942) dealing with the Aeglidae, and Griffin (1970) treating *Lyreidus.* There are, thus, adequate comparative studies for the fossil, as well as the recent, decapod fauna of the southern hemisphere.

2.4 *Problems of sample size*

An important remaining question is whether a sample, particularly related to a group of organisms such as the decapod crustaceans, is adequate in terms of numbers of individuals to permit a reasonable base for interpreting paleobiogeographic distribution. Traditionally, paleobiogeography has focused on examination of taxa embracing more numerous and perhaps more readily identifiable forms; for example, the molluscs used by Zinsmeister (1982). Species of molluscs can very commonly be represented by hundreds, or thousands, of individuals owing to the durability of their hard parts and to the original large number of individuals in most populations. By contrast, fossil decapods tend to be rare.

When one assesses the scarcity of decapod remains, a number of interesting observations, some of them relatively recently made, can be offered to explain this paucity. These include the observation that they are typically not as numerous as are individuals occupying low-level filter feeding niches; the exoskeleton of most decapods tends to be relatively fragile and disarticulates readily upon death or moulting so that a very high proportion of individuals, even if compared to other fossilized organisms, are destroyed by mechanical processes; and, finally, because decapods are easily eaten (compared to a heavy-shelled mollusc) they are often sought by other animals, so that scavenging and predation upon decapod remains is widespread. This, in turn, leads to destruction of skeletal material because in order to get at the edible remains, it is necessary to fragment the integument. A recent observation recorded by Ward & Wicksten (1980), in studying the ecology of *Nautilus*, may have profound implication, when one considers the fate of fossil decapod remains. The pearly nautilus has been observed to selectively ingest carapace material of palinurid lobsters, probably to increase calcium intake for shell production. Ward & Wicksten observed that, in both *Nautilus macromphalus* and *N.pompilius*, etched and pitted shell fragments of the spiny lobster, *Panulirus longipes*, were retained in the caecum for extended time periods, probably to be dissolved as a source for calcium in secretion of their own shells. Visual observations of the feeding behavior of nautiloids confirmed that they seek palinurids as a food resource. Although this manner of comminution of skeletal material of decapods would have to be considered a minimal factor in the Holocene, owing to the paucity of shelled cephalopods, it might very well have been a significant factor in destruction of decapod remains during the Mesozoic and early Cenozoic when these fierce predators were far more numerous and widespread in marine habitats.

Ward & Wicksten observed, further, that the predation on palinurids was focused primarily on the abdominal regions of the organisms. This suggests that the relative proportion of cephalothoracic regions left behind, potentially to contribute to the fossil record, would be high. Although I have not tested this quantitatively, I am almost certain that the majority of fossil lobsters, homarids and palinurids, described from North America were based on descriptions of either claw material or cephalothoracic material, because relatively fewer abdomina are preserved. Whether this is correct or not, the desirability of crustacean tissue as a food resource is only one of the factors that may result in diminution in numbers of fossil arthropods.

This combination of attributes, viz. the desirability of the organisms as a food resource, the delicate nature of skeletal material, and the original numbers of individuals, must be weighed against the factors favoring preservation and discovery if one is to develop a measure of the completeness of our sample of decapods and, therefore, the appropriateness of these organisms as paleobiogeographic indices.

Yet, there is some evidence, again qualitative, to document the opinion that decapod remains are likely candidates for discovery and study. The remains of decapods tend to be relatively distinctive and readily noted during collecting. The color of decapod fossils often tends to be rather dark and quite different from that of the surrounding matrix. Thus the specimens stand out and are more easily recognizable than their invertebrate associates with calcitic shells. Additionally, and in the subjective human sense, there is very little doubt that arthropod remains are considered prized specimens. An examination of the collections of the Smithsonian Institution, for example, or examination of published records of fossil arthropods (see Rathbun 1935), testifies to the fact that every bit and piece, down to tips of claws, is to be saved, curated, and often named. The tiniest fragments of decapods are set aside, because of their rarity, whereas comparable sized fragments of molluscs would never be thought worth collecting. I suspect that, if rock units that have been studied in any detail contain any arthropod remains at all, they would be so noted.

2.5 *Summary*

The point of this brief analysis is to demonstrate that we have a reasonably good paleogeographic framework with which to work, we have powerful tools for determining geologic age and for demonstrating correlation, and the taxonomic framework of decapods is well enough known. Therefore, although there are relatively few organisms with which to work, the sample is a reasonable reflection of the organisms preserved in the fossil record so that it is possible to draw some meaningful conclusions about paleobiogeography from this group. As an example, this conclusion can be reinforced by looking at specimens from two groups of decapods in the southern hemisphere and by presenting information from new descriptions of fossil forms. The biogeographic distribution and dispersal style of these creatures, as compared to the described paleobiogeographic model of Zinsmeister (1982) using the more commonly studied molluscan elements of fossil faunas, is remarkably similar.

3 THE AEGLIDAE

3.1 *Recent distribution*

The Aeglidae, a family of fresh and brackish water anomurans from South America, have been known, previously, to embrace only a few species within a single genus, *Aegla.* These distinctive creatures were studied in detail by Schmitt in a monograph published in 1942, in which he catalogued and described 20 species. Subsequently, the number has risen to 34 recognized species and subspecies (Hobbs 1979).

Schmitt (1942:443) made the reasonable suggestion that, because aeglids were known only from South America, their point of origin must have occurred there. He further concluded (p.443) that their ancestry had undoubtedly included marine forms but, in the absence of tangible evidence, could say relatively little about their geologic history. The present day distribution of species of *Aegla* is confined to drainages on both the eastern and western coasts of temperate South America. Based on the notion that the evolution of *Aegla* would have proceeded from relatively simple, unornamented forms to more complexly ornamented forms, Schmitt (1942:442) suggested that the origin of the Aeglidae was on the Atlantic side of the continent, because these forms were relatively less ornamented, and that populations in the Pacific drainages were derived. This interpretation ran counter to the opinion of Ortmann (1902a), who suggested that the Pacific populations embraced the more primitive forms. Ortmann's conclusions were apparently based on the distribution of *Aegla* compared to the distribution of *Parastacus* in South America. He suggested that, because the distributions were very much the same, the organisms had enjoyed a common history (1902:389). No justification, on morphologic grounds, was forwarded. In the absence of fossil evidence, of course, all these conclusions were speculative. Fossil evidence, presented below, suggests that Ortmann may have been correct, but for the wrong reason. The only fossil form now known is heavily ornamented.

3.2 *Fossil distribution*

Recent examination of a splendid collection of fossil decapods from New Zealand, turned up three specimens which are, without question, a fossil form closely allied to the modern *Aegla* (Fig.4). The specimens, however, were collected from *marine* rocks in New Zealand, dateable as Maastrichtian, or Late Cretaceous, in age. They have been assigned to a new genus and species, *Haumuriaegla glaessneri* Feldmann, 1984. The assignment of these specimens to the Aeglidae is based on their close similarity to species of *Aegla* in respect to carapace conformation, groove patterns, and claw morphology. Even the diagnostic flabellate expansion of the propodus is present in *Haumuriaegla.* The surface of the carapace is covered with coarse scales, similar to, but better developed than, the ornamentation on some species of *Aegla.*

3.3 *Synthesis*

With discovery of these few specimens, it is now possible to suggest that the origin of this unusual group of freshwater decapods was not in South America, but was half-way around the globe in the region of New Zealand (Fig.6). Dispersal of aeglids from this point of origin occurred, perhaps, by dispersal of larvae in the main counter-clockwise gyre of ocean circulation in the southern Pacific Ocean.

This dispersal probably occurred prior to the close of the Oligocene. Prior to the break-up of the continental land mass bounding the southern circum-Pacific (Fig.2), New Zealand, Antarctica, and southern South America would have experienced similar cool-temperate coastal conditions. Dispersal, either as a single chance event or in a series of events as yet undocumented in the fossil record, would have been more likely than in post-Oligocene time. Separation of Australia and New Zealand from the Antarctic continent, concomitant northward movement of Australia, and development of the circum-Antarctic current system would have resulted in more complete isolation of New Zealand from South America. The most parsimonious interpretation, therefore, would suggest dispersal at the time of, and perhaps triggered by, the environmental changes associated with the breakup.

Here, then, is a clear indication that a modern paleogeogrphic interpretation of the southern hemisphere aids considerably in interpreting a dispersal pattern. A clear understanding of the continental positions during the Cenozoic and Late Mesozoic as well as the ability to accurately date thermal and biological events, permits drawing quite a reasonable conclusion regarding the paleobiogeography of the Aeglidae, even though the fossil record is sparse.

4 LYREIDUS

4.1 *Recent distribution*

The second group of organisms that displays an interpretable paleobiogeographical

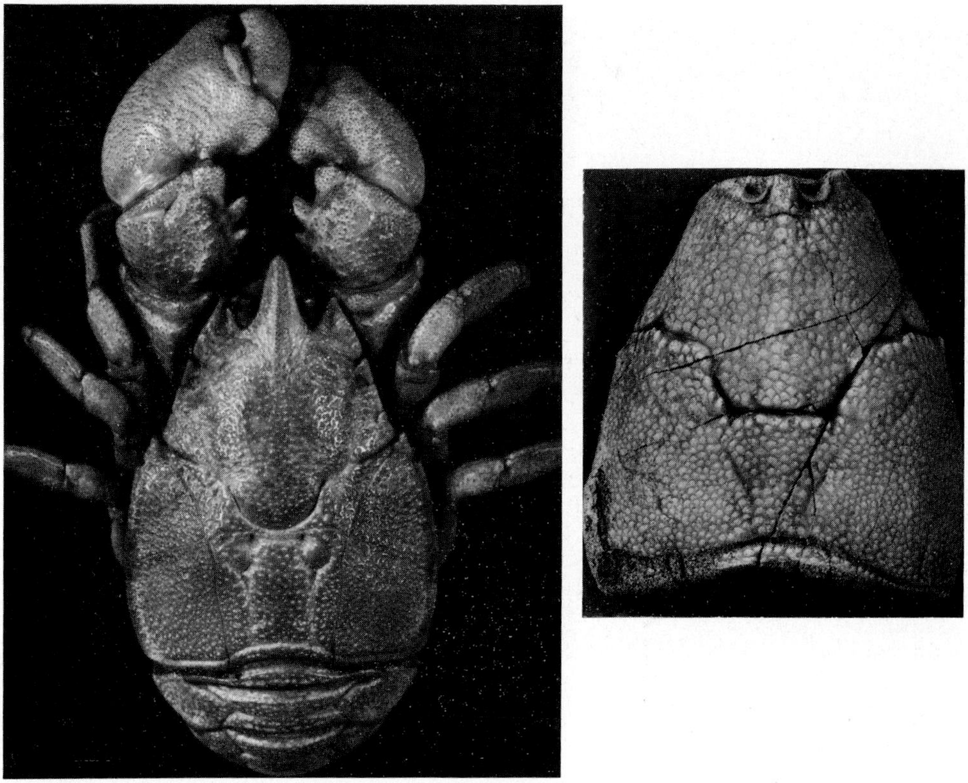

Figure 4. Representatives of the Aeglidae. *Aegla affinis* Schmitt, USNM Catalogue 176807, × 1.5, is typical of the Recent forms (left). *Haumuriaegla glaessneri* Feldmann, New Zealand Geological Survey AR915, × 1.25, is the sole fossil representative of the family (right).

Figure 5. Representatives of *Lyreidus*. *Lyreidus tridentatus*, (approx. × 1), the type species of the genus (left). *Lyreidus antarcticus*, USNM 365444, × 1.5, represents one of the earliest occurrences of the genus (right).

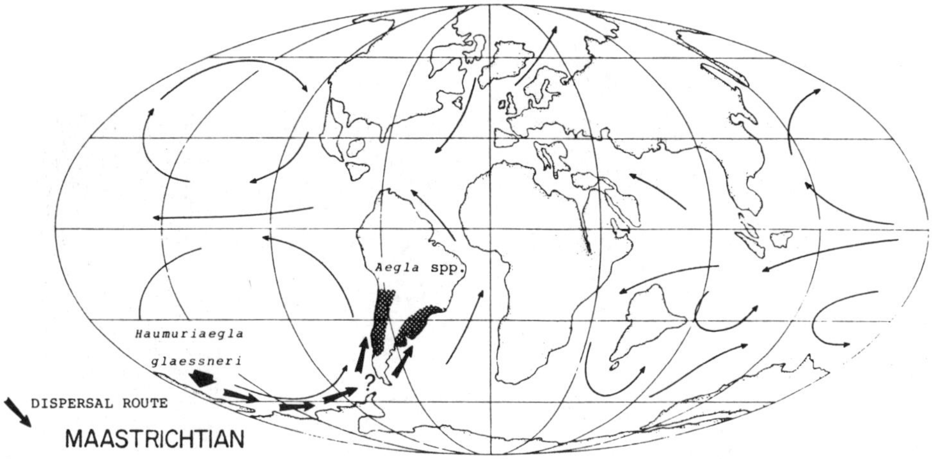

Figure 6. Paleogeographic map (from Scotese 1979) showing generalized current circulation patterns in the Maastrichtian (Late Cretaceous) as well as the site of the probable origin of the Aeglidae determined by the location of *Haumuriaegla glaessneri*, in New Zealand and the modern distribution (stipple) of the Aeglidae. Primary introduction along the west coast of South America is favored, although there is little fossil evidence to support the contention. Distribution of modern Aeglidae from Schmitt (1942).

distributional pattern in the southern hemisphere, is the genus *Lyreidus*, a raninid crab (Fig.5).

The living species of *Lyreidus* have a broad, disjunct distributional pattern, recently summarized by Griffin (1970) (Fig.7). The type species, *Lyreidus tridentatus*, is distributed throughout the western Pacific and into the area of Australia and New Zealand. It is also present along the Australian coast of the Indian Ocean. Three other species of the genus are distributed in what is almost universally an overlapping range. Two of the species have ranges extending into the Indian Ocean. *Lyreidus brevifrons* has been collected from the South China Sea and also from the African coast of the Indian Ocean. *Lyreidus channeri*, also known from the South China Sea, has also been collected in the Bay of Bengal. The range of *Lyreidus stenops* closely overlaps the distribution of the type species. The fifth living species, however, has quite a different distributional pattern. *Lyreidus bairdi* is known only from the Atlantic and Caribbean coastal regions of North America (Rathbun 1937).

4.2 *Fossil distribution*

The distribution of species of *Lyreidus* can, I think, be adequately explained by examining the geologic history of the group. Recent representatives of the genus are creatures of moderately deep-water. Most records of their occurrence are in temperate latitudes and most specimens have been collected from outer shelf and slope habitats (Griffin 1970). Although temperature data are not normally available for described specimens, I would suspect that these are organisms that prefer cool water habitats.

Fossils of *Lyreidus* have been found in some relatively unlikely places (Fig.8). Several specimens of this species have been collected from late Eocene rocks on Seymour Island

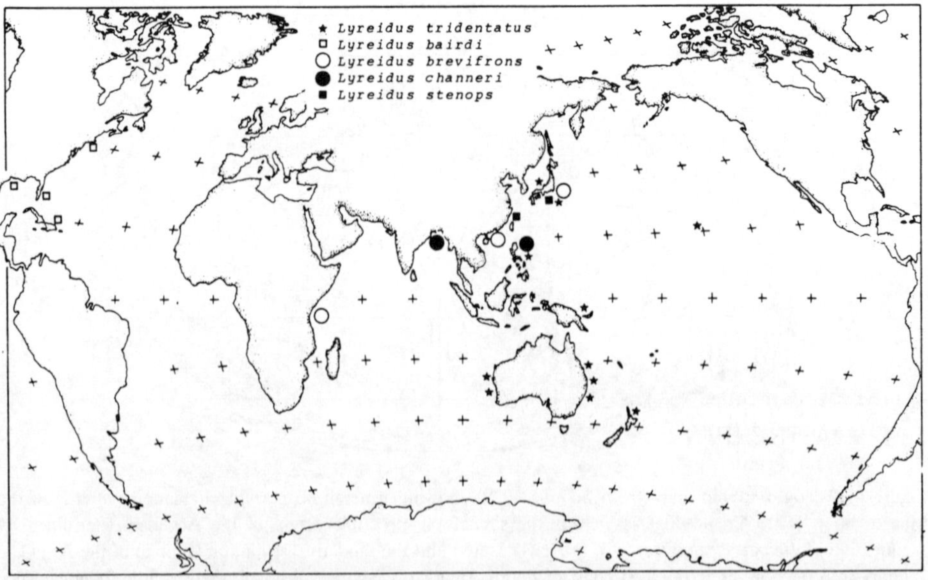

Figure 7. Map showing the generalized distribution of Recent species of *Lyreidus.* Data from Griffin (1970) and Rathbun (1937).

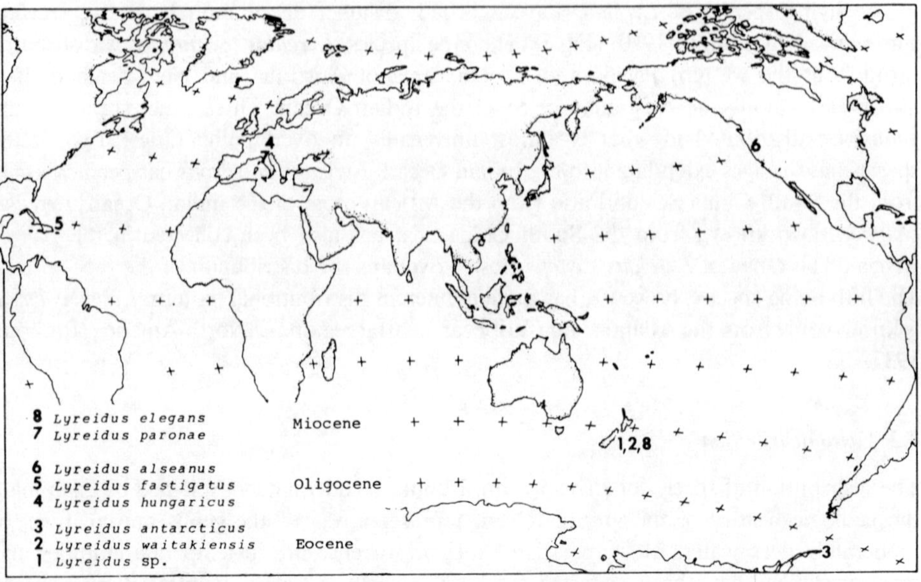

Figure 8. Distribution of all known fossil species of *Lyreidus.* Sources of data are cited in the text.

along the Antarctic Peninsula. This species has recently been described as *L.antarcticus* Feldmann & Zinsmeister (1984a). The abdomina of these organisms do not seem to extend dorsally in the same fashion as do the recent forms (Fig.6) and the fifth pereiopod is not nearly as reduced as it is in Recent organisms. Finally, the claws of these organisms are not nearly as specialized as are those on the Recent forms. This morphologic state is more generalized than that of modern representatives of the genus so that it is possible to suggest that these animals are rather near the root-stock of the taxon.

Species of *Lyreidus* have been collected from two other Eocene exposures in the southern hemisphere, both in New Zealand. *Lyreidus waitakiensis* has been described (Glaessner 1980) from middle to late Eocene rocks and another species, yet to be described, has also been noted in late Eocene rocks of South Island. These occurrences, coupled with the single species known from the Palmer Peninsula in Antarctica, represent probably the best documented fossil species known for the genus. Furthermore, they represent the oldest occurrences of *Lyreidus* known to date. Examining the paleogeographic setting in the southern hemisphere during the Eocene, and nothing that the circum-Antarctic current has yet to be established, the distribution of species of *Lyreidus* along the New Zealand and Antarctic coasts does not seem to be particularly surprising. Early representatives of the genus were endemic to the Wedellian Province.

Subsequent to the Eocene, five other notices of the genus, four in the northern hemisphere, are noteworthy. A species, described from relatively well-preserved carapace material from Hungary, *Lyreidus hungaricus* Beurlen (1939:146), formed one of the earlier references to the genus in the fossil record. This species was collected from Oligocene rocks, of approximately the same age as units yielding two other specimens referred to the genus. One, *Lyreidus alseanus*, was collected from Oregon (Rathbun 1932:239) and another *L.fastigatus*, was described (Rathbun 1919:168) from Anguilla in the West Indies. I have examined the specimens of Rathbun, in the Smithsonian Institution, and only the eye of faith would permit their assignment to the genus. The Oregon specimen has been so severely crushed and distorted that it could be assigned to almost any genus of decapods and the remains from the Caribbean consists only of a badly preserved fragment of carpus; again, not assignable to the genus with any degree of confidence.

In the Miocene, moderately well-documented occurrences of the genus are known from New Zealand and Italy. *Lyreidus elegans* was described (Glaessner 1960) from two poorly preserved carapaces from South Island, New Zealand. Additionally, Crema (1895:671) described *Lyreidus paronae* from the Tertiary of northeastern Italy. Subsequently, Sacco (1907) documented the Miocene age of the occurrence.

4.3 *Synthesis*

It is interesting to note that, historically, the European and North American occurrences were the first fossil representatives of the genus described. For that reason, and because of a kind of general sentiment that many dispersal patterns reflect radiation from the high diversity regions of the Tethys into more remote regions, such as the southern Pacific region, one might logically postulate an origin for *Lyreidus* in the Tethyan region. Contrary to this interpretation, however, the oldest fossils of the genus are found in New Zealand and Antarctica, suggesting that routes of dispersal (Fig.9) may have been counter-clockwise through the south Pacific region. The timing for the dispersal was early enough to result in introduction of the genus into the Oligocene rocks of Hungary. This would permit

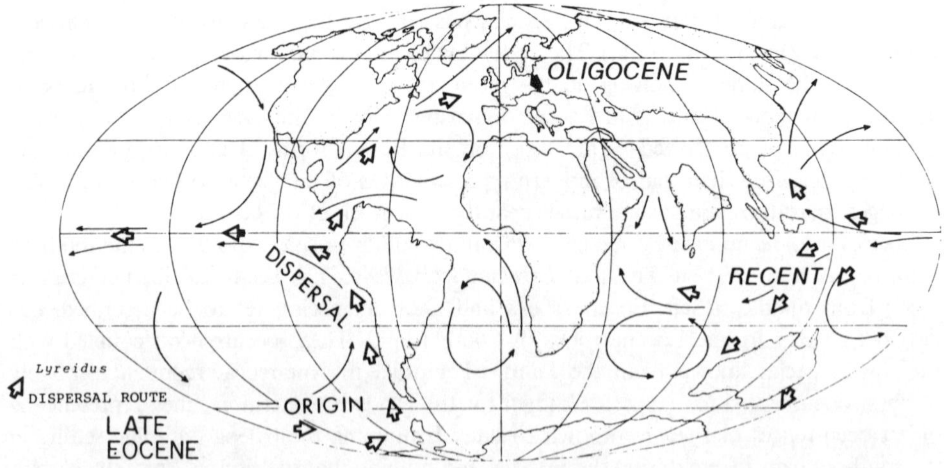

Figure 9. Paleogeographic map (from Scotese 1979) showing generalized current circulation patterns in the Late Eocene as well as the suggested direction of dispersal. The origin of the genus was in high latitudes, either in the region of New Zealand or Antarctica. Subsequent dispersal was counterclockwise in the Pacific basin with introduction into the Atlantic basin through the region of Central America.

a reasonable model for early Oligocene radiation of *Lyreidus* into habitats in the Pacific Ocean basin as well as the Indian Ocean basin and into the north Atlantic, as well.

Whether, in fact, *Lyreidus* ever did disperse northward along the west coast of North America is doubtful. The only evidence for this would be the one, inadequately preserved specimen known from Oregon and, certainly, this is not enough documentation to consider significant.

It is interesting to note that by the time of occurrence of fossil *Lyreidus* in the northern hemisphere in the Oligocene, the paleogeographic setting of the southern hemisphere had begun to change markedly. Australia had migrated northward away from the Antarctic continent (Fig.2). For the first time, the possibility of development of a deep ocean separation between Antarctica and Australia existed; a condition which is required for the development of a circum-Antarctic current.

5 SUMMARY AND CONCLUSIONS

Zinsmeister's summary of molluscan paleobiogeography of the southern hemisphere concluded that, because of isolation during the Cretaceous and early Tertiary, a distinctive, southern molluscan assemblage with a high degree of endemism developed. He referred to this association as the Wedellian Province. This province existed until the Oligocene, during which time the development of the circum-Antarctic current and breakup of the large southern continental mass began to decrease opportunities for dispersal of organisms into the Indo-Pacific region and the Northern Hemisphere. Separation of the southern continents and formation of a distinctive circum-Antarctic current system was associated with a general decline of sea surface temperatures. These events might very well have been the triggering mechanisms for dispersal of the arthropods. If so, the timing of the dispersal would be rather precisely defined as the early part of the Oligocene. Although Zinsmeister & Feldmann (1984) demonstrated that several taxa, endemic to the Wedellian Province in their early history, dispersed northward from the Antarctic region, there is no evidence to support application of a vicariance model to the data. Rather, all present evidence supports an origin of the groups in the southern circum-Pacific and subsequent radiation into lower latitudes in a variety of styles and, probably, at different times.

Data regarding the geographic distribution of the decapods fit nicely with the conclusions drawn by Zinsmeister. This is, indeed, reassuring and permits me to conclude that, although the sample of decapod crustaceans is relatively small, we are on the right track in interpreting their paleobiogeogrphic distribution.

Early representatives of the Aeglidae and *Lyreidus* did, in fact, occupy shallow water habitats in the southern circum-Pacific region (Feldmann 1984, Feldmann & Zinsmeister 1984a). At least in the case of *Lyreidus*, it can be demonstrated that the taxon had radiated into regions stretching from New Zealand to peninsular Antarctica. By the close of Oligocene time, *Lyreidus* had radiated into the northern hemisphere. This early widespread dispersal was probably paralleled by dispersal of the Aeglidae, although no specific data document that interpretation. Certainly, the dispersal would have been faciliated by the absence of the circum-Antarctic current system. About the time of the onset of the Oligocene, this was to change markedly with the northward migration of Australia away from the Antarctic continent and the establishment of the circum-Antarctic current. Thus, Antarctica was separated for the first time, ecologically, from Australia, New Zealand, and southern South America. I suggest, therefore, that any dispersal of organisms from the New Zealand region eastwards into Antarctica, in fact, must have been initiated prior to the onset of the Oligocene. Subsequent dispersal into lower latitudes was probably triggered by the early Oligocene climatic change associated with the breakup of Gondwana. Evidence for the development of the circum-Antarctic current during the early Oligocene is well documented by oxygen isotope analyses of deep sea sediments in the Palmer Peninsula region (Fig.3). A drop in the sea surface temperature from 10°C to almost 5°C occurred over a span of time as brief as 100 000 years. This same dramatic temperature decline has been recorded in a number of places in the circum-Antarctic region and provides a powerful piece of evidence that coastal habitats in Antarctica changed dramatically beginning in the Oligocene, from tolerably cool, temperate environments to the polar conditions that exist today.

Certainly, description of the paleobiogeography of two groups of decapods does not establish a pattern that can be applied to all forms. Not all decapods in the southern hemisphere had their origins in the high latitudes of that hemisphere and, certainly, there are numerous groups for which there is no known fossil record. There is, however, evidence to suggest that these are not unique examples. The present day distribution of *Nephrops* is primarily Indo-Pacific with the single species, *N.norvegicus*, found in the North Atlantic. My guess would be that, eventually, we might very well be able to demonstrate through the examination of the fossil record, an origin of this genus in the Wedellian Province. As with any major region, the assemblage of organisms currently inhabiting the area forms a composite of distributional types ranging from strongly endemic to cosmopolitan. The South Pacific region is no exception but, until recently, very little evidence had been assembled to document the contribution of endemic forms in the southern circum-Pacific.

ACKNOWLEDGEMENTS

The manuscript was read and reviewed by Kenneth L.Heck, Jr, Director, Patrick Center for Environmental Research and Robert H.Gore, Research Associate, The Academy of Natural Sciences of Philadelphia. Richard Bambach provided enlarged base maps from Scotese (1979). My thanks to these gentlemen.

REFERENCES

Ball, H.W. 1960. Upper Cretaceous Decapoda and Serpulidae from James Ross Island, Graham Land. *Falkland Islands Dependencies Survey Scientific Reports* 24.

Beurlen, K. 1939. Neue Decapoden-Krebse aus dem ungarischen Tertiar. *Palaeontol. Zeit.* 21: 135-161.

Crema, C. 1895. Sopra alcuni decapodi terziarii del Piedmonte. *Atti della Roy. Acad. delle Sci. di Torino* 30: 671-677.

Feldmann, R.M. 1984. *Haumuriaegla glaessneri* n.gen. and sp. (Decapoda; Anomura; Aeglidae) from Haumurian (Late Cretaceous) rocks near Cheviot, New Zealand. *New Zealand J. Geol. and Geophys.*

Feldmann, R.M. & W.J.Zinsmeister 1983. New crabs from the La Meseta Formation (Eocene) of Antarctica: biogeography and paleoecology. *Geol. Soc. Amer. Abs. with Prog.* 15(6): 571.

Feldmann, R.M. & W.J.Zinsmeister 1984a. First occurrence of fossil decapod crustaceans (Callianassidae) from the McMurdo Sound region, Antarctica. *J. Paleontol.* 58: 1041-1045.

Feldmann, R.M. & W.J.Zinsmeister 1984b. New fossil crabs (Decapoda: Brachyura) from the La Meseta Formation (Eocene) of Antarctica: paleogeographic and biogeographic implications. *J. Paleontol.* 58: 1046-1061.

Firstbrook, P.L., B.M.Funnell, A.M.Hurley & A.G.Smith 1979. *Paleoceanic reconstructions 160-0 Ma.* Washington: National Science Foundation.

Flemming, C.A. 1962. New Zealand biogeography. A paleontologist's view. *Tuatara* 20: 53-108.

Flemming, C.A. 1963. The nomenclature of biogeographic elements in the New Zealand biota. Trans. Roy. Soc. New Zealand (General) 1: 13-22.

Flemming, C.A. 1967. Cenozoic history of Indo-Pacific and other warm-water elements in the marine Mollusca of New Zealand. *Venus* 25: 105-117.

Glaessner, M.F. 1960. The fossil decapod Crustacea of New Zealand and the evolution of the Order Decapoda. *New Zealand Geol. Surv. Paleontol. Bull.* 31.

Glaessner, M.F. 1980. New Cretaceous and Tertiary crabs (Crustacea: Brachyura) from Australia and New Zealand. *Trans. Roy. Soc. S. Australia* 104: 171-192.

Griffin, D.J.G. 1970. A revision of the Recent Indo-Pacific species of the genus *Lyreidus* de Haan (Crustacea, dDecapoda, Raninidae). *Trans. Roy. Soc. New Zealand* 12: 89-112.

Hickey, L.J., R.M.West, M.R.Dawson & D.K.Choi 1983. Arctic terrestrial biota: paleomagnetic evidence of age disparity with mid-northern latitudes during the Late Cretaceous and early Tertiary. *Science* 221: 1153-1156.

Hobbs, H.H. III 1979. A new species of the endemic South American genus *Aegla* from Parana, Brazil (Crustacea: Anomura: Aeglidae). *Proc. Biol. Soc. Wash.* 91: 982-988.

Kennett, J.P. 1978. Cainozoic evolution of circumantarctic palaeoceanography. In E.M.van Zinderen Bakker (ed.), *Antarctic glacial history and world palaeoenvironments.* Rotterdam: Balkema.

Opdyke, N.D., E.Lindsay, G.D.Johnson, N.Johnson, R.A.K.Tahirkheli & M.A.Mirza 1979. Magnetic polarity stratigraphy and vertebrate paleontology of the Upper Siwalik Subgroup of Northern Pakistan. *Palaeogeog., Palaeclimat., Palaeoecol.* 27: 1-34.

Ortmann, A.E. 1902a. The geographical distribution of freshwater decapods and its bearing upon ancient geography. *Proc. Amer. Philos. Soc.* 41: 267-400.

Ortmann, A.E. 1902b. Part II = Tertiary invertebrates. In W.B.Scott (ed.), *Reports of the Princeton University Expeditions to Patagonia, 1896-1899.* Stuttgart: Princeton Univ. Press.

Rathbun, M.J. 1919. West Indian Tertiary decapod crustaceans. *Carnegie Inst. of Washington Pub.* 291: 159-184.

Rathbun, M.J. 1932. New species of fossil Raninidae from Oregon. *Washington Acad. Sci. J.* 22: 239-242.

Rathbun, M.J. 1935. Fossil Crustacea of the Atlantic and Gulf coastal plain. *Geol. Soc. America Spec. Paper* 2: 1-160.

Rathbun, M.J. 1937. The oxystomatous and allied crabs of America. *U.S. Nat. Mus. Bull.* 166: 1-278.

Sacco, F. 1907. Se facies faunistiche del Miocene Torinese. *Revista Italiana di Paleontologia* 13: 116-119.

Schmitt, W.L. 1942. The species of *Aegla*, endemic South American fresh-water crustaceans. *Proc. U.S. Nat. Mus.* 91: 431-520, 4 pls.

Schopf, T.J.M. 1980. *Paleoceanography.* Cambridge: Harvard Univ. Press.

Scotese, C.A. 1979. *Continental Drift*, 2nd ed. Published privately.

Secretan, S. 1964. Les crustaces decapodes du Jurassique superieur et du Cretace de Madagascar. *Mem. Mus. National d'Hist. Natur.* 14: 1-226, 20 pls.

Shakleton, N.J. & J.P.Kennett 1973. Paleotemperature history of the Cenozoic and the initiation of Antarctic glaciation: oxygen and carbon isotope analyses in DSDP sites 277, 279, and 281. *Intial Repts. of Deep Sea Drilling Proj.* 29: 743-755.

Smith, A.G. & J.C.Briden 1977. *Mesozoic and Cenozoic paleocontinental maps.* Cambridge: Cambridge Univ. Press.

Taylor, B.J. 1979. Macrurous Decapoda from the Lower Cretaceous of south-eastern Alexander Island. *British Antarctic Surv. Scient. Repts.* 81: 1-39.

Ward, P. & M.K.Wicksten 1980. Food sources and feeding behavior of *Nautilus macromphalus. Veliger* 23: 119-125.

Zinsmeister, W.J. 1982. Late Cretaceous-early Tertiary molluscan biogeography of the southern circum-Pacific. *J. Paleontol.* 56: 84-102.

Zinsmeister, W.J. & R.M.Feldmann 1984. Cenozoic high latitude heterochroneity of Southern Hemisphere marine faunas. *Science* 224: 281-283.

Origin of the Hawaiian marine fauna: Dispersal and vicariance as indicated by barnacles and other organisms

WILLIAM A. NEWMAN
Scripps Institution of Oceanography, La Jolla, California, USA

ABSTRACT

The Hawaiian Archipelago is situated in relative isolation, especially from other high islands, in the subtropical fringe of the North Pacific between 20 and 29°N. The isolation from high islands as well as low islands is emphasized because there are some notable differences between the biotas of the two types. Most of the biota inhabiting the Hawaiian Islands represents a depauperate extension or attenuation of the Indo-West Pacific biota and most marine forms are wide ranging eurytopic species. There are also some cosmopolitan species as well as a variety of endemics. The Hawaiian shallow water fauna is relatively young (Tertiary), at least as far as the balanomorph barnacles and hermatypic corals are concerned, and within the latter there has been a marked turnover at the generic level since the Miocene. Therefore, I agree with most previous authors that long range dispersal accounts for the general nature of the biota.

A number of authors have noted a marked similarity, including shared species, between some elements of the Hawaiian biota with that of the high island reaches of the eastern South Pacific. This observation is inconsistent with the general Indo-West Pacific attenuation pattern. Representatives should be found in between, and to explain their absence 'extinction' and 'island integration' hypotheses have been proposed. So far, island integration has not been corroborated on geological grounds, and it is biologically unsound because of the time frame involved. However, there is historical evidence favoring the extinction hypothesis. The distributional anomaly is primarily an amphitropical pattern, comparable to that observed in the zooplankton. The latter has been explained on oceanographic grounds, by invoking Pleistocene events in the Eastern Pacific, with dispersal between the two regions from south to north, the same direction as for the insular biotas. This explanation becomes especially attractive for the insular biotas when Pleistocene eustatic sea level changes, as well as concomitant cooler sea surface temperatures across the tropics, are considered. During low stands of the sea, low islands such as the Line Islands, which range across the equator towards Hawaii from the south, became high islands, whereby the notable isolation of the Hawaiian Archipelago from high islands of the South Pacific became dramatically reduced. It also appears that the isolated islands of the Eastern Pacific were involved as stepping stones for insular forms between these two regions.

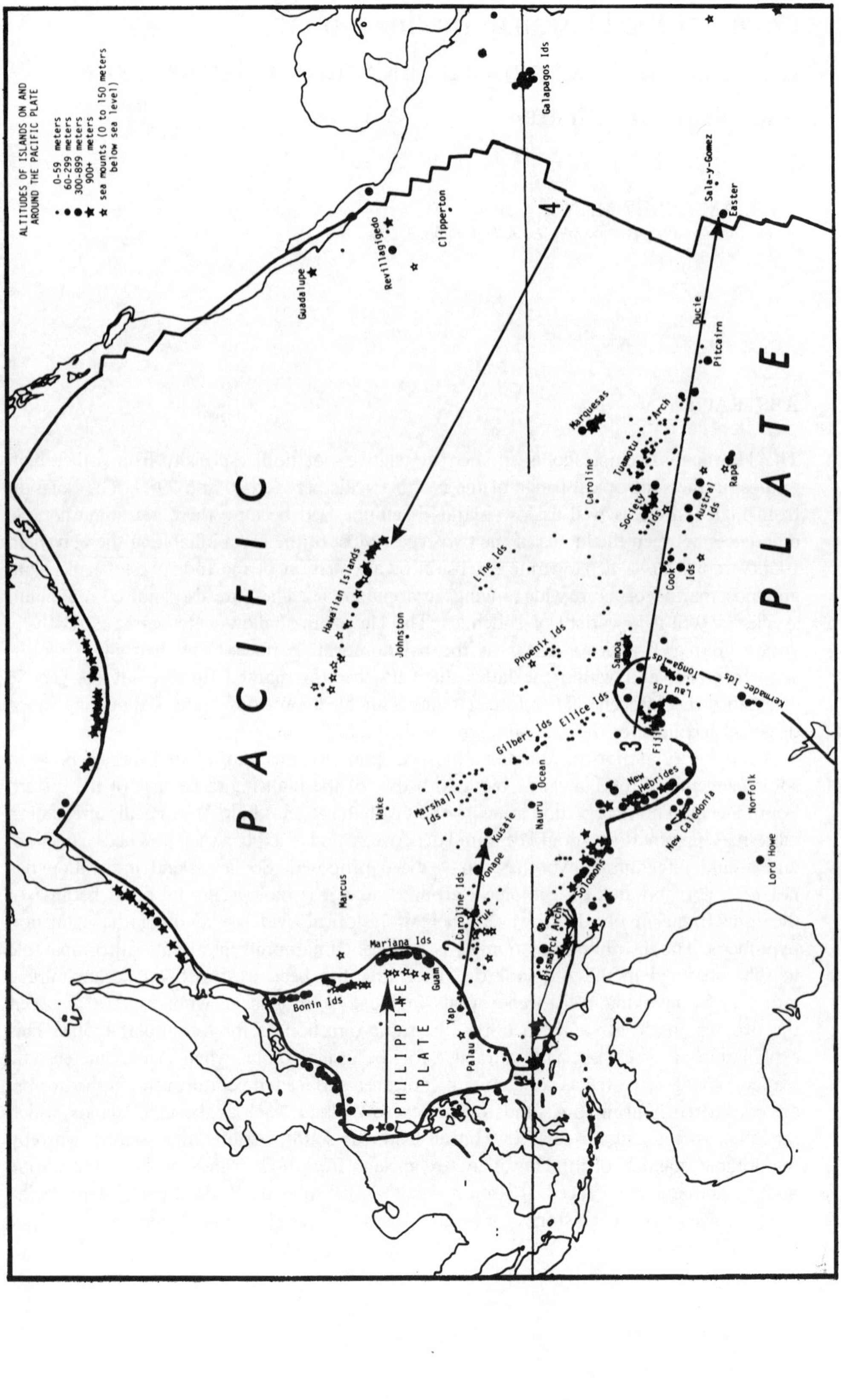
ALTITUDES OF ISLANDS ON AND AROUND THE PACIFIC PLATE
0-59 meters
60-299 meters
300-899 meters
900+ meters
sea mounts (0 to 150 meters below sea level)
PACIFIC
PLATE
PHILIPPINE PLATE
Galapagos Ids
Clipperton
Revillagigedo
Guadalupe
Hawaiian Islands
Johnston
Line Ids
Phoenix Ids
Marquesas
Caroline
Tuamotu Arch
Society Ids
Austral Ids
Rapa
Cook Ids
Pitcairn
Ducie
Sala-y-Gomez
Easter
Samoa
Tonga Ids
Lau Ids
Kermadec Ids
Fiji
Ellice Ids
Gilbert Ids
Marshall Ids
Wake
Marcus
Ocean
Nauru
Kusaie
Ponape
Truk
Caroline Ids
Mariana Ids
Guam
Bonin Ids
Yap
Palau
Bismarck Arch
Solomons
New Hebrides
New Caledonia
Norfolk
Lord Howe
1
2
3
4

1 INTRODUCTION

This paper investigates the origin of the shore fauna of the isolated oceanic islands of the Hawaiian Archipelago (Fig.1), as deduced from barnacles and other organisms such as the hermatypic corals. It therefore impinges on the current quasi-controversy in biogeography involving the relative importance of vicariance versus dispersal. The principles involved pose a particularly difficult mix where the marine faunas of oceanic islands are involved (Kay 1980). However, I think it regrettable that two separate disciplines, dispersal and vicariance, seem to have arisen in the minds of some. Historically the two aspects were and should remain simply facets, rather than factions in biogeography.

Long before the advent of 'Vicariance biogeography', barriers to dispersal were sought as explanations for disjunct or interrupted distributional patterns in related animals or plants. One need only recall Wallace's Line in the Indo-Malayan Region. Another important example was the similarity between the freshwater fishes of Africa and South America, and the geological evidence, utilized by Wegener (1915), in the best known early attempt to document continental drift. At about the same time the similarities among the southern portions of southern continents were noted and proposed as the basis for the then hypothetical protocontinent, Gondwanaland, before its break-up at the end of the Mesozoic (Suess 1904-1909). Particularly relevant to the present paper was the development of our understanding of the final break-up of the Tethys Sea and the concomitant development of tropical provincialism in the Cenozoic, primarily owing to barriers erected at Suez and the Isthmus of Panama, and to the barrier formed by the expanse of open water between the Americas and the Pacific islands to the west (Darwin 1859, Ekman 1953). Barriers to dispersal are at the very heart of our understanding of the biogeography of the Pacific Basin, beginning with the ideas of such classical biogeographers as Wallace in the west and Darwin in the east.

Biogeographers concerned with the Hawaiian Archipelago have long been puzzled by taxa found there that seemed too distant from parent populations to be readily accounted for by chance dispersal. They have looked for alternative explanations such as changes in the strength and/or direction of oceanic current systems (Ekman 1953) or the drowning of island stepping stones from the west (Zimmerman 1963). Therefore, while some more sophisticated methods and criteria have been developed and espoused in recent years, under the aegis of 'Vicariance biogeography' (cf. Nelson & Rosen 1981) I personally do

---

Figure 1. Islands and shallow seamounts of the Pacific Basin (Springer 1982), illustrating the relative isolation of the Hawaiian Archipelago from other islands. Springer's (1982) principal biogeographical observations include: (1) An islandless barrier to dispersal across the Philippine Plate; (2) An abrupt drop in diversity when moving off the Philippine Plate and on to islands of the Pacific Plate; (3) As compared to the gradual decline in diversity across the south Pacific; and (4) The integration of Cretaceous islands from the southeast Pacific into the Hawaiian Archipelago. (Heavy line contours indicate plate margins.) Springer's (1982) original chart included Wentworth Seamount and Kōkō Guyot with summits within 150 m of the sea surface. Actually Wentworth resides at approximately 1800 m and there is no evidence that it was ever an island (Newman & Foster 1983). Kōkō, on the other hand, is capped by reefal material forming a flat central depression at about 335 m (Dalrymple et al. 1980) and peripheral shallows up to 261 m (Clague et al. 1980). Kōkō was apparently an island in the Eocene, but it is presently too deep to have been at the sea surface during the last Pleistocene eustatic low stand of the sea (cf. Figs.8 and 9). Therefore these seamounts have been deleted.

not see that there is any real change in the philosophy an eclectic biogeographer needs to make.

The present paper will propose that the barnacles and corals of the Hawaiian Archipelago are by and large not remnants of an ancient biota isolated by vicariant events but rather Cenozoic invaders. Evidently they became established by relatively long range dispersal, for the most part from the west, following the extinction of the Mid-Pacific shallow water fauna at the end of the Mesozoic and local extinctions following the Miocene. Even though most of what one finds among the shallow water marine invertebrates and fish of the Hawaiian Archipelago is explained by long range dispersal, that which is not explained may require alternative hypotheses, and needs to be explored.

One of the fascinating exceptions not readily explained by ordinary dispersal involves elements of the Hawaiian marine fauna that have their closest affinities with the eastern Southern Pacific. Actually this pattern includes land snails and plants, and therefore could be the result of long-range, chance dispersal across the deep tropics, at least in part (Peake 1981, Carlquist 1981). But it also includes certain marine water striders and zooplankton, as well as shore fishes and invertebrates. Therefore, it apparently represents, in good part, an amphitropical pattern resulting from some general vicariant event(s). Rehder (1980) offers an extinction hypothesis (Théel 1886), whereby the ancestral stocks in the intervening region were replaced by modern forms. Springer (1982), on the other hand, offers 'Island integration' (Rotondo et al. 1981), whereby one or more southeast Pacific Cretaceous islands, and their biotas, were incorporated into the Hawaiian Archipelago. Newman & Foster (1983) favor the extinction hypothesis for a number of reasons including lack of geological support for island integration. On the other hand, amphitropical distributions have long been considered related to climatic cycles of the Pleistocene (Darwin 1859, Berg 1933, Hubbs 1952, Randall 1981). Therefore, if Théel's extinction hypothesis could be integrated with knowledge of Pleistocene eustatic sea level and concomitant insular geological phenomena, as well as climatic change, it would form a plausible hypothesis explaining at least the marine if not some of the terrestrial aspects of Hawaiian-South Pacific amphitropicality.

1.1 *Preamble*

Nothing has been published concerning the nature, affinities and origins of the Hawaiian shore barnacles, although Edmondson (1946) and earlier workers noted Indo-West Pacific affinities for the shore fauna in general. My own interests began with the demonstration that: 1) a new species of *Octolasmis*, the only obligate barnacle commensal in the gill chambers of certain decapod crustaceans from California, was distinct from a species referred to as *O.lowei* from Hawaii (Newman 1960a); 2) *Paralepas*, another commensal of large decapods such as lobsters from Kapingamarangi and Hawaii, had South African affinities (Newman 1960b); and 3) *Balanus eburneus*, a North Atlantic estuarine barnacle, had been introduced there (Matsui et al. 1964). Subsequently I was asked to write a chapter on Hawaiian cirripeds for a revision of Edmondson's (1946) *Reef and Shore Fauna of Hawaii.* The manuscript, first submitted in July of 1967 and since revised, included remarks on the apparent 'oldness' of the barnacle fauna concomitant with the notion that the dispersal route by which most had arrived in Hawaii had since been lost. To date, the crustacean volume of the *Reef and Shore Fauna of Hawaii* has not been published.

One factor that made the Hawaiian shore barnacles appear old was the lack of such modern forms as the coral balanoids (Pyrgomatidae), an otherwise widely distributed family in the Indo-West Pacific and Caribbean which first appeared in the Miocene (Newman & Ladd 1974a, b). It seemed obvious at first glance; the appropriate reef corals, which the barnacles inhabit elsewhere, must have gotten to Hawaii via island stepping stones that disappeared before the modern coral barnacles appeared in the Miocene. This in turn was corroborated by the fact that the average generic age of the hermatypic corals of Hawaii was among the oldest in the Indo-West Pacific (Stehli & Wells 1971).

Where were the stepping stones to Hawaii that had disappeared before the Miocene? They could have been the Mid-Pacific Mountains (Hamilton 1956). Following the suggestion of Menard & Hamilton (1963) to that effect, an expedition (SIO STYX-7, 1967) was mounted to dredge the drowned Cretaceous reefs, on the margins and slopes of the Mid-Pacific Mountains of the Marcus-Necker Ridge, to look for evidence that they had actually acted as stepping stones relative to the origin of the contemporary Hawaiian reef fauna. Although numerous interesting findings were made, including the Pacific's 'oldest atoll' (Ladd et al. 1974), the primary goal failed; it was evident from the abundant rudist-scleractinian reef materials recovered that the Cretaceous shallow-water fauna of the Central Pacific had nothing obvious to do with the contemporary reefs of the Hawaiian Archipelago. The fauna had become extinct and the islands supporting it had drowned (Hamilton 1956, Ladd et al. 1974, cf. Alvarez et al 1984 regarding worldwide abruptness of some of these extinctions).

There was only a vague idea, during the 1960s and early 1970s, of the age of the Hawaiian Archipelago and sea floor spreading and hotspot models were just beginning to appear. Only in recent years have we begun to understand the history of the Pacific Basin since the end of the Mesozoic, and only Nelson (1975: 501), Kay (1980), Rotondo et al. (1981) and Springer (1982) have applied it to Pacific Basin biogeography. Their principal conclusions are summarized below.

1.2 *Some of the latest views on Pacific Basin biogeography*

Kay (1980), in considering generalizations about the distribution of the biota of the Pacific Basin, prudently separates observations from explanations. She denotes three principal generalizations: 1) the affinities of the Basin flora and fauna are, by and large, with the west; 2) there is a decline in numbers of species and genera; 3) and this decline is concomitant with a gradual elimination of major groups of animals and plants from west to east. To these she adds four subsidiary generalizations: 1) distributions within the Pacific Basin are often disjunct and/or patchy; 2) certain marine as well as terrestrial biotas are frequently correlated with island type (high versus low island; cf. Figs.1, 2); and 3) while there is a recognizable Pacific Island biota, as far as marine invertebrates are concerned, 4) there has been relatively little radiation (species swarms, etc.).

To the first group of generalizations one can add that, where investigated (corals, Stehli & Wells 1971, coral barnacles, Newman et al. 1976; and shore barnacles — herein) there is an increase in average generic age away from the Indo-West Pacific center of distribution, and this increase is particularly evident from west to east.

Following the generalizations, Kay (1980) looks to explanations. Courtesy is paid to early 19th century thoughts on 'centers of creation' before moving on to Darwin/Wallace and dispersal, as did MacArthur & Wilson (1967), including the inference that the endemism associated with peripheral islands, such as those of the Hawaiian Chain, is a result of speciation after occasional chance dispersal (Mayr 1942, Gosline 1968, Carlquist 1981). Thus the primary three generalizations, and some of the secondary four, are attributed to dispersal and its eventual attenuation as a result of distance from western sources. However, there are aspects of the distributional patterns that are not explained by dispersal and Kay (1980) identifies three: 1) a recognizable Pacific Basin biota as well as 2) areas of endemism other than simply in peripheral regions, and 3) disjunct distributions within the Basin. Some of these she relates to ecological differences between high and low islands and the possible effects of changes in climate and sea level, but no specific examples are explored.

Kay (1980) finds vicariance theory (viz. the study of how present distributional patterns can be explained by barriers to dispersal interrupting once widespread distributions), difficult to apply in the Pacific where islands are the basis for distributional patterns. She goes on to provide a framework into which the general distribution of both the past and present can be fitted and concludes that there were two major phases in the history of the Pacific Basin biota: 1) the Mesozoic Pacific inhabited by a cosmopolitan Tethyan biota that essentially disappeared from the late Cretaceous through early Eocene, and 2) its replacement in the Eocene through late Miocene by a new biota, many elements of which occurred in the Eocene through Miocene of Europe (Fleming 1979). To my knowledge these two phases have not been generally understood by Pacific Basin biogeographers.

From this frame of reference and the primary generalities discussed above, Kay concludes that long-range dispersal is essential in accounting for most of the contemporary biota. She also mentions the possibility of 'island integration' (Rotondo et al. 1981), whereby Cretaceous islands and their biotas are carried by the northwesterly motion of the Pacific Plate over the Hawaiian Hotspot, thus explaining biotic similarities between Hawaii and the eastern South Pacific. But the possibility is not looked at critically nor are other explanations explored (see Kay 1984).

1.3 *Pacific Plate biogeography*

Springer (1982), in his masterful treatise, distinguishes islands of the Pacific Plate from Pacific Islands in general. He reviews much of what is known about the distribution of molluscs, arthropods, tetrapods, and plants and adds substantially to our knowledge concerning the distribution of the shore fishes. An avowed 'vicariographer', he nonetheless takes an eclectic approach by ignoring for the most part what is readily explained by dispersal. He focuses on endemics and defines three types: 1) endemics in numerous and widely separated island groups; 2) endemics limited to a few islands or island groups within a restricted area and found primarily on southern islands; and 3) single island or island group endemics (which he notes is the most common type of endemism and which predominates at high islands rather than low islands). He also considers, as did Kay (1980), the history of the Pacific Basin beginning 290 million years before present, prior to the breakup of Gondwanaland, but in explicit detail especially in the Cenozoic. He then explains the occurrence of the principal types of endemics by invoking vicariant events that were or may have been in progress during this critical time (Fig.1).

Springer (1982) considers the development of the Philippine portion of the Asiatic Plate as instrumental in creating an islandless void between the Philippines and the Mariana Islands to the east. This void created a barrier to dispersal that would account for 1)

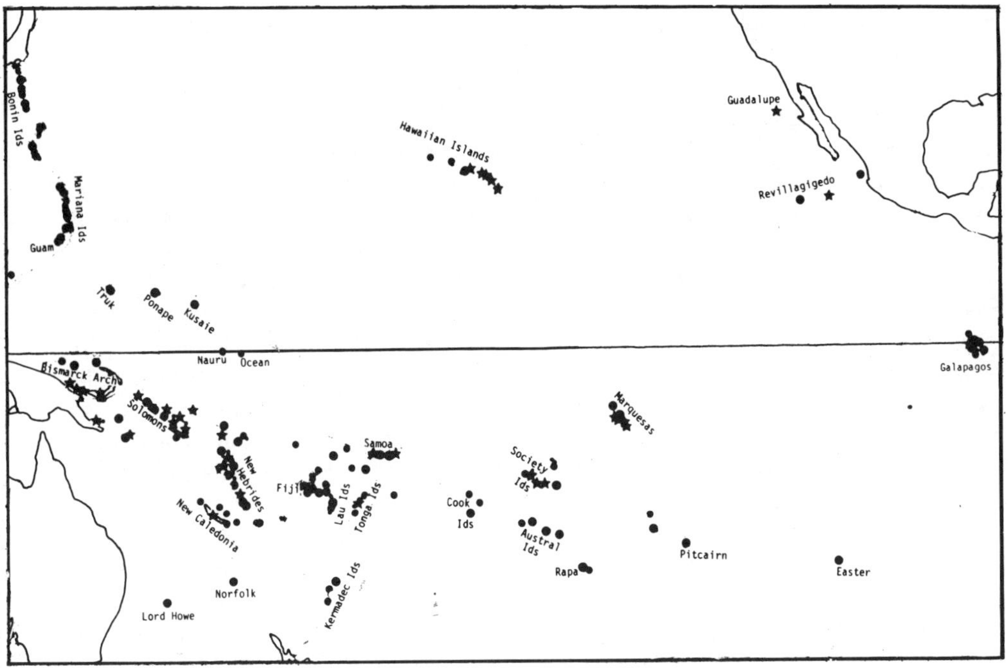

Figure 2. Relative isolation of the high islands of the Hawaiian Archipelago from other high islands of the Pacific Basin illustrated by the deletion of shallow banks and islands of less than 60 m elevation seen in Figure 1 (modified from Springer 1982).

the abrupt reduction seen in the biota as one moves east onto the primarily atoll-bearing portion of the Pacific Plate, and 2) the isolation considered necessary for the formation of the type 1 endemics. He then explores the more southern region where the reduction in numbers of taxa is gradual rather than abrupt as one progresses from New Guinea and Australia through the high islands to the east and onto the high islands of the Pacific Plate where most of the type 2 endemics are to be found. The changes in the configuration of the region and the ages of the islands involved are brought to bear and our attention is drawn to significant tectonic alterations. For example, Fiji was closer to the New Hebrides 9 to 10 million years ago than today, and the Tonga Island Arc was further from the Samoa Islands in the Miocene than today, if the Samoas were then extant.

For an explanation of the type 3 endemics, such as those found in Hawaii and Easter Island, Springer looks to 'island integration' (Rotondo et al. 1981). This hypothesis states that islands formed in the Cretaceous on the southeast Pacific margin of the Pacific Plate were carried northwest to the Hawaiian Hotspot where they and their biotas are further hypothesized to have become integrated into the Hawaiian Archipelago. Area cladograms are developed to determine whether competing hypothess can be excluded but, as is frequently the case in such attempts, Springer's results are ambiguous. Various explanations for the peculiarities of the Marquesas and their type 3 endemics are also explored, including limited dispersal (founder principle), but some sort of a vicariant explanation ends up being favored. A plausible historical oceanographic scenario is developed, but the problem, which includes the lack of well developed coral reefs there, remains unresolved.

Following this extensive and in-depth treatment, Springer (1982: 147) comes to the surprising (and disgruntling) conclusion that the Cenozoic history he has reviewed, and interpreted relative to the endemics of the Pacific Plate, seems to be of little consequence in explaining the endemism found there. Despite some reluctance, owing to the apparent Cenozoic nature of extant fish genera and the fact that Plate-associated endemics are mostly at the species level, Springer looks to earlier events that might bear on their origin. The scenarios developed are intriguing, but because they do not bear directly on Springer's fishes nor on the organisms I will be dealing with shortly, they will not be pursued here.

Springer (1982), mistakenly I believe, considers Kay (1980) to be a 'vicariance biogeographer' of the sort that considers that distributional patterns can be explained by events interrupting previously widely distributed biotas. But as noted above, although Kay was concerned over facets of Pacific Basin biogeography not readily explained by dispersal, dispersal was her primary explanation for most present day distributional patterns. Furthermore Springer does not consider the massive Cretaceous extinctions emphasized by Kay, even though he observes on several occasions the relative youth of fishes at the generic level. What is really puzzling is Springer's summary dismissal of the facts, and to some extent the demonstrated importance of climatic change and eustatic changes in sea level, especially in view of his recognition of: 1) the biotic differences between low and high islands; 2) the paleontological as well as geological evidence that, subsequent to having been volcanic islands, some contemporary low islands have at times been high limestone islands; and 3) the present marked isolation of the Hawaiian Archipelago from high island sources (Fig.2). There were substantial eustatic lowerings of sea level and concomitant extinctions and reinvasions during the Miocene and the Pleistocene, which also need to be considered, and these will be returned to shortly.

2 HAWAIIAN SHORE BARNACLES

A check list of Hawaiian shore barnacles is provided in Table 1. It includes the known distribution of the species, their larval forms, and the generic age (known or inferred) of the genus to which each belongs. The majority of the species are ordinary barnacles or thoracicans. There are a few acrothoracicans (cf. Tomlinson 1969), small relatives of ordinary barnacles found burrowing primarily in calcareous substrates and known to occur in the Paleozoic (Devonian; Rodriguez & Gutschick 1977). Other than the disconcerting fact that most species of acrothoracicans have wide geographic ranges despite any obvious means of dispersal, they will not concern us much in the following discussion. Suffice it to say that the Hawaiian acrothoracicans have their affinities with the Indo-West Pacific, as noted in Table 1.

In general the thoracican cirriped fauna includes: 1) one to three introduced species (6.25-18.75%); 2) one Pacific Plate species (6.25%) not counting Hawaiian endemics; 3) five Indo-West Pacific species (32.25%); 4) five cosmopolitan species (31.25%); and 5) two endemic species (12.5%). Thus, by a simple majority (3 + 4; 63.5%) the Hawaiian cirripeds are wide ranging eurytopic forms. The Pacific Plate species, including the two endemics, total 18.75%. With regard to genera, 11 (78%) of 14 are cosmopolitan while three (22%) are Indo-West Pacific.

Table 1. Shore and shallow water barnacles (Cirripedia) of the Hawaiian Archipelago. * Obligate commensals; ** probably introduced; *** introduced. 1. *K.tinkeri* is either a sister species or the same species as that found in Japan; 2. *O.indubia* has its closest relative in Japan; 3. *P.hawaiense* may equal *P.amygdalum* known from as far west as Madagascar; 4. *V.cookei* is either a sister species or the same species as that found in the Line Islands; and 5. *M.tanagrae* has its closest relatives at Clipperton I and the Galapagos (from Newman, in prep.).

	Distribution of sp.	Larval form	Generic age	My BP
THORACICA				
Lepadomorpha				
* *Paralepas palinuri* (Barnard)	S.Africa - Line Is.	Prob. naup.	?Mesozoic	200 (Palinura Trias.)
* *Microlepas diademae* Hoek	Western Pacific	?	?Mesozoic	200 (Diadematoids Trias.)
* *Koleolepas tinkeri* Edmondson	Endemic[1]	?	?Mesozoic	150 (Hermit crabs Juras.)
* *Octolasmis lowei* Darwin	E.Atlan. Indo - W.Pac.	Nauplius	Fam. U.Eo.	45
* *Octolasmis indubia* Newman	Endemic[2]	Prob. naup.	Fam. U.Eo.	45
* *Poecilasma hawaiense* Pilsbry	Endemic[3]	Prob. naup.	Gen. U.Eo.	45
Verrucomorpha				
Verruca cookei Pilsbry	Endemic[4]	Prob. naup.	Cretaceous	100
Balanomorpha				
Euraphia hembeli Conrad	Ceylon - Pitcairn I	Prob. naup.	?Eo.	45
Euraphia intertexta (Darwin)	W.Pac. - Pitcairn I	Prob. naup.	?Eo.	45
Tetraclitella divisa (Nilsson-Cantell)	Cosmo. except E.Pac.	Cyprid	?Oligo.	32
Tesseropora cf. *atlantica* Newman & Ross	Cosmo. except E.Pac.	Cyprid	Oligo.	32
Balanus trigonus Darwin	Cosmo.	Nauplius	Oligo.	32
** *Balanus amphitrite* Darwin	Cosmo.	Nauplius	Oligo.	32
** *Balanus reticulatus* Utinomi	Cosmo. except E.Pac.	Nauplius	Oligo.	32
*** *Balanus eburneus* Gould	W.Atlantic	Nauplius	Oligo.	32
Megabalanus tanagrae (Pilsbry)	Endemic[5]	Prob. naup.	Oligo.	32
ACROTHORACICA				
* *Weltneria hirsuta* (Tomlinson)	Japan	Cyprid	Fam. Dev.	–
* *Lithoglyptes mitis* Tomlinson	W.Pac.	Cyprid	Fam. Dev.	–
* *Cryptophialus unguiculus* Tomlinson	W.Pac.	Cyprid	?Mesozoic	–

2.1 *Lepadomorph/Balanomorph ratio*

A measure of balance in a cirriped fauna is the Lepadomorph/Balanomorph ratio and its correlation with latitude (Newman & Ross 1971). Observations for 10 localities (Spivey 1981) indicate that the ratio is generally greater than 1 at high latitudes and less than 1 at low latitudes, and the additional localities to be added here fit this trend. Quite simply, there is a greater diversity of balanomorph barnacles relative to lepadomorphs in shallow waters at low latitudes as compared to high latitudes, and the same trend holds when the comparison is made with deep water.

The ratio for the Hawaiian cirripeds (Table 1) ranges from .66, when taken at face value (15 spp.), to as much as 1 when the potentially introduced, as well as the introduced species are excluded. The ratio for Fiji is .33, but the species list (16 spp., Foster 1974) does not include coral balanomorphs known to be present there, which, if included, would render the ratio substantially lower. The cirripeds of Hong Kong are better documented (38 spp., Foster 1982) and yield a ratio of .23, as would be expected for a fauna from near the Indo-Malayan center of distribution.

Taking these data at face value, Hawaii is in keeping with the expected trend at the high latitude or peripheral extremes of the tropics. Not only is the number of species low, but some groups are conspicuous by their absence. However, because there are no obvious habitat restrictions, one can ask whether this is a result of extinction, isolation, or both, and we will return to this aspect of the problem shortly.

2.2 *Average generic age*

Average generic age (AGA) has been shown to increase with distance from centers of distribution; for corals (Stehli & Wells 1971) and for coral inhabiting balanomorphs (Newman et al. 1976). The same trend apparently holds for shore barnacles. The AGA for Hong Kong and Fiji is 47 and 56 million years (My) respectively, whereas that for Hawaii can be inferred to be in the order of 99 My (as calculated from the lists of Foster [1974, 1982] and Table 1 herein).

The high AGA for Hawaii, compared to the deep tropics, is a result of the higher proportion of lepadomorphs present (nine, compared to six balanomorph genera) and the absence of such common forms as the coral inhabiting barnacles (Pyrgomatidae).

Another reason for the high AGA for Hawaiian barnacles is the presence of a shallow water, indeed intertidal, verrucomorph species. The verucomorphs consist of the predominately deep-water genus *Verruca*, species of which are known from elsewhere in shallow water only in the Northeast Atlantic (Nilsson-Cantell 1978) and the southern Southeast Pacific (Nilsson-Cantell 1957). A shallow water to intertidal species in the tropics is therefore remarkable and because the genus is Cretaceous in age (Schram & Newman 1980), its presence in Hawaii substantially increases the AGA for thoracican barnacles found there.

It is noteworthy that all of the lepadomorphs in Hawaii are obligate commensals (*Microlepas* on diadematoid sea urchin spines; *Koleolepas* between the pedal disc of an anemone and a gastropod shell inhabited by a hermit crab; *Poecilasma, Octolasmis* and *Paralepas* variously distributed on large decapod crustaceans). If we exclude the lepadomorphs, because their presence is dependent upon that of their hosts, and the verrucomorph because it has no counterpart in the tropics, from consideration and concentrate on the balanomorphs, the AGA falls to 34 My. This will allow us to discuss them in connection with the hermatypic corals.

2.3 *Dispersal*

Dispersal capabilities in shore barnacles need to be discussed even though I agree with van Balgooy (1971) and others that many insular organisms are widely distributed despite any obvious long range dispersal capabilities. The barnacles are no exception in this regard. Their life history generally includes six naupliar stages and a cyprid larva. The nauplii generally feed in the plankton for two weeks or so before metamorphosing into the non-feeding cyprid whose primary function is to find a place to settle down permanently. The cyprid is not the sustained swimmer the nauplius is, and it must maintain itself on food reserves. Thus its tenure in the plankton is relatively short. In the tropics the cyprid's capacity to settle and metamorphose into the adult usually lasts but a few days to a week following its metamorphosis from a nauplius.

Given three weeks in the plankton and a one knot current (currents in the open Pacific are generally less, J.Reid, pers. comm.), the range for ordinary barnacles would be in the order of 500 nautical miles. A three knot current provides nearly a 1500 mile range but, other than at the beginning of the Kuroshio, such strong currents are not presently known to impinge on Hawaii (cf. Grigg 1981).

More enigmatically, at least five of the species found in Hawaii release their larvae as cyprids rather than nauplii (*Tetraclitella divisa, Tesseropora* cf. *atlantica, Weltneria hirsuta, Lithoglyptes mitis* and *Cryptophialus unguiculus*). If these species were endemic, the loss of larval dispersal capabilities would make sense (cf. Scheltema & Williams 1983, for discussion). However, they are wide ranging and therefore long-range dispersal by adult propagules must have recently, and may still be, taking place.

Some adult shore barnacles can disperse as adults. Fouling species are easily dispersed and if the two possibly introduced species of *Balanus* eventually prove to have reached Hawaii by natural means it will not be surprising. In fact *Balanus trigonus*, which is now known to have been in Hawaii since the Pleistocene (Resig 1969), has been found on the whale barnacle *Coronula* on whales, which effectively acting as ships (Newman & Abbott 1980), migrate great distances to Hawaiian waters. Therefore fouling species can apparently take advantage of active forms of transport. As for passive transport, I have found *Tetraclitella divisa* on the pelagic barnacle *Lepas* which, in turn attaches to wood, pumice and other floating objects such as the shells of *Nautilus.* The tendency of *Tetraclitella* to settle on *Lepas* as well as rocky shores may well account for its nearly cosmopolitan distribution as well as its occurrence in Hawaiian waters. Mollusks are also known to raft, an example being a marine gastropod native to the Juan Fernandez Islands that arrived at Easter Island on a piece of pumice (Rehder 1980). It is by way of their molluskan hosts that many of the shell-burrowing acrothoracicans may have been dispersed. Thus, while long range dispersal of shore barnacles by larval means seems remote, adults apparently have a number of means of being transported great distances.

2.4 *Endemic species*

As Springer (1982) has so aptly pointed out, endemics of oceanic islands are of special interest in biogeographical studies because, though they or their immediate ancestors once dispersed, the dispersal events were either very unusual or have been drastically curtailed. There are several Hawaiian endemics among the Lepadomorph barnacles. These include *Octolasmis indubia*, a species of a large genus apparently having its only close relative, *O.clavula*, in Japan (Newman 1961); *Koleolepas tinkeri*, belonging to a small genus having a representative in the Loyalty Islands and another in Japan (Edmondson 1951); and *Poecilasma hawaiense* which may actually be a junior synonym of an Indo-West Pacific species (Foster 1974); *Verruca cookei*, which may be represented by the same species or a sister species, in the Line Islands; and finally, the balanomorph, *Megabalanus tanagrae*, having its closest affinities with *M.clippertonensis* and *M.galapaganus* from islands of the Eastern Pacific, as indicated by their specific names (cf. Zullo 1969). *Megabalanus tanagrae*, therefore, appears to belong to a complex of Eastern Pacific peripheral species, but it is unknown whether the distribution of this group of *Megabalanus* also includes the eastern South Pacific.

2.5 *Conclusions regarding shore barnacles*

The majority (63.5%) of the Hawaiian shore cirripeds under consideration are wide ranging cosmopolitan or Indo-West Pacific species. There are three Pacific Plate species (18.75%), two of which are Hawaiian endemics. The non-endemic Plate species, known to occur elsewhere only in the Line Islands, is the only member of the genus *Verruca* known from intertidal waters of the tropics. One of the endemics, *Megabalanus*, has its closest relatives in the Eastern Pacific while the others have their closest affinities with the Indo-West Pacific. Western Pacific, South Pacific and Eastern Pacific contributions to the biota of the Hawaiian Archipelago have been noted by many authors (e.g. Zimmerman 1948, Croizat 1958, Carlquist 1981, Heads 1983), so it should not be surprising to find similar elements among the barnacles.

The Hawaiian shore barnacle fauna is depauperate. Its AGA is greater than that for Indo-West Pacific faunas near the Indo-Malayan center of distribution. Nonetheless, the Hawaiian shore barnacle fauna is, according to the so-called Lepadomorph/Balanomorph ratio, comparable in balance to other tropical localities in general. However, all of the lepadomorphs present are obligate commensals; free-living species of genera such as *Capitulum, Ibla* and especially *Lithotrya* found elsewhere in the Indo-West Pacific are absent. *Capitulum* (sensu Foster 1978) is known to extend onto the Pacific plate as far as Samoa, *Ibla* as far as the Marshall Islands and *Lithotrya* as far as the Line Islands. Therefore, only the absence of the last from Hawaii is notable.

The AGA of the Hawaiian balanomorph barnacles is higher than expected owing in good part to the absence of the coral-inhabiting balanomorphs (Pyrgomatidae) known from elsewhere in the tropics since the Miocene. The Hawaiian shore barnacle fauna is thus not as balanced as the Lepadomorph/Balanomorph ratio would suggest, unless there are habitat restrictions that would prevent the missing elements from existing there. However there has been at least one introduction (*Balanus eburneus*) and possibly two more (*B.amphitrite* and *B.reticulatus*), suggesting that open niches are relatively common. The openness of niches could be the result of a climatic amelioration rather than, or in addition to, under-saturation resulting from isolation from a source fauna.

Shore barnacles are generally not well equipped for long-range dispersal as larval forms and a number of widely distributed species found in Hawaii have abbreviated larval series in which the planktonic nauplius is deleted. Such forms hatch out as cyprid larvae which, as far as is known, have poor dispersal capabilities at best. However, various methods of rafting have been observed and are available to adults, especially in the so-called fouling forms, and in obligate commensals that can be carried about by their hosts. Thus some of the Hawaiian shore barnacle fauna can be explained by long range dispersal. However, the chances for such dispersal seem slim for some species, especially the obligate commensals of decapod crustaceans because the hosts generally disperse as larvae rather than adults. Therefore, any vicariant explanation consistent with these facts would be welcome. The same problem is encountered, although perhaps not as severely, with the Hawaiian corals, and it will be returned to below.

Table 2. Hermatypic coral genera of the Hawaiian Archipelago. * *Acropora* apparently reinvading Hawaiian waters; ** *Acropora* cf. *valida* recently discovered at Gorgona I, Colombia (Wells & Glynn, pers. comm.). [1] = extinct genera. Data from Heck & McCoy 1978, Griggs 1983, Maragos 1977, Wells 1982, 1983.

	Generic age	(My)	Distribution
Extant			
*Acropora**	Eo.	47	[1]Europ.; W.Atlan. & Indo-Pac.**
Cyphastrea	Oligo.	32	W.Atlan. & Indo-W.Pac.
Fungia (Scutaria)	Mio.	12	Indo-W.Pac.
Leptastrea	Eo.	47	[1]E.Pac. Indo-W.Pac.
Leptoseris	Eo.	47	Cosmopolitan
Montipora	Paleo.	60	[1]W.Atlan.; Indo-Pac.
Pavona	Oligo.	32	[1]Europ. & W.Atlan.; Indo-Pac.
Pocillopora	Paleo.	60	Indo-Pac.
Porites	Eo.	47	Cosmopolitan
Porites (Synaraea)	Oligo.	32	Cosmopolitan
Psamnocora (Stephanaria)	Mio.	12	[1]W.Atlan.; Indo-Pac.
Extinct			
Acropora	Eo.	47	see *Acropora* above
Favia	L.Cret.	120	[1]E.Pac.; Atl. & Indo-W.Pac.
Favities	U.Cret.	70	[1]Atlan. & E.Pac.; Indo-W.Pac.
Galaxea	Mio.	12	[1]N.Amer.; Indo-W.Pac.
Gardeneroseris	Pleisto.	2	Indo-Pac.
Platygyra	Eo.	47	[1]Eur., W.Atlan. & E.Pac.; Indo-W.Pac.
Plesiastrea	Mio.	12	[1]Eur., W.Atlan. & E.Pac.; Indo-W.Pac.
Seriatopora	Mio.	12	Indo-W.Pac.
Stylophora	Eo.	47	[1]Eur., W.Atlan. & E.Pac.; Indo-W.Pac.

3 HERMATYPIC CORALS

It has been demonstrated that the AGA of the hermatypic corals of the world increases from centers of distribution to the periphery and that concomitantly the AGA of the Hawaiian hermatyic corals is higher than that of the Indo-Malayan region (Stehli & Wells 1971). The AGA, approximately 39 My, is similar to but somewhat greater than that for the balanomorph barnacles of Hawaii (34 My).

Hawaiian reefs were better developed as geomorphic features in the Pleistocene than they are today (Menard et al. 1962, Maragos 1977); the northwestern atolls (Midway & Kure) are presently approaching the 'Darwin point' (the point where growth rate can no longer keep pace with erosion and subsidence under prevailing climatic conditions, Grigg 1982, 1983), and only 11 of 19 coral genera recorded from Hawaii are extant (Wells 1982, pers. comm., cf. Table 2). The Midway extinctions were rather abrupt, stratographically, and were presumably a result of climatic deterioration. The Hawaiian Hotspot has not moved appreciably, certainly since the Oligocene if not the Jurassic (cf. Rotondo et al. 1981) and therefore new islands of the chain have been forming at the same latitude. That climate has improved is attested to by the apparent reinvasion, perhaps from Johnston Island, of three species in one of the genera that went extinct (*Acropora*) (Grigg 1981, Grigg et al. 1981). However, the evidence suggests that the climatic situation is still suboptimal because the species of *Acropora* are apparently incapable of breeding in the cooler, leeward reaches of the Hawaii Archipelago.

It is instructive that four of nine coral genera that went extinct in Hawaii (*Favites, Favia, Stylophora* and *Plesiastrea*) also went extinct in the Eastern Pacific. However the Cretaceous genera, *Favites* and *Favia*, went extinct in the Eastern Pacific during the Cretaceous, while the Eocene *Stylophora* and Miocene *Plesiastrea* remained until the Pleistocene (Heck & McCoy 1978). On the other hand, *Gardeneroseris*, extant in the Indo-West Pacific and Eastern Pacific but not Hawaii, has a fossil record nowhere else but Hawaii where it appeared and disappeared in the Pleistocene (Wells, pers. comm.). Also instructive is the fact that *Acropora*, extinct at Midway following the Miocene and Oahu in the Pleistocene, but presently represented by three species in the central reaches of the archipelago, has been found in the Eastern Pacific for the first time (Gorgona Island, Colombia, Wells, pers. comm.). There is no fossil record *Acropora*, an Eocene genus, in the Eastern Pacific (Heck & McCoy 1978, Wells 1983). The Colombian species is probably *A.valida* (Glynn, pers. comm.), an Indo-West Pacific species and one of the three presently apparently invading the central reaches of the Hawaiian Archipelago from the southwest (Grigg 1981). It would be very interesting to know whether *A.valida* has long been resident on Gorgona Island or has recently arrived there following climatic amelioration since the Pleistocene.

The important thing to note is that contemporary hermatypic corals of Hawaii are, for the most part, a replacement rather than an attenuated remainder of the Miocene fauna (Wells 1982). The principal hermatypic species (about half the total) include all genera present in the archipelago and they inhabit the entire range of islands from the small atolls of Kure and Midway in the northwest to the high islands of Oahu and Hawaii to the southeast. These species, except for the three species of *Acropora* restricted to the central reaches mentioned above, are the dominant reef forms encountered on all reefs and are therefore considered wide ranging eurytopic forms (Grigg 1981). Grigg's evidence indicates that the species of *Acropora* in Hawaii do not produce larval propagules and therefore are presently incapable of spreading throughout the archipelago.

There are no endemic hermatypic coral genera in Hawaii. Of the 11 extant genera, eight (73%) are cosmopolitan and three (27%) are only known from the Indo-Pacific. The proportion for the extinct genera, of which there are eight, is the same. These proportions are also similar to those for the barnacles although no genera of the latter are yet known to have become extinct in Hawaii.

3.1 *Average generic age*

The AGA of the hermatypic corals of Hawaii is approximately 39 My. The cosmopolitan genera average a little older than the Indo-Pacific genera; i.e. 37.6 and 34.3 My respectively. Although this trend is in keeping with the general observation that older genera are generally more wide ranging than younger genera, the difference here is hardly significant. However, there is a substantial number of extinct coral genera in Hawaii that are extant elsewhere, and therefore an additional exercise concerning age and area is possible. But before approaching it, there is an aspect of AGA regarding hermatypic corals of the world that has not been pointed out.

It has been observed that 1) wide ranging genera generally have a greater AGA than short range ones and 2) that the AGA of Atlantic corals is greater than those in the Pacific (Stehli & Wells 1971). We know that the Atlantic reefs have undergone considerable distributional contraction since the Oligocene (Newell 1971), as have the reef corals of the Eastern Pacific (Heck & McCoy 1979) and of the Hawaiian Islands (Menard et al. 1962, Maragos 1977, Wells 1983, herein). This is apparently a result of climatic deterioration an concomitant events such as the break-up of Tethys. Under such conditions, wide ranging species (and the genera they represent) would, on the average, be less susceptible to extinction than short range species because they are more eurytopic. Thus, up to a point, the AGA of the communities involved should increase as the geographic area of suitable habitat decreases. Because the tropical Atlantic has deteriorated proportionately more than the Indo-West Pacific, there should have been more extinctions and therefore the AGA there should be greater, which it is. An alternative hypothesis would be that despite deterioration in both oceans, there has still been the evolution of more young genera in the Pacific than in the Atlantic and this accounts for its general, overall lower AGA. This may be part of the story, but as we shall see, it is not the primary one as far as the Neogene history of Hawaii is concerned.

We can now return to Hawaii where a well-documented change in coral diversity took place a few million years ago, apparently owing to its marginal high latitude situation and the climatic deterioration during the Plio-Pleistocene. The species most commonly found on Hawaiian reefs range throughout the entire archipelago, from 20°N to 29°N, and they represent all genera present except *Acropora* (Grigg 1982, 1983). We have already noted that the AGA of extant hermatypic coral genera found there is approximately 39 My. Despite the fact that *Favia* and *Favites*, the only Cretaceous genera known from Hawaii, are among those that have become extinct, the AGA for the extinct genera is only 37.4 My (Table 2). As already noted, *Favia* and *Favites* also went extinct in the Eastern Pacific (Heck & McCoy 1978); the former, with several species in the Miocene but only one today, is presently doing relatively poorly in the Western Atlantic (Wells, pers. comm.) while the latter also became extinct in the Western Atlantic. Thus these two genera are apparently on the verge of becoming Indo-West Pacific Cretaceous relicts and therefore one can ask what the effect would be if they were removed from the calculation? Removal brings the AGA of the extinct Hawaiian genera down to 25.6 My, reflecting the fact that three Miocene genera present became extinct. That the extinction was primarily among the younger genera is supported by the fact that the only fossil record for the Indo-Pacific genus *Gardeneroseris* is the Pleistocene of Hawaii (Wells, pers. comm.).

3.2 *Dispersal*

Corals apparently are well equipped for dispersal by larvae; as noted by Grigg (1981) for three species of *Acropora* from Johnston Island or possibly Wake Island, and by Richmond (1981) concerning long range dispersal capabilities for *Pocillopora damicornis* from laboratory experiments. These species are widely distributed in nature. Furthermore, corals apparently are not limited to dispersal by larvae: I have seen small coral colonies on seeds, pumice and glass fishing floats that have washed ashore in the Caroline and Marshall Islands, and Jokiel (1984) reports on pumice of known origin that had travelled thousands of kilometers before arriving in Hawaii carrying substantial colonies of *Pocillopora.* Thus, corals that 'foul' are potentially capable of being transported great distances as adults.

Dispersal must have played an important part in the formation of the contemporary Hawaiian hermatypic coral fauna, as attested to by corals taken on Midway (Wells 1982). As Wells notes, six of the eight genera present were replaced by nine contemporary genera. Apparently they arrived from elsewhere by long range dispersal.

If the marine climate of the Hawaiian archipelago is presently comparable to what it was before the Pleistocene, the hermatypic coral fauna is but one portion of an undersaturated marine biota, as suggested by short species lists and attested to by the recent reinvasion by three species of *Acropora* as well as by introductions such as barnacles, mangroves and the mangrove-associated crab *Scylla serrata.*

4 CORAL INHABITING BARNACLES

It was noted earlier that the coral inhabiting balanomorphs (family Pyrgomatidae) are not present in Hawaii. There is one species known from the Line Islands and the number of genera and species increases dramatically progressing westward on the Pacific Plate (Ross & Newman 1973). These barnacles have a well-documented fossil record going back into the Miocene in both the Western Atlantic and the Indo-West Pacific (Newman & Ladd 1979a, b). Were they never present in Hawaii or did they become extinct?

There are four genera of coral balanoids in the Indo-West Pacific that occur on four of the coral genera known to have become extinct in Hawaii (cf. Ross & Newman 1973, Table 3 herein). *Acropora* is indicated as extinct rather than extant, in Table 3, because it is believed to have reinvaded Hawaii since its extinction in the Plio-Pleistocene (Grigg 1981). Thus *Savignium*, the only coral barnacle genus known to range as far east as the Line Islands, lost two out of three of its potential coral hosts following the Miocene. Therefore it is possible that extinction rather than a lack of dispersal capability accounts for its absence in Hawaii. This possibility is heightened by the fact that the present coral fauna of Hawaii is evidently a replacement of a previous fauna rather than a depauperate remnant from the Miocene (Wells 1982). Yet, the relatively poor larval dispersal capabilities of barnacles in general, the very low probability of coral balanoids being dispersed as adults by corals on which they are obligate commensals, and the rapid attenuation in the number of genera and species from west to east on the Pacific Plate, suggests they never reached Hawaii. Coral balanoids were commonly encountered, as far back as the

Table 3. Hermatypic coral genera of the Hawaiian Archipelago known to host coral balanoids (Pyrgomatidae) elsewhere in the Indo-West Pacific, illustrating the possibility of extinction of all coral balanoids in Hawaii except *Megatrema* (data from Ross & Newman 1973). (*Acropora* shown as extinct but not extant because it is considered to have recently reinvaded Hawaii – cf. Table 2).

	Extant								Extinct			
	Cyphastrea	*Leptastrea*	*Leptoseris*	*Montipora*	*Pavona*	*Pocillopora*	*Porites*	*Psamnocora*	*Acropora*	*Favia*	*Galaxea*	*Stylophora*
Megatrema						●		●				
Savignium						●			●	●		
Cantellius	●		●	●	●	●	●	●	●			●
Nobia	●						●				●	
Creusia		●				●				●	●	

Miocene, in the cores drilled at Enewetak (Newman & Ladd 1974a). However, none were recovered from the drillings of Midway in which advanced techniques were used and core recovery was excellent (Gross et al. 1969). A single sample of a coral balanoid in the fossil record of Hawaii is all that would be needed to falsify the latter hypothesis.

5 VICARIANT AND DISPERSAL EVENTS

From the foregoing discussion of balanomorph barnacles, corals, and generic age considerations, and from what Springer (1982) has said about fish genera, it is apparent that we are dealing with primarily a widely distributed Indo-Pacific fauna predominantly Tertiary in age. This fauna, in order to get to Hawaii must have arrived by substantial dispersal because it is too young to have been left over from the Cretaceous. However the story is by no means as simple as this because as Kay (1980), Rehder (1980) and Springer (1982) have noted with regard to marine animals, and van Balgooy (1971) with regard to terrestrial plants, the Hawaiian biota shares marked affinities with the South Pacific; in a general way, the connection seems to be or have been through the Eastern rather than the Western Pacific. Carlquist (1981) has detailed the likelihood that birds have been involved in the long-range dispersal of many terrestrial plants, but some other explanation is needed for marine organisms. Rehder (1980) utilized an extinction hypothesis; namely that the Central Pacific relatives of the two faunal extremes have been replaced by modern invaders. Springer (1982), on the other hand, looked to 'Island integration', whereby Cretaceous islands that formed near the East Pacific Rise to the southeast of Hawaii were subsequently carried by the Pacific Plate into the vicinity of the Hawaiian Hotspot where they and their biotas were integrated into the Hawaiian Archipelago.

Newman & Foster (1983) reject island integration because there is no evidence that Wentworth Seamount or the Cretaceous portion of Necker Island had ever been islands. There are no early Tertiary, much less Cretaceous, high islands on the Pacific Plate today; subaerial erosion and subsidence over the time period involved precludes their existence. Furthermore, there are presently no islands of any sort on the Pacific Plate in a position to be integrated with the Hawaiian Archipelago in the next 60 My or so (Fig.1). Those we know were high islands in the Cretaceous and early Tertiary are presently guyots (e.g. Mid-Pacific Mountains; Emperor Seamounts: there is no evidence for Tertiary islands in the Mid-Pacific Mountains; Grigg's (1983) mention of them was a slip of the pen; Grigg, pers. comm.) or atolls (e.g. Line Islands; Marshall Islands). Therefore if an ancient island had been integrated into the Hawaiian Archipelago in the past 30 My or so it would have to have been an atoll and as such it should have survived to the present. Necker and Wentworth fail on this count, and Johnston is the only island in the region we know of that even came close. Nonetheless, as improbable as it may be, it is still possible that one or more integrated atolls lie buried under the low leeward islands of the Hawaiian Archipelago and this could be tested by drilling.

The most damaging argument against island integration is that it is not sound on biological grounds. The balanomorph barnacles and extant coral genera in the Hawaiian Archipelago belong to Tertiary, not Cretaceous genera and therefore could not have become part of the Hawaiian fauna by integration of Cretaceous island from the eastern South Pacific. Springer (1982) notes that his fish genera are, for the most part, Eocene if not more likely Miocene in age and that the similarities with the eastern South Pacific are

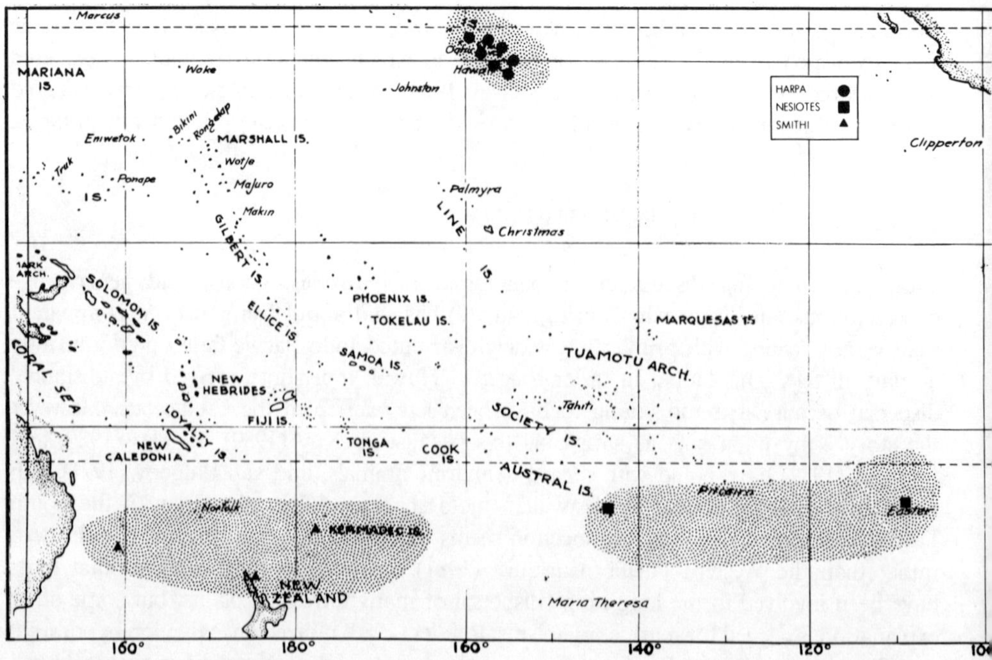

Figure 3. The amphitropical distribution of the three species of the marine gastropod genus *Neothais* (modified from Rehder 1980).

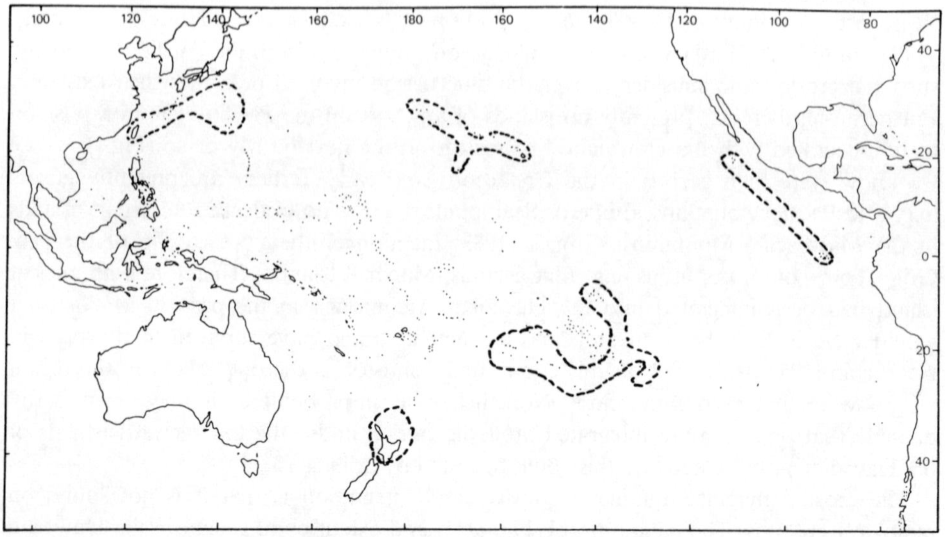

Figure 4. The amphitropical distribution of pulmonate land snail genus *Tornatellides* (modified from Peake 1981). Note the Eastern Pacific component. Possible occurrence at Johnston Island without Peake's confirmation.

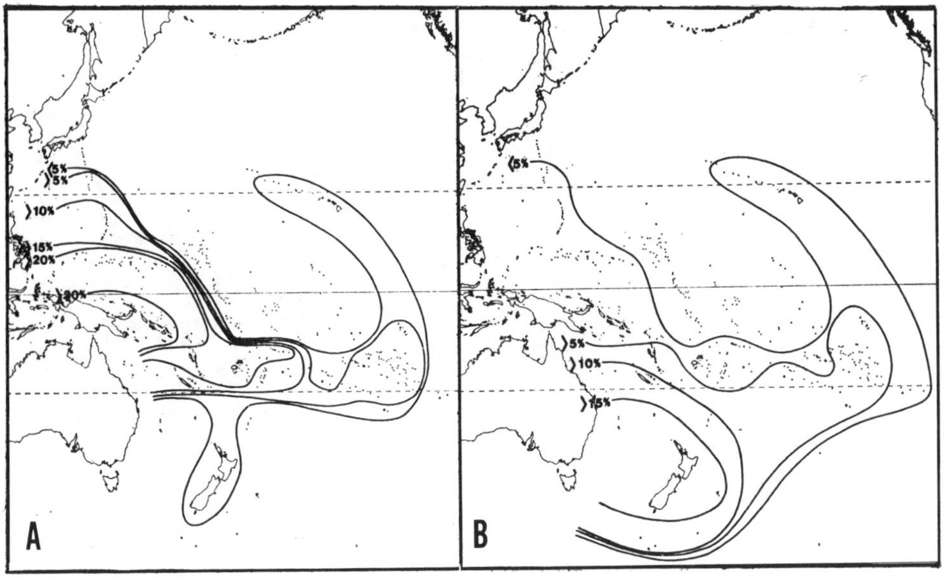

Figure 5. The amphitropical distribution of certain vascular plant genera (from van Balgooy 1971).

primarily at specific rather than generic or higher taxonomic levels. Therefore he cannot utilize island integration either. Even if island integration could be demonstrated, it doesn't matter because it doesn't explain the biology of the situation. But the biotic similarities between the South Pacific and Hawaii exist, and we can't ignore the question of how they came to be.

The extinction hypothesis, utilized by Rehder (1980), could help explain the high degree of endemism in the South Pacific, but at first glance it does not readily explain the similarities between the eastern South Pacific and Hawaii because of the notable isolation of the latter. I think the answer to this demise has already been touched upon by Springer (1982) but not applied because he did not fully evaluate eustatic sea level changes and concomitant climatic and oceanographic events; they impinge upon the problem to a significant degree.

5.1 *Affinities of the South Pacific with the Hawaiian Archipelago*

Springer (1982) has stressed certain affinities between the faunas of the eastern South Pacific and the Hawaiian Archipelago. The affinity actually involves Southern Oceania, including Australia and New Zealand, as far as the Pacific is concerned. This is well illustrated by a number of fish (Randall 1981), by the gastropod mollusk *Neothais* (Fig.3, Rehder 1980), certain marine fungi (Kohlmeyer 1983), land snails (Fig.4, Peake 1981), and the terrestrial plants (Fig.5, van Balgooy 1971). The monotypic fish genus *Sectator* (cf. Rosenblatt 1967, Springer 1982) as well as the land snail *Tornatellides*, also illustrate an Eastern Pacific component of what is otherwise an amphitropical situation, one which will be returned to shortly. The amphitropical aspect of the problem is strikingly illustrated by certain oceanic zooplankton (Fig.6, Brinton 1962, 1975) and by an oceanic water strider

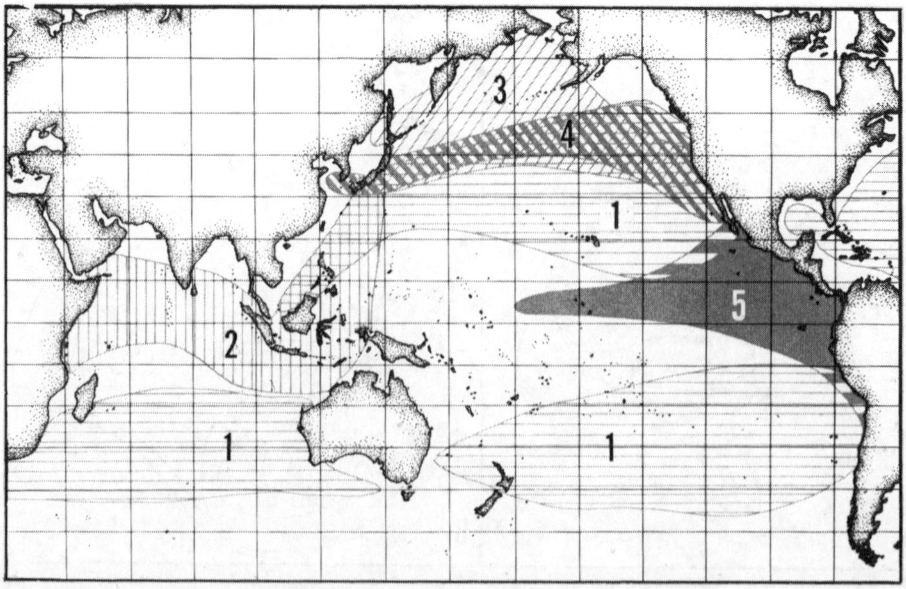

Figure 6. The distribution of five species of the planktonic copepod genus *Eucalanus* (*E.hyalinus* group). Note the amphitropical distribution of *E.hyalinus* in the Pacific and the close proximity of the two populations in the region of overlap with the Eastern Pacific species, *E.inermis*(after Fleminger & Hulsemann 1973, from van der Spoel & Heyman 1983). (1) *E.hyalinus*; (2) *E.elongatus*; (3) *E.bungi bungi*; (4) *E.b.californicus*; (5) *E.inermis.*

(Fig.7, Cheng & Shulenberger 1980), both examples noted by Springer (1982, 'antiequatorial').

I mention the planktonic situation not just for the sake of completeness, but rather with two words of caution. The first is that for certain zooplankton and water striders, as for fish and marine fungi, the amphitropical taxa are considered to be the same species and, if so, communication between the two regions was either relatively recent or is still going on. The second is that the following discussion concerns vicariant events relative to insular biotas and I can see no reason to suspect, if there had been no islands, that the distribution of the oceanic amphitropical zooplankton and water striders would be much (if any) different that it is today. Nonetheless, the amphitropical pattern for the insular biotas and the plankton must be coupled because, in addition to some having shared or sharing the same current systems in getting about, they share the same marine climates.

In explaining the amphitropical pattern in the plankton Brinton (1962, 1975) and Reid et al. (1978) look primarily to events in the Eastern Pacific, with the direction of exchange from south to north during the Pleistocene. Springer (1982) looks to exchange in the same direction for the insular shore fauna, as does van Balgooy (1971) for the insular flora.

6 AN INTEGRATED HYPOTHESIS

There is considerable merit in the 'oceanographic' explanation provided by the zooplankton

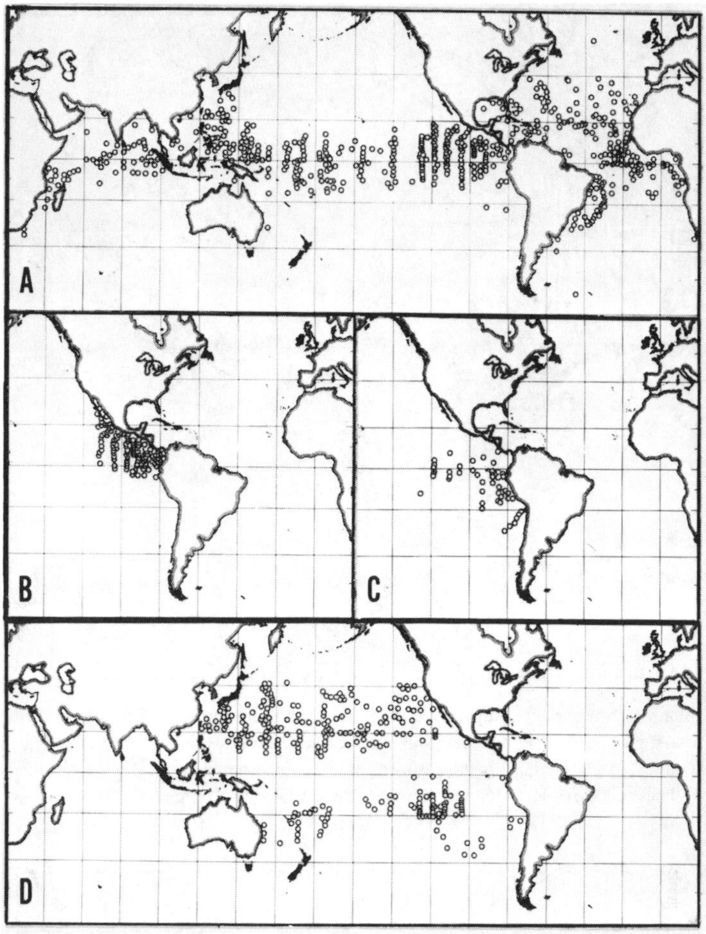

Figure 7. The distribution of four species of the marine water strider genus *Halobates*. Note the trans-pacific, amphitropical distribution of *H.sericeus* (D), the Eastern Pacific amphitropical distribution of *H.sobrinius* (B) and *H.splendens* (C), and the tropical distribution of *H.micans* (A) (modified from Cheng & Shulenberger 1980).

and physical oceanography; namely, the present amphitropical pattern was established during the Pleistocene, and perhaps also in the Miocene, via south to north oceanographic connections in the Eastern Pacific. However, the situation should be explored further because as satisfying as this explanation might be, much of the shore fauna and certainly the terrestrial biotas do not have the dispersal capabilities of the plankton, and the distances that must presently be spanned are immense. If the oceanographic explanation could be supplemented with some related phenomenon favoring the transport of southern insular species to the north, it would be more acceptable. Because there is no possibility of appreciably increasing the number of islands previously or presently existing between the two regions in the time frame involved, some alternative explanation is necessary.

6.1 *The Pleistocene*

The strengthening of circulation and concomitant reductions in sea-surface temperature in the eastern boundary and equatorial currents (Fig.8, Moore et al. 1980) apparently

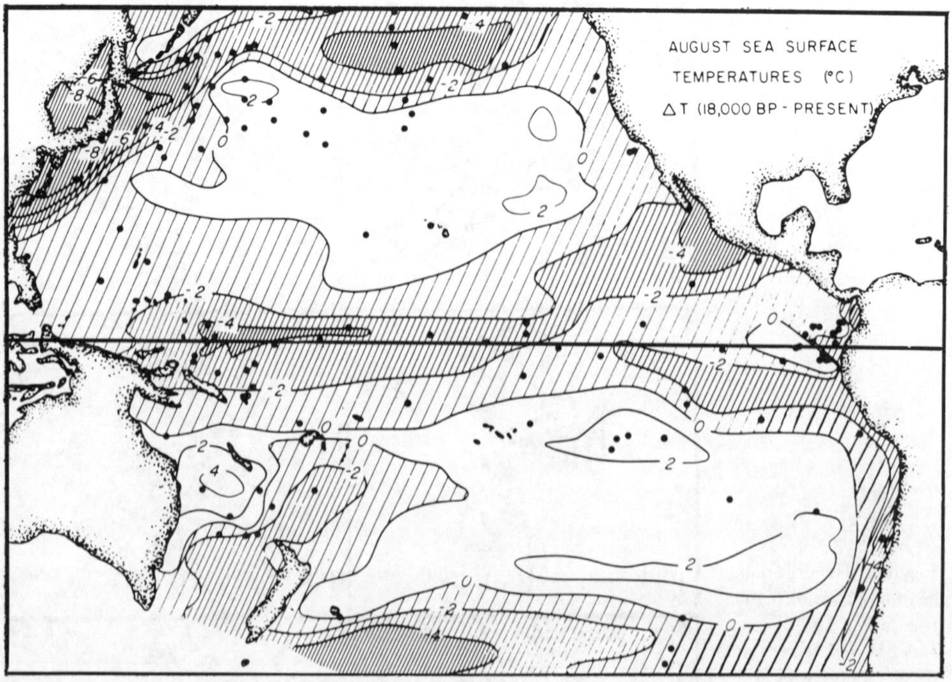

Figure 8. Sea surface temperature differences between August, 18 000 years before present and the averages for August in the present. Contour interval for isotherms is 2°C, light shading indicates 0-2°C, intermediate shading 2-4°C, and dark shading more than 4°C cooler than present. (Dots indicate sampling locations; continental outlines reflect sea-level 150 m below present level; from Moore et al. 1980).

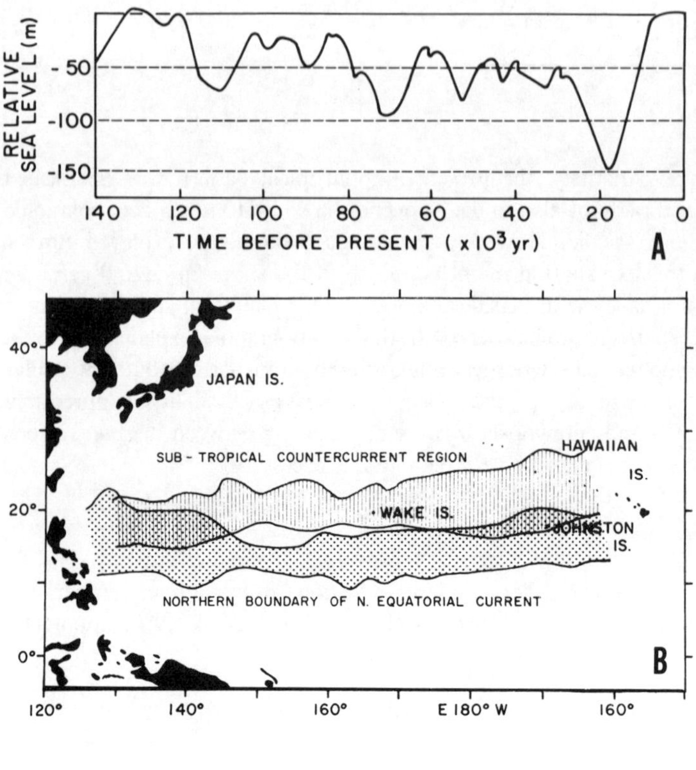

Figure 9. (A) Eustatic sea-level fluctuations since the last interglacial illustrating the low stand of approximately 150 m 18 000 years before present and the rapid rise to present level a few thousand years ago (from Potts 1983). (B) Geographical location of the Subtropical Countercurrent and the North Equatorial Current in the western Pacific (from Grigg 1981 after Robinson 1969).

afforded a mechanism for the establishment of amphitropical distributions for the zooplankton just noted. In addition, the Pleistocene saw several dramatic lowerings of sea level. Are these important factors in explaining the insular distributional patterns in the Central Pacific, especially those that are amphitropical?

6.2 *Sea level*

Substantial eustatic lowerings of sea level in the Cenozoic were apparently not limited to the Pleistocene (Vail et al. 1977), but because those of the Pleistocene are the best known and of a time frame sufficient to include speciation events, the Pleistocene low stands will suffice for the moment. Eustatic changes in sea level can be relatively rapid compared to wide ranging isostatic ones, and they impinge on shores around the world at the same time. Springer (1982) apparently did not appreciate this or he would not have so summarily dismissed Briggs (1974) on the matter.

The last eustatic low stand was but 18 000 years ago and was in the order of 150 m below the present level (Fig.9, Stoddart 1971, Potts 1983). The important point to note, in addition to the fact that present day low islands were then high islands, is the rapidity of the subsequent rise during which these high limestone islands underwent

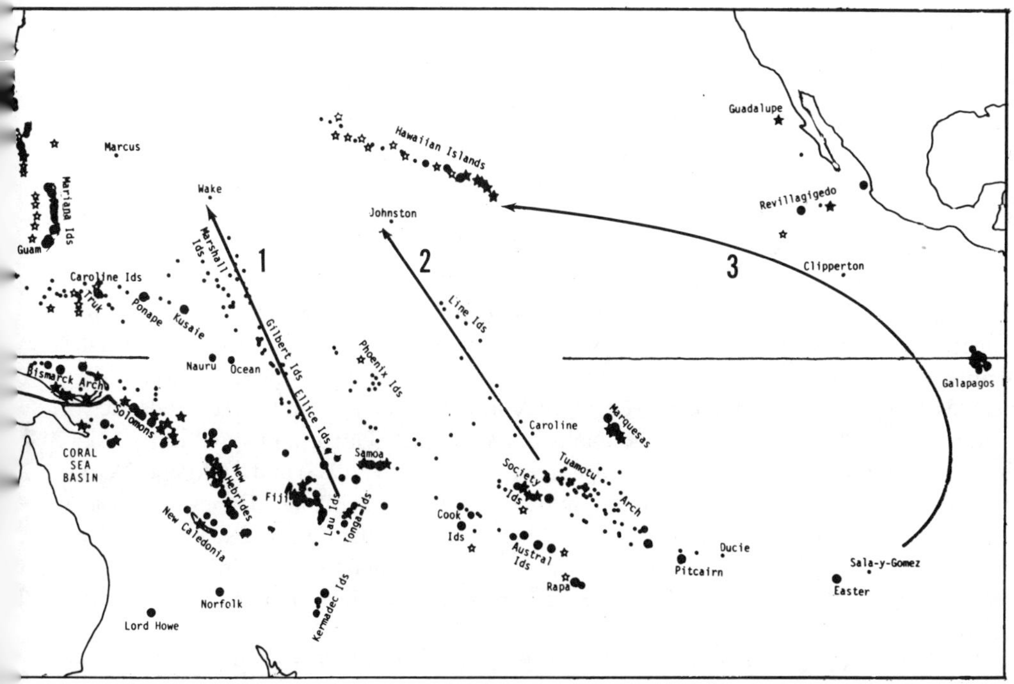

Figure 10. Transequatorial high island routes between the South Pacific and the Hawaiian Archipelago during late Cenozoic, eustatic low stands of the sea and concomitantly cooler sea surface temperatures across the tropics (cf. Figs.8 and 9a): (1) the Ellice, Gilbert an Marshall route to Wake and the region of the North Subtropical Countercurrent; (2) the Society-Tuamotu and Line route toward Johnston; and (3) via the Eastern Pacific and the North Equatorial Current.

Table 4. Longevity of 270 Western Pacific island marine gastropod mollusk species having a fossil record, illustrating that 1. most species (approximately 85 %) are apparently Pleistocene in age and that 2. significant Plio-Pleistocene extinctions occurred (data from Ladd 1982).

144 living species having a fossil record:		
Holocene	3	2.1 %
Pleistocene - Holocene	122	84.7 %
Pliocene - Holocene	14	0.7 %
L.Miocene - Holocene	1	
E.Miocene - Holocene	4	
126 species that went extinct:		
Holocene	1	0.8 %
Pleistocene	56	44.4 %
Pliocene	43	34.1 %
Miocene	23	18.3 %
L.Eocene	3	2.4 %

submergence in less than 10 000 years, and those that did not keep pace were drowned. It is also important to note that this was a period of extinction and speciation. We have already mentioned the partial, if not total, extinction of the Midway Miocene hermatypic coral fauna and its replacement by predominantly Indo-West Pacific species. It is further instructive that for the 270 Western Pacific marine gastropod mollusks having a fossil record studied by Ladd (1982, Table 4), 56 of the 126 extinctions (44.4%) known to have occurred since the Eocene were in the Pleistocene, and of the 144 extant species, 122 (84.7%) appeared in the Pleistocene. Only five (3.5%) of the 144 extant species had survived from the Miocene. Furthermore, Ladd found no extant species that had survived since the Eocene, although three species (2.4% of the total) are known to have become extinct during this epoch. The absence of any Oligocene records may be additionally instructive because the Oligocene, as well as the Miocene, was also a period of substantial lowerings of sea level (Vail et al. 1977).

A *contemporary* eustatic lowering of sea level of 150 m would: 1) change atolls into rimmed elevated limestone islands with no or negligible marine lagoons but with substantial opportunities for estuarine and fresh water conditions comparable to but better developed than seen on Fanning, Washington an Palmyra Islands today; 2) change 'drowned atolls', such as Mogami and Greyfeather west of Truk in the Caroline Islands, into moderately elevated table islands; and 3) change banks, such as Shoal Guyot and some other seamounts of the Sala y Gómez-Nazca Ridge (Newman & Foster 1983) and Ladd and Nero Banks just north and south of Midway respectively, into low islands. The event would not only dramatically increase the number of high islands in the Central Pacific, it would significantly increase land area, and provide or expand habitats suitable for high island and estuarine biotas over vast reaches of the Central Pacific where primarily atolls occur today.

The first substantial lowering early in the Pleistocene, nearly 2 My ago, would have been even more dramatic because limestone islands were for the most part table reefs rather than atolls (Tayama 1935) (most contemporary atolls are subaerial erosion forms developed during the Pleistocene (MacNeil 1954), rather than the product of gradual subsidence as Darwin (1842) had postulated). This also would have been the situation during the Oligocene and Miocene lowerings of sea-level detailed by Vail et al. (1977).

Curiously, the Oligocene is missing from the fossil record over much of the Pacific, but land snails and rain forest vegetation were present on Bikini in the Miocene (Solem 1981) as were land snails on Midway at the same time, as noted by Springer (1982).

During such low stands, elevated limestone structures are modified and degraded primarily by subaerial erosion (MacNeil 1954, Purdy 1974, Steers & Stoddart 1977). The rate of subaerial erosion will depend upon the morphology and composition of an exposed limestone island, as well as upon annual rain fall (Hutchinson 1950). The distribution of rain fall can differ by an order of magnitude over relatively short distances in the open ocean (e.g. approximately 600 mm/yr in the vicinity of Christmas versus approximately 5000 mm/yr in the vicinity of Palmyra, Line Islands, cf. Taylor 1973). The rate of subaerial erosion will also depend on the degree to which atmospheric circulation and concomitant adiabatic processes are altered by the elevation and form of individual islands. Therefore, reefs elevated by a eustatic lowering of sea level will undergo varying degrees of subaerial erosion unless sufficient time elapses for their complete truncation. Because atoll form is for the most part the result of subaerial erosion and because atolls are the predominant reef form today, it is evident that few reefs were completely truncated during the Pleistocene. Those that were failed to keep pace with the following rise of sea level were temporarily or permanently drowned (e.g. many present day atolls and all of the aforementioned 'drowned atolls' and banks). Concomitant with these events extinctions must have occurred, being in some cases total on land and to lesser degrees intertidally and in shallow water.

6.3 *The biota of Hawaii and the South Pacific*

The biological impact of changes in sea level was noted by Springer (1982) but not brought to bear on the biogeographic problem at hand; namely, concerning the similarities between some biotic elements of Hawaii and the eastern South Pacific. Springer relied heavily on a study by Frakes (1979), which was concerned primarily with climate and only touched briefly upon eustatic sea level. However, Frakes did not delve into submarine terraces and karst, lagoon depths, and related matters that provide ample evidence for the simultaneous subaerial exposure of reefs during the low stand of 18 000 years ago and earlier in the Pleistocene (cf. Wiens 1962, Stoddart 1971, Adey 1978).

An important point is that the generally considered notable feature of the Hawaiian Archipelago is its isolation from other islands in the region: Wake to the west, Clipperton to the east, and Johnston and the Line Islands to the south. All of these are presently low islands (Fig.1). More noteworthy is the Archipelago's isolation from other high islands, the closest being the Marianas to the west, the high islands of the Carolines (Kusaie, Ponape & Truk) to the west-southwest, the few high islands of the Eastern Pacific (Revilla Gigedo, Galápagos, Easter-Sala y Gómez, Juan Fernandez) and the numerous high islands of the South Pacific (Fig.2).

With a substantial lowering of sea level, the Carolines become a string of high islands reaching from the west to the Marshall Islands (thereby improving the 'Caroline Conduit' of Springer 1982). The Ellice and Gilbert Islands, now also high islands, extend from the Marshall Islands across the equator to the high islands in the south (Fig.10). Thus, a chain of high islands would range northward nearly to Wake Island, and Wake lies in line with the now high islands of the central reaches of the Hawaiian chain via the sub-tropical counter current (Fig.9b, cf. Grigg 1981). Similarly the Line Islands, including Johnston as well as Kingman Reef, extend as high islands across the equator from the

Tuamotu, Society and Cook Islands, and other previously high islands of the South Pacific, directly toward the central reaches of the Hawaiian Archipelago and into the sub-tropical counter current. Thus while the isolation of the Hawaiian Archipelago from islands remained virtually unchanged, its isolation from other high islands, especially those of the eastern South Pacific, was dramatically reduced. When sea level rose and the elevated, limestone islands returned to being low islands, the Hawaiian Archipelago again became isolated from high island sources.

This is all well and good for understanding the possibilities for dispersal and extinction during eustatic sea level fluctuations, across the presently low islands of the Central Pacific. But in the present scenario an 'Eastern Pacific connection' between the eastern South Pacific and Hawaii is also required. The oceanographic connection is used to explain the amphitropical zooplankton situation, in conjunction with the islands of the Eastern Pacific, especially the Revilla Gigedo, Clipperton, Galápagos and Easter-Sala y Gomez. Indeed, the present distribution of the land snail genus *Tornatellides* (Fig.4) indicates an Eastern Pacific connection between the otherwise amphitropical populations of the genus, as does the monotypic fish genus *Sectator* and the group to which the barnacle *Megabalanus tanagrae* of Hawaii belongs.

ACKNOWLEDGEMENTS

Many thanks are due to R.Gore and K.Heck, for convening this symposium as well as for critiques and editorial expertise on the manuscript for the present paper. I am particularly grateful to E.A.Kay and V.Springer for critical reviews. I would also like to thank Kay, for a preprint of her paper on 'Patterns of speciation in the Indo-Pacific' (1984) as well as for pointing out Randall's (1981) paper; and also Springer, not only for his spirited discussions (in litt.), but for being reasonably patient with me concerning dispersal hypotheses and other matters over which we are likely to continue to disagree. I thank L.Cheng for pointing out Kohlmeyer's paper (1983), as well as for providing material for, and approving of the modification of an original figure (Fig.7, herein). I regret that J.F.Peake was apparently unable to comment on my interpretation and modificatin of his figure (Fig.4, herein). While indebted to editors and colleagues, responsibility for errors and other shortcomings falls on me. This research was supported, in part, by the National Science Foundation: in the field by grants G-21905 (Fanning Islands Expedition — 1963), GA-729 (SIO-CARMARSEL, Caroline & Marshall Island Sea Level Expedition — 1967), GB-7596 (SIO STYX-7, Mid Pacific Mountain Expedition — 1968), and DES 7100559 (SIO SOTW-1, Eastern Tropical Pacific Transequatorial Expedition to Pitcairn — 1971) under the able leadership of S.A.Wainwright; R.Curray & F.Shepard; H.Ladd & R.Rosenblatt; and R.Hessler respectively; and, in the laboratory, by grant DEB 78-15052 (systematics of the Cirripedia).

REFERENCES

Adey, W.H. 1978. Coral reef morphogenesis: A multidimensional model. *Science* 202: 831-837.

Alvarez, W., E.G.Kauffman, F.Surlyk, L.W.Alvarez, F.Asaro & H.V.Michel 1984. Impact theory of mass extinctions and the invertebrate fossil record. *Science* 223: 1135-1140.

Berg, L.S. 1933. Die bipolare Verbrietung der Organismen und die Eiszeit. *Zoogeographica* 1: 444-484.

Briggs, J.C. 1974. *Marine Zoogeography.* New York: McGraw-Hill.

Brinton, E.B. 1962. The distribution of Pacific euphausiids. *Bull. Scripps Inst. Oceanogr.* 8: 51-270.

Brinton, E.B. 1975. Euphausiids of Southeast Asian waters. *NAGA Report* 4(5): 1-287.

Carlquist, S. 1981. Chance dispersal. *Amer. Sci.* 69: 509-516.

Cheng, L. & E.Shulenberger 1980. Distribution and abundance of *Halobates* species (Insecta: Heteroptera) in the Eastern Tropical Pacific. *Fish. Bull.* 78: 579-591.

Clague, D.A., G.B.Dalrymple, H.G.Greene, D.Wald, M.Kono & L.W.Kroenke 1980. Bathymetry of the Emperor Seamounts. In J.Shambach (ed.), *Initial Reports of the Deep Sea Drilling Project* 55: 845-849. Washington: National Science Foundation.

Croizat, L. 1958. *Panbiogeography.* Published by the author, Caracas.

Dalrymple, G.B., H.G.Greene, B.D.Rupple, T.E.Bear & D.A.Clague 1980. Pre-leg 55 site survey geophysical data from R/V *S.P.Lee* cruise Lee8-76.NP. In J.Shambach (ed.), *Initial Reports of the Deep Sea Drilling Project* 55: 801-843.

Darwin, C. 1842. *The structure and distribution of coral reefs.* London: Smith, Elder.

Darwin, C. 1859 (1872). *The origin of species by means of natural selection.* (Reprint of 6th edition.) New York: The Modern Library.

Edmondson, C.H. 1946. *Reef and Shore Fauna of Hawaii.* Honolulu: Bishop Museum.

Edmondson, C.H. 1951. Some central Pacific crustaceans. *Occ. Papers, Bishop Museum* 20(13): 183-243.

Ekman, S. 1953. *Zoogeography of the Sea.* London: Sidgwick & Jackson.

Fleming, C.A. 1979. Evolution of the South Pacific marine biota: The expanding fossil record. *Proceedings, International Symposium on Marine Biogeography and Evolution in the Southern Hemisphere.* New Zealand DSIR Information Series 137: 5-42.

Foster, B.A. 1974. The barnacles of Fiji, with observations on the ecology of barnacles on tropical shores. Pac. Sci. 28: 35-56.

Foster, B.A. 1978. The Marine Fauna of New Zealand: Barnacles (Cirripedia: Thoracica). *Mem. N.Z. Oceanogr. Inst., Wellington* 69: 1-160.

Foster, B.A. 1982. Shallow water barnacles from Hong Kong. In B.S.Morton & C.K.Tseng (eds.). *The Marine Flora and Fauna of Hong Kong and Southern China*: 207-232. Hong Kong: Hong Kong University Press.

Frakes, L.A. 1979. *Climates through geologic time.* North-Holland: Elsevier.

Gosline, W.A. 1968. Considerations regarding the evolution of Hawaiian animals. *Pac. Sci.* 22: 267-273.

Grigg, R.W. 1981. *Acropora* in Hawaii. Part 2. Zoogeography. *Pac. Sci.* 35: 15-24.

Grigg, R.W. 1982. Darwin Point: A threshold for atoll formation. *Coral Reefs* 1: 29-34.

Grigg, R.W. 1983. Community structure, succession and development of coral reefs in Hawaii. *Mar. Ecol. Prog. Series* 11: 1-14.

Grigg, R.W., J.W.Wells & C.Wallace 1981. *Acropora* in Hawaii. Part 1. History of the scientific record, systematics and ecology. *Pac. Sci.* 35: 1-13.

Gross, M.G., J.D.Milliman, J.I.Tracey Jr & H.S.Ladd 1969. Marine Geology of Kure and Midway Atolls, Hawaii: a preliminary report. *Pac. Sci.* 23: 17-25.

Hamilton, E.L. 1956. Sunken islands of the Mid-Pacific mountains. *Geol. Soc. Amer. Mem.* 64.

Heads, M. 1983. Pacific plate biogeography, with special reference to shorefishes. *J. Biogeogr.* 10: 543-548.

Heck, K.L. Jr & E.D.McCoy 1978. Long-distance dispersal and reef-building corals of the Eastern Pacific. *Mar. Biol.* 48: 349-356.

Hubbs, C.L. 1952. Antitropical distribution of fishes and other organisms. *Seventh Pacific Science Congress* 3: 324-329.

Hutchinson, G.E. 1950. The biogeochemistry of vertebrate excretion. In *Survey of Existing Knowledge of Biogeochemistry, Bull. Amer. Mus. Nat. Hist.* 96: 160-164, 481-482.

Jokiel, P.L. 1984. Long distance dispersal of reef corals by rafting. *Coral Reefs.* 3: 113-116.

Kay, E.A. 1980. Little worlds of the Pacific: An essay on Pacific Basin biogeography. *University of Hawaii, Harold L.Lyon Arborteum Lecture* 9: 1-40.

Kay, E.A. 1984. Patterns of speciation in the Indo-West Pacific. In F.J.Radovsky, P.H.Raven & S.H.Sohmer (eds.) *Biogeography of the Tropical Pacific.* Bishop Mus. Sp. Publ. 72: 15-31. Honolulu.

Kohlmeyer, J. 1983. Geography of Marine Fungi. *Aust. J. Bot. Suppl. Ser.* 10: 67-76.

Ladd, H.S. 1982. Cenozoic fossil mollusks from western Pacific islands; Gastropods (Eulimidae and Volutidae through Terebridae). *Geol. Surv. Prof. Paper* 1171: 1-100 + Pls.1-41.

Ladd, H.S., W.A.Newman & N.F.Sohl 1974. Darwin Guyot, the Pacific's oldest atoll. *Proceedings of the Second International Coral Reef Symposium* 2: 513-522.

MacArthur, R.H. & E.O.Wilson 1967. *The Theory of Island Biogeography.* Princeton: Princeton University Press.

MacNeil, F.S. 1954. The shape of atolls: an inheritance from subaerial erosion forms. *Amer. J. Sci.* 252: 402-427.

Maragos, J. 1977. Order Scleractinia: stony corals. In D.M.Devaney & L.G.Eldredge (eds.), *Reef and Shore Fauna of Hawaii.* Sect.1: Protozoa through Ctenophore: 158-241. Honolulu: Bishop Museum Press.

Matsui, T., W.A.Newman & G.Shane 1964. On *Balanus eburneus* Gould (Cirripedia, Thoracica) in Hawaii. *Crustaceana* 7: 141-145.

Mayr, E. 1942. *Systematics and the Origin of Species.* New York: Columbia University Press.

Menard, H.W. & E.L.Hamilton 1963. Paleogeography of the tropical Pacific. *10th Pac. Sci. Congr., Pac. Sci. Assoc.* (1961).

Menard, H.W., E.C.Allison & J.W.Durham 1962. A drowned Miocene terrace in the Hawaiian Islands. *Science* 138: 896-897.

Moore, T.C. Jr, L.H.Burckle, K.Geitzenauer, B.Luz, A.Molina-Cruz, J.H.Robertson, H.Sachs, C.Sancetta J.Thiede, P.Thompson & C.Wenkam 1980. The reconstruction of sea surface temperatures in the Pacific Ocean of 18,000 B.P. *Mar. Micropaleo.* 5: 215-247.

Nelson, G. 1975. Biogeography, the vicariance paradigm, and continenta drift. Syst. Zool. 24: 490-504.

Nelson, G. & D.E.Rosen (eds.) 1981. *Vicariance Biogeography: A Critique.* New York: Columbia University Press.

Newell, N.D. 1971. An outline history of tropical organic reefs. *Amer. Mus. Novit.* 2465: 1-37.

Newman, W.A. 1960a. *Octolasmis californiana* spec.nov., a pedunculate barnacle from the gills of the California Spiny Lobster. *Veliger* 3: 9-11, pl.2.

Newman, W.A. 1960b. Five pedunculate cirripeds from the Western Pacific, including two new forms. *Crustaceana* 1: 100-116.

Newman, W.A. 1961. On certain littoral species of *Octolasmis* (Cirripedia, Thoracica) symbiotic with decapod Crustacea from Australia, Hawaii and Japan. *Veliger* 4: 99-107, pls.21-23.

Newman, W.A. 1979. On the biogeography of Balanomorph barnacles of the Southern ocean including new balanid taxa: A sub-family, two genera and three species. *Proc. of the Int. Symp. of Mar. Biogeo. and Evol. in the S.Hemisphere 1978* 1: 279-306.

Newman, W.A. & D.P.Abbott 1980. Cirripedia: the barnacles. In R.H.Morris, D.P.Abbott & E.C.Haderlie (eds.), *Intertidal Invertebrates of California*: 504-535. Stanford: Stanford University Press.

Newman, W.A. & B.A.Foster 1983. The Rapanuian faunal district (Easter and Sala y Gómez): In search of ancient archipelagos. *Bull. Mar. Sci.* 33: 633-644.

Newman, W.A. & H.S.Ladd 1974a. Occurrences of modern coral-inhabiting barnacles in the Miocene of Fiji and the Marshall Islands (Cirripedia, Balanomorpha). *International Symposium on Indo-Pacific Reef Biology*: 29. University of Guam.

Newman, W.A. & H.S.Ladd 1974b. Origin of coral-inhabiting balanids (Cirripedia, Thoracica). *Verhandl. Natürf. Ges. Basel* 84: 381-396, 2 pls.

Newman, W.A. & A.Ross 1971. Antarctic Cirripedia. *Antarctic Res. Ser.* 14: 1-257.

Newman, W.A., P.A.Jumars & A.Ross 1976. Diversity trends in coral-inhabiting barnacles (Cirripedia, Pyrgomatinae). *Micronesica* 12: 69-82.

Nilsson-Cantell, C.A. 1957. Thoracic cirripeds from Chile. *Lunds Univ. Arsskr. N.F.* (2) 53(9): 1-25.

Nilsson-Cantell, C.A. 1978. Cirripedia Thoracica and Acrothoracica. *Marine Invertebrates of Scandinavia* No.5: 1-135. Oslo: Universitetesforlaget.

Peake, J.F. 1981. The land snails of islands — a dispersalist's viewpoint. In P.L.Forey (ed.), *The Evolving Biosphere.* London: British Museum (Nat. Hist.).

Potts, D.C. 1983. Evolutionary disequilibrium among Indo-Pacific corals. *Bull. Mar. Sci.* 33: 619-632.

Purdy, E.G. 1974. Reef Configurations: Cause and effect. In L.F.Laporte (ed.), *Reefs in Time and Space. Soc. Econ. Paleo. Miner. Spec. Publ.* 18: 9-76.

Randall, J.E. 1981. Examples of antitropical and antiequatorial distribution of Indo-West Pacific fishes. *Pac. Sci.* 35: 197-209.

Rehder, R.A. 1980. The marine mollusks of Easter Island (Isla de Pascua) and Sala y Gómez. *Smith. Contr. Zool.* 289: 1-167.

Reid, J.L., E.Brinton, A.Fleminger, E.L.Venrick & J.A.McGowan 1978. Ocean circulation and marine life. In H.Charnock & Sir G.Deacon (eds.), *Advances in Oceanography.* New York: Plenum.

Resig, J.M. 1969 Paleontological investigations of deep borings on Ewa Plain, Oahu, Hawaii. *Hawaii Inst. Geophys.* 69(2): 1-99.

Richmond, R. 1981. Energetic considerations in the dispersal of *Pocillopora damicornis* (Linnaeus) planulae. *Proc. Fourth Int. Coral Reef Symp.* Manila 2: 261-264.

Rodríguez, J. & R.C.Gutschick 1977. Barnacle borings in live and dead hosts from the Louisiana Limestone (Famennian) of Missouri. *J. Paleont.* 51: 718-724.

Rosenblatt, R.H. 1967. The zoogeographic relationships of marine shore fishes of tropical America. In *Proceedings of the International Conference on Tropical Oceanography. Studies in Tropical Oceanography* 5: 579-592. Coral Gables: University of Miami.

Ross, A. & W.A.Newman 1973. Revision of the coral-inhabiting barnacles (Cirripedia: Balanidae). *Trans. San Diego Soc. Nat. Hist.* 17: 137-174.

Rotondo, G.M., V.G.Springer, G.A.J.Scott & S.O.Schlanger 1981. Plate movement and islands integration — a possible mechanism in the formation of endemic biotas with special reference to the Hawaiian Islands. *Syst. Zool.* 30: 12-21.

Scheltema, R.S. & I.P.Williams 1983. Long-distance dispersal of planktonic larvae and the biogeography and evolution of some Polynesian and Western Pacific mollusks. *Bull. Mar. Sci.* 33: 545-565.

Schram, F.R. & W.A.Newman 1980. *Verruca withersi* n.sp. (Crustacea: Cirripedia) from the middle of the Cretaceous of Colombia. *J. Paleo.* 54: 229-233.

Solem, A. 1981. Land-snail biogeography: A true snail's pace of change. In G.Nelson & D.E.Rosen (eds.), *Vicariance Biogeography: A Critique*: 197-221. New York: Columbia Univ. Press.

Spivey, H. 1981. Origins, distributions and zoogeographic affinities of the Cirripedia (Crustacea) of the Gulf of Mexico. *J. Biogeogr.* 8: 153-176.

Springer, V.G. 1982. Pacific plate biogeography, with special reference to shorefishes. *Smith. Contr. Zool.* 367: 1-182.

Steers, J.A. & D.R.Stoddart 1977. The origin of fringing reefs, barrier reefs and atolls. In O.A.Jones & R.Endean (eds.), *Biology and Geology of Coral Reefs 4, Geology 2.* New York: Academic Press.

Stehli, F.G. & J.W.Wells 1971. Diversity and age patterns in hermatypic corals. *Syst. Zool.* 20: 115-126.

Stoddart, D.R. 1971. Environment and history in Indian Ocean reef morphology. In D.R.Stoddart & C.M.Yonge (eds.), *Regional variation in Indian Ocean coral reefs. Symp. Zool. Soc. Lond.* 28: 3-38.

Suess, E. 1904-1909. *The Face of the Earth* Vol.1-5. Oxford: Clarendon.

Tayama, R. 1935. Table reefs, a particular type of coral reef. *Proc. Imperial Acad. Japan* 11: 268-270.

Taylor, R.C. 1973. An atlas of Pacific Island rainfall. *Hawaii Inst. Geophys.* Data Report 25: 1-5 + 13 figs. and appendices A-E.

Théel, Hj. 1886. Report on the holothuroidea dredges by H.M.S. Challenger; Part 2, *Challenger reports. Zoology* 44: 1-290.

Tomlinson, J.T. 1969. The burrowing barnacles (Cirripedia: Order Acrothoracica). *U.S. Natl. Mus. Bull.* 296: 1-162.

Udvardy, M.D.F. 1969. *Dynamic Zoogeography with special reference to land animals.* New York: Van Nostrand, Reinhold.

Vail, P.R., R.M.Mitchum & S.Thompson 1977. Seismic stratigraphy and global sea level. Part 4: Global cycles of relative changes of sea level. *Amer. Assoc. Petrol. Geol. Mem.* 26: 83-97.

Van Balgooy, M.M.J. 1971 Plant-geography of the Pacific. *Blumea Suppl.* 6: 1-222.

Van der Spoel, S. & R.P.Heyman 1983. *A comparative Atlas of Zooplankton.* Springer-Verlag.

Wegner, A. 1915. *Die Entstehung der Kontienente und Ozeane.* Sammlung Vieweg. Braunschweig.

Weins, H.J. 1962. *Atoll Environment and Ecology.* New Haven: Yale University Press.

Wells, J.W. 1982. Fossil corals from Midway Atoll. *Geol. Surv. Prof. Pap.* 680-G: G1-G7.

Wells, J.W. 1983. Annotated List of the Scleractinian Corals of the Galapagos. In P.W.Glynn & G.M.Wellington, *Corals and Coral Reefs of the Galapagos Islands*: 213-318. Berkeley: University of California Press.

Zimmerman, E.C. 1948. *Insects of Hawaii* Vol.1. Honolulu: University of Hawaii Press.

Zimmerman, E.C. 1963. Pacific Basin Biogeography: A summary discussion. In J.L.Gressitt (ed.), *Pacific Basin Biogeography*: 477-481. Honolulu: Bishop Museum.

Zullo, V.A. 1969. A new subspecies of *Balanus tintinnabulum* (Linnaeus, 1758) (Cirripedia, Thoracica) from Clipperton Island, Eastern Pacific. *Proc. Calif. Acad. Sci.* (4) 36: 501-510.

Centers of radiation of freshwater crabs in the Neotropics

GILBERTO RODRIGUEZ
Instituto Venezolano de Investigaciones Cientificas, Caracas, Venezuela

ABSTRACT

Presence of shared synapomorphies in the respiratory structures of freshwater crabs strongly suggests a monophyletic origin for all groups. In the Trichodactylidae and other relatives from Africa plesiomorphy is more apparent, while in Pseudothelphusidae and other subgroups from West Africa and India, adaptation to air breathing has progressed further. The present distribution of both groups suggests a primitive Gondwanian distribution made disjunct by a mid-Cretaceous vicariant event. The most primitive Pseudothelphusidae categorized as such by the morphology of the third maxilliped, has a disjunct distribution the Greater Antilles and the Northern Andes of Colombia, explainable by a Pliocene vicariant event that isolated the Greater Antillean biota from that of Central and South America. The somatic and gonopodal characters form three distinct chorological series that radiate from northern Colombia to Central America and Mexico, to northern South America and the Guianas, and to the Southern Andes. Sympatries and the presence of geographic morphoclines indicate that this radiation of characters occurred in a series of dispersion events from a dispersion center in northern Colombia. The disjunct distribution observed in the Lesser Antilles and Guianas can only be attributable to chance dispersal after the Miocene. Consequently the present distribution of freshwater crabs in the neotropics cannot be understood simply in terms of either vicariance or dispersal, but as the result of a complex series of events.

1 INTRODUCTION

Recently the preeminence of the concept of center of origin in the Darwinian sense as a conceptual model fundamental to historic zoogeographical analysis has been challenged (Croizat et al. 1974, Rosen 1976, McCoy & Heck 1976). As an alternative the importance of 'generalized tracks' has been stressed, i.e. generalized patterns of biotic distribution, as estimates of ancestral biotas that owing to changing geography, have become subdivided into descendent biotas in localized areas. This vicariance hypothesis is clearly not incompatible with dispersal, because in its most modern sense dispersal has become attributable to a general cause: i.e. removal of preexisting barriers (Brundin 1981). This idea is not far removed from the point of view of vicariance biogeography, as stated by Nelson (1981: 526): 'In Croizat's terms, the removal of barriers allows "means of survival" to become "means of dispersal", and marks the shift from a period of "immobilism" to a period of "mobilism"'.

In its original form, vicariance biogeography deals with whole biotas, or at least with information derived from as many groups of organisms as possible. However, as Nelson & Platnick (1980) discuss, it is useful to test this hypothesis on monophyletic groups of organisms whose synapomorphies may be clearly established. The neotropical freshwater crabs seem to be ideal biogeographical tools for this purpose. The two existing families, Trichodactylidae and Pseudothelphusidae, comprise numerous species, possibly more than 200, many of them located at the junction of the North and South American land masses. They are strictly freshwater organisms, without a free swimming larva, and cannot stand desiccation on dry land. Furthermore, the species of Pseudothelphusidae have aerial respiration, and hence do not tolerate prolonged submersion in deep water such as might occur in large rivers. Consequently, while the species of Trichodactylidae inhabit large rivers in the flat lands an have extensive specific ranges, the Pseudothelphusidae usually have very small ranges in streams, between 100 and 3000 m altitude. This pattern of non-overlapping distributions operates throughout the range of the two families.

On closer examination, these neotropical crabs present several problems. The few fossil remains are younger than 4000 a BP (Rodríguez & Díaz 1977). In the Pseudothelphusidae, the somatic morphology is almost completely uniform and the species can only be distinguished by the structure of the male gonopod. In both families their relationship with species from the Old World are not well understood. Ortmann (1902), who was the first to deal in a systematic way with the biogeography of these crabs, envisaged a close relationship between the American Pseudothelphusidae and the African Potamidae, particularly with *Acanthothelphusa*, and suggested a connection of the respective parts. He considered the Trihodactylidae as morphologically isolated and sharply defined and thus an ancient group, but with a remarkably recent character in its distribution. Bouvier (1917) considered all freshwater crabs as having originated in the Tethyan Sea from a marine ancestor which gave rise to two groups: (1) A primitive group comprised of the American Trichodactylidae and the African Potamonidae. 'Even in Tertiary times, when the Brazil and neighbouring areas formed with Africa a Brasilo-ethiopian continent, the evolution of some Potamoninae into Trichodactylinae had already begun in the occidental region; this evolution was completed from the time when the sinking of the Atlantide created an enormous oceanic barrier between America and the African continent' (Bouvier 1917: 753). (2) A more advanced group comprising the Pseudothelpusidae and the Gecarcinucidae occupied America and the Indo-Australian region, respectively.

Sharply contrasting with the ideas of these two authors are those of Colosi (1921). He compared the presumed 'antiquity' of each group of freshwater crabs with the geological age of the areas that eah occupies. As has been done by more recent vicariant biogeographers, he denies the possibility of migrations of these organisms,[1] and concluded that there are no centers of origin, but that each group of freshwater crabs had an ancestor in the respective adjacent sea. An equally polyphyletic origin has been postulated by Pretzmann (1973), who suggested that the Pseudothelphusidae originated from the Xanthidae, during the Upper Eocene and the Oligocene, at a time when Northern South America was fragmented into numerous islands. Each group of Pseudothelphusidae colonized an 'island' through the brackish waters: the Kingsleyini, Northern South America; *Epilobocera*, the Greater Antilles; *Guinotia*, the Lesser Antilles; *Hypolobocera*, the Cauca Valley in Colombia; *Potamocarcinus*, an island in Eastern Panama and so on. Of course, this extremely polyphyletic hypothesis ignores any links between the paleotropical and the neotropical crabs.

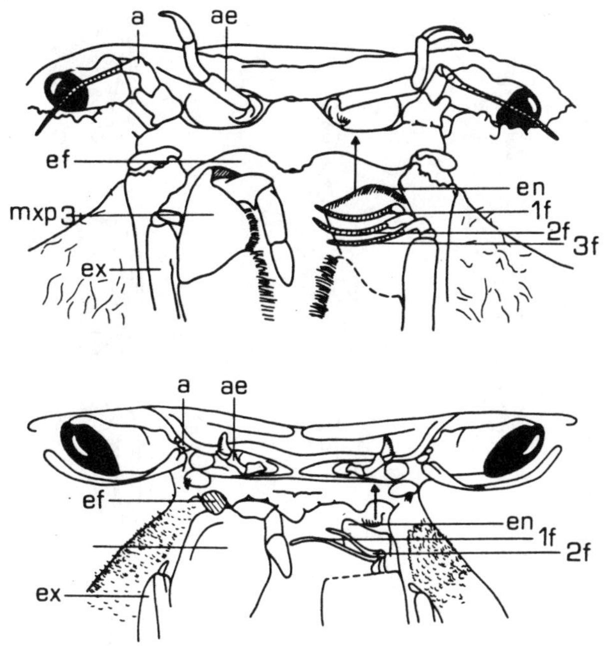

Figure 1. Semidiagrammatic representation of the buccal area in the Trichodactylidae (above) and the Pseudothelphusidae (below). a – antenna; ae – antennulae; ef – opening of efferent channel; en – endopod of 1st maxilliped; ex – exognath of 3rd maxilliped; 1f, 2f, 3f – flagella of 1st, 2nd and 3rd maxillipeds; mxp3 – 3rd maxilliped.

2 SYNAPOMORPHIES RELATED TO RESPIRATORY STRUCTURES

One of the conditions for the reconstruction of phylogenetic patterns (Cracraft 1981) is the formulation of hypotheses about homology (plesiomorphic) and those at which they are derived (apomorphic). In the freshwater crabs these characters can best be studied in the structures related to semi-terrestrial adaptation, i.e. the reaccommodation of mouth parts for respiration and water conservation.

A general condition in many terrestrial and semi-terrestrial crabs is the possession of a highly vascularized lining in the branchial chamber, presumably for gaseous interchange. The chamber itself is partly filled with air and partly filled with water. Air flow within the chamber apparently occurs by beating of the scaphognatite (Cameron 1975). Renewal of this air may take place only through the opening of the efferent channels located inside the endostome.

A distinguishing feature of the Pseudothelphusidae is the possession of a specialized area in the lining of the branchial chamber which acts as a lung for aerial respiration (Díaz & Rodríguez 1977). This organ was acquired early in the evolution of the group, because all members of the family possess it, and so is a strong argument in favor of monophyly of the Pseudothelphusidae. Development of this organ required the coadaptation of other structures to ensure renewal of the air supply. Figures 1 and 2 compare the plesiomorphic conditions in Trichodactylidae with the synapomorphies found in the Pseudothelphusidae. As summarized in Table 1, the transformations found in Pseudothelphusidae are as follows: (a) the openings of the efferent channels are located outside the endostome, in the lateral angle of the buccal frame, and are closed on their lower margin by the endopodite of the first maxilliped (Fig.1b), thus isolating the channels from the mouth and insuring a free communication; (b) the flagella of the first and second

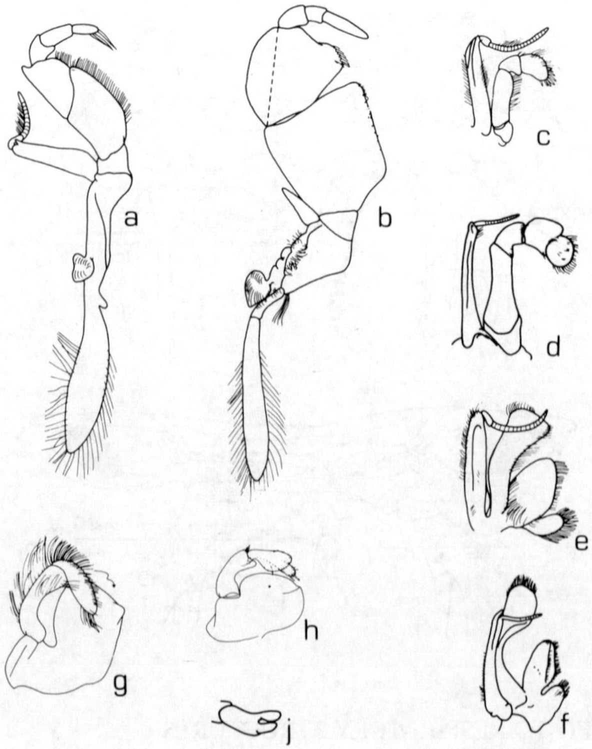

Figure 2. Buccal appendages of Trichodactylidae and Pseudothelphusidae. (a), (c), (e), (g) 3rd maxilliped, 2nd maxilliped, 1st maxilliped and mandible of Trichodactylidae. (b), (d), (f), (h), (j) 3rd maxilliped, 2nd maxilliped, 1st maxilliped, mandible, and mandibular palp in dorsal view of Pseudothelphusidae.

Table 1. Differences in the buccal region of neotropical freshwater crabs.

Character	Trichodactylidae	Pseudothelphusidae
Efferent channel opening	Inside endostome	Outside endostome
1st maxilliped endopod	Wide and short	Narrow and long
1st maxilliped flagellum	Strong	Weak
2nd maxilliped flagellum	Strong	Weak
3rd maxilliped flagellum	Present	Absent
3rd maxilliped exognath	Developed	Mostly reduced
3rd maxilliped merus	Narrow	Wide (with 'wing')

maxillipeds are weak (Fig.2d,f); (c) the flagellum of the third maxilliped is lost and the exognath is correspondingly reduced (Fig.2b); and (d) the merus of the third maxilliped is expanded, forming a 'wing' (delimited in Fig.2b by a dashed line) that covers the endostome and delimits the jugal border of the efferent openings.[2]

The resulting condition of the migration of the efferent channel openings in the Pseudothelphusidae is presented schematically in Figure 3. The air in the branchial chamber can be in free communication with the external air, even if the mouth is tightly closed by the third maxilliped. Movement of the scaphognathite will then provide a measure of 'air breathing' for the pseudolung. The reduction of the antennulae and antenna is

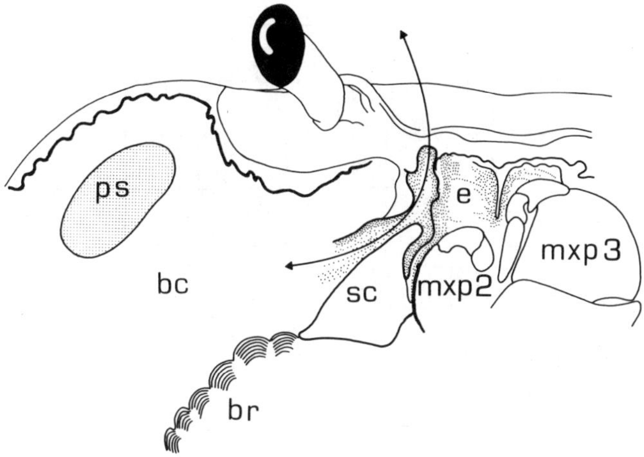

Figure 3. Schematic cut-away representation of air admission to the branchial chamber of Pseudothelphusidae. bc – branchial chamber; br – branchiae; e – endostome; ps – pseudolung; sc – scaphognathite; mxp2, mxp3 – right 2nd and left 3rd maxillipeds, respectively.

perhaps a secondary condition associated with the position of the aerial openings on the epistome.

There is no evidence for the presence of lungs in Paleotropical freshwater crabs, but a reduction of the exognath flagellum of the third maxilliped similar to the one found in Pseudothelphusidae, does occur in 16 taxa (species and subspecies) belonging to 10 genera of paleotropical Gecarcinucidae and 10 species of four genera of Asiatic Potamidae (Bott 1970). The upper margin of the endostome in some of these crabs is deeply indented at both external angles, a condition perhaps more primitive than the complete exclusion of efferent channel openings found in Pseudothelphusidae. The species with reduced flagella present a particular distribution: the Pseudothelphusidae in America, the genera *Globonautes* and the *Afrithelphusa* in West Africa, and *Liothelphusa* and other genera in India, Indochina and Malaysia. Associated with these transformations of the external maxillipeds are some characters of the mandibular palp. In a first group of crabs, the Eupotamonea which comprises the Trihodactylidae and the palaeotropical superfamily Potamoidea, the first article of th mandibular palp is simple, elongated and lamellar, placed in front or sometimes behind the anterior border of the mandible (Fig.2g). This character represents the generalized brachyuran condition. The species in the second group, the Parapotamonea which comprises the Pseudothelphusidae and the palaeotropical superfamily Parathephusidea, have the last article of the mandibular palp deeply cleft into two lamellar lobes which cover the anterior cutting surface of the mandible (Fig.2h,j). A transitional condition can be found in a third group comprising some Hydrothelphusida of Madagascar, in which the last article of the mandibular palp has a small lobe at the base (Bott 1965).

3 THE GEOGRAPHICAL DISTRIBUTION OF SOMATIC SYNAPHOMORPHIES IN PSEUDOTHELPHUSIDAE

3.1 *Centers of high taxa density*

The genera and species of Pseudothelphusidae show an unequal distribution of taxa through the Caribbean, Central and South America (Rodríguez 1982). In order to assess the density

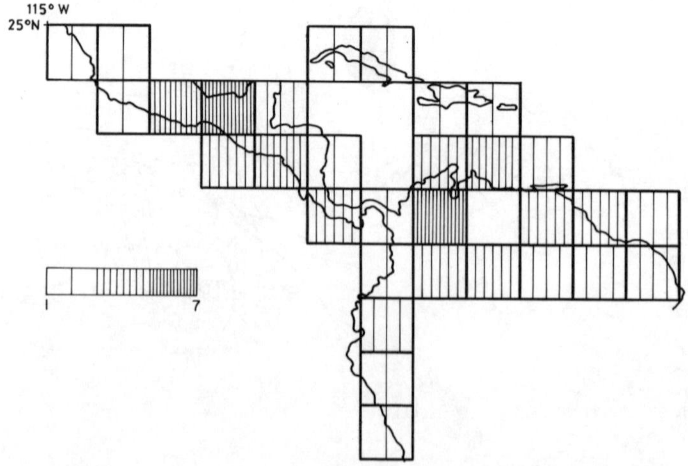

Figure 4. Density of genera of Pseudothelphusidae in the New World. Vertical hatching indicate number of species per quadrant in a scale of 1 to 7 genera.

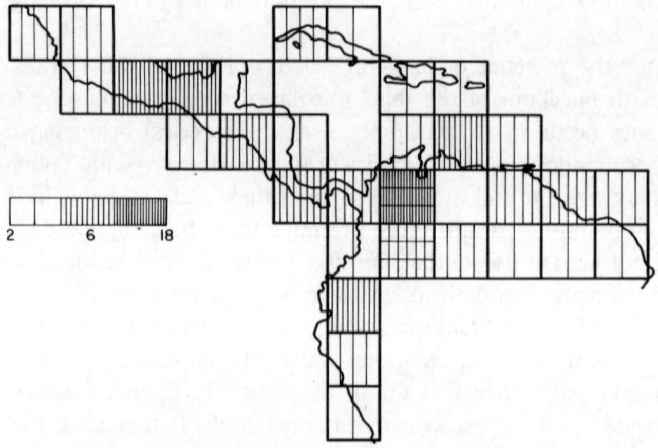

Figure 5. Density of species of Pseudothelphusidae in the New World. Vertical hatching indicate number of species per quadrant in a scale of 2 to 18 species.

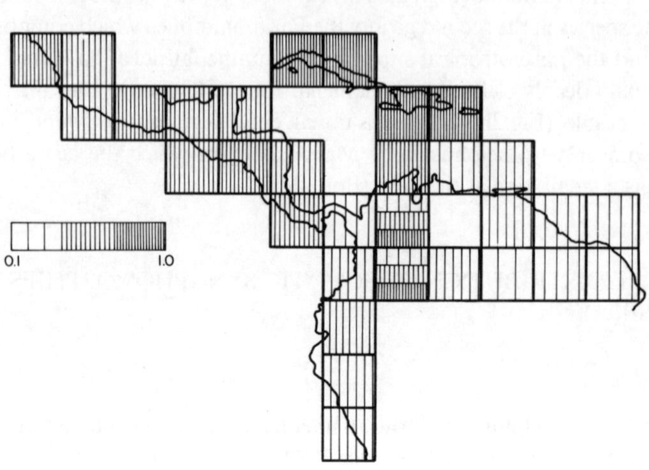

Figure 6. Geographical distribution of the relative length of the exognath of the 3rd maxilliped in Pseudothelphusidae. This parameter fluctuates from species with a stump-like exognath (0.1) to those in which the exognath is as long as the ischium of endognath (1.0).

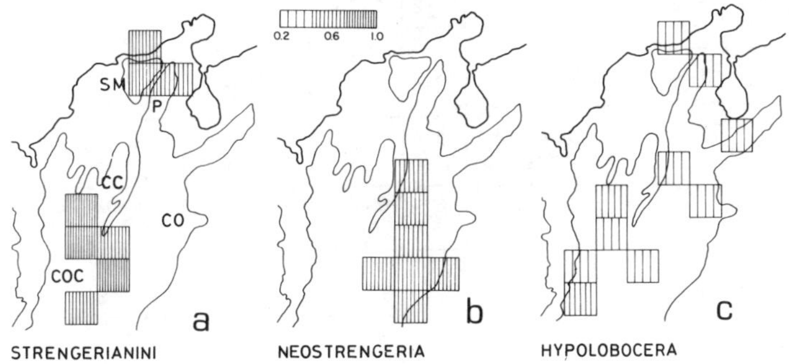

Figure 7. Sympatric distribution of groups with different relative lengths of the exognath of 3rd maxilliped in the Pseudothelphusidae of northern Colombia. Scale as in Figure 6. CO – Cordillera Oriental; CC – Cordillera Central; COC – Cordillera Occidental; SM – Sierre de Santa Marta: P – Sierra de Perijá.

of taxa the whole area was divided in arbitrary quadrats of 5° latitude by 5° longitude, progressing southward from parallel 25°N, and eastward from meridian 115°W (Figs.4, 5). The concentration of species found by this method is very similar to the concentration of genera and it showed in both cases maximum values in two areas, one near the isthmus of Tehuantepec with seven genera and 11 species, and another in the northern Andes of Colombia with seven genera and 18 species. From these two centers the number of taxa decreases in all directions.

3.2 *Relative length of the exognath*

As stated before, one distinguishing feature of the Pseudothelphusidae is the loss of the flagella of the third maxillipeds and the consequent degeneration of the exognath. This, however, is not uniform for all species. The genera in two areas, the Greater Antilles (*Epilobocera*) and the Northern Andes of Colombia (*Neostrengeria*) have the exognath slightly larger than the ischium of the endognath and one species in Puerto Rico. *Epilobocera sinuatifrons*, even has a rudimentary palp. The proportions of the exognath of the species in the rest of the distributional area of the family is presented in Figure 6. The species from Mexico and Central America have relatively long exognaths, while in most of those from South America it is very short or even reduced to a stump with the exception of some Colombian species which are commented upon in the following paragraph.

3.3 *Sympatric distribution of characters in the Northern Andes of Colombia*

Species with different types of exognath occur only in the Northern Andes of Colombia, including the Sierras of Santa Maria and Perijá (Fig.7). In that area are found (a) the already mentioned group of species with a complete or almost complete exognath (Tribe Strengerianini), (b) a group with exognath approximately 0.5 the length of the endognath (genus *Neostrengeria*) and (c) a group with a very short exognath (genus *Hypolobocera*).

3.4 *Lateral extension of the merus of the third maxilliped*

One of the changes in the process of reaccommodation of the mouth parts in Pseudothelphusidae, is supposedly the enlargement of the external margin of the merus, which at the same time changed its shape. To study these two characters, drawings of this area

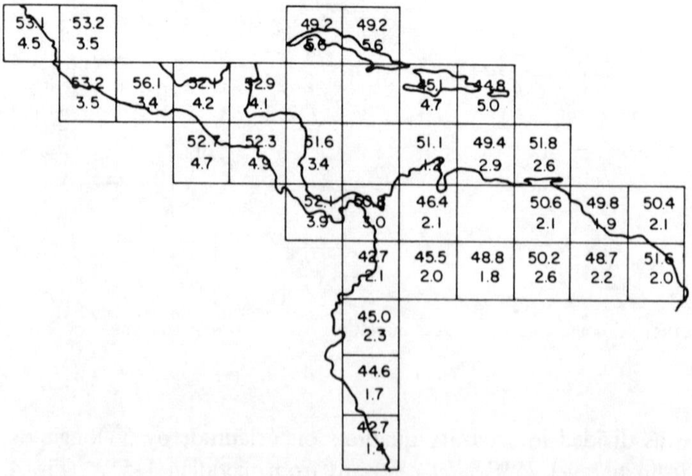

Figure 8. Geographical distribution of the shape (upper values) and extension (low values) of the lateral projection of merus of 3rd maxilliped in Pseudothelphusidae. Values in arbitrary units (see text).

of the merus in all species available were made with a camera lucida and the curve representing the external margin was digitized. From these data two parameters were calculated (a) the distance (hatched line in Fig.2b) at which 50% of the integation of the 'wing' surface is achieved, and (b) the total area of the 'wing' (i.e. the area above the hatched line in Fig.2b). The first parameter is a measure of the shape of the 'wing'. Species with low values have this area advanced distally whereas in those with higher values this area is more regularly rounded. Both parameters were measured in 25 specimens of *Guinotia garmani* ranging from 22.4 to 99.0 mm of carapace breadth to test for allometric growth. The slope of the regression line of the shape of the merus (parameter a) on the carapace breadth does not significantly differ from zero ($t_{0.05}$ = 1.29), but the total area of the 'wing' (parameter b) shows a clear allometric growth (coefficients for linear and exponential regressions = 0.869 and 0.940, respectively). Consequently, the total areas in different species of Pseudothelphusidae were normalized by photographic reduction of the drawings to make horizontal (x-axis) distances equal. The normalized areas thus obtained represent the relative extension of the merus in antero-lateral direction perpendicular to the hatched line in Figure 2b.

The distribution of the normalized areas of the merus within the geographical range of the Pseudothelphusidae (Fig.8) shows a clear discrimination of three regions (a) in the Greater Antilles the merus is very extended anterolaterally (4.7-5.6), (b) in Mexico and Central America it has intermediate values (3.0-4.5), and (c) in South America it is extended very little (1.4-2.6). The shape of the merus follows a similar, although less well-defined geographical pattern. In the species from Mexico and Central America it is regularly rounded (50.8-56.1), while in the Antilles and South America it is expanded distally (44.8-49.2 and 42.7-51.8, respectively).

4 CHOROLOGICAL SERIES BASED ON GONOPOD STRUCTURE

4.1 *Intergroup continuity of gonood characteristics*

If the gonopods of the Pseudothelphusidae are examined for homologies it is possible

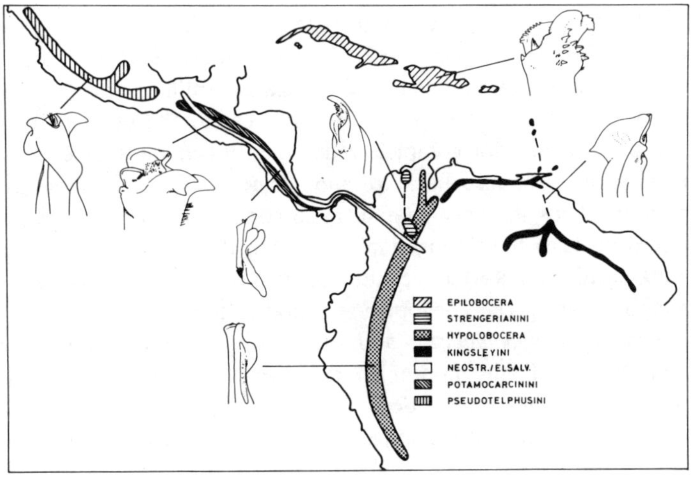

Figure 9. Geographical distribution of the gonopod patterns in Pseudothelphusidae.

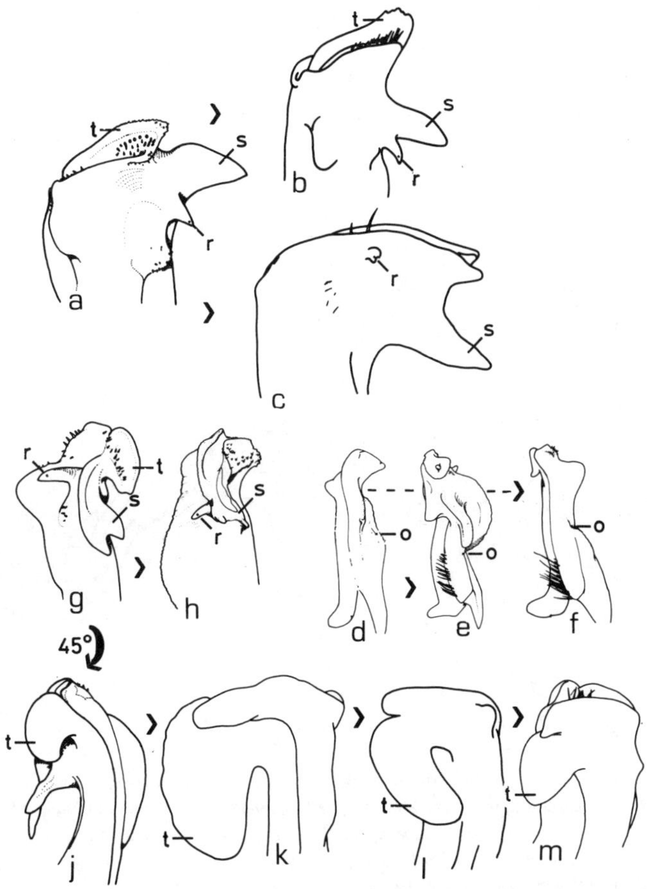

Figure 10. Homologies in the gonopods of Central American and Mexican Pseudothelphusidae. (a-c) the *Potamocarcinus* phyletic line; (d-m) the *Elsalvadoria* phyletic line. (a) *Potamocarcinus aspoekorum*; (b) *Typhlopseudothelphusa mocinoi*; (c) *Odonthothelphusa maxillipes*; (d) *Ptychophallus paraxanthusi*; (e) *Spirothelphusa verticalis*; (f,g,j) *Elsalvadoria zurstrasseni*, (j) is turned 45° with respect to (g); (h) *Lobithelphusa mexicana*; (k) *Epithelphusa mixtepensis*; (l) *Tehuana veracruzana*; (m) *Pseudothelphusa guerreroensis*. o – notch of the lateral lobe; r – cephalic spine; s – mesial spine; t – caudal process that gives rise to the hatchet-shaped mesial process of *Elsalvadoria* and Pseudothelphusini. All figures represent the apex of the gonopod, except (d,e,f), which are views of the total appendage.

to distinguish seven structural patterns within the family, each with a well-defined geographical distribution (Fig.9). The first pattern (*Epilobocera*) is morphologically isolated from the rest and confined to the Greater Antilles. The second pattern (Strengerianini) is restricted to northern Colombia and cannot be derived from any other. The other five patterns are not completely discrete, but in most cases characters of one gonopod pattern can be traced back to some character from another supposedly more primitive pattern.[3] Thus the very characteristic gonopod of *Hyolobocera* from the Andes can be traced back to the Strengerianini through the genus *Martiana*, and the gonopod of the Kingsleyini from Northern South America can also be traced to the Strengerianini through the genus *Strengeria*. The fourth gonopod pattern, also from South America (*Neostrengeria*) is related to *Hypolobocera* and their characters are continued in species from Central America and finally appear in the Pseudothelphusini from Mexico. The last pattern (Potamocarcinini) can be derived as a lateral branch of some of the Central American relatives of the *Neostrengeria* pattern, possibly through the genus *Elsalvadoria*.

4.2 *Multiplicity of phyletic lines in the Isthmus of Tehuantepec*

As indicated above, the Isthmus of Tehuantepec is one of the two areas of high taxa density. Through a study of the homologies of the gonopods it is possible to discriminate several series of species which seem to change morphologically in a westward direction. Three genera reach the isthmus: *Potamocarcinus, Ptychophallus* and *Elsalvadoria*. The presumed homologies are presented in Footnote 4 and illustrated in Figure 10. The continuity

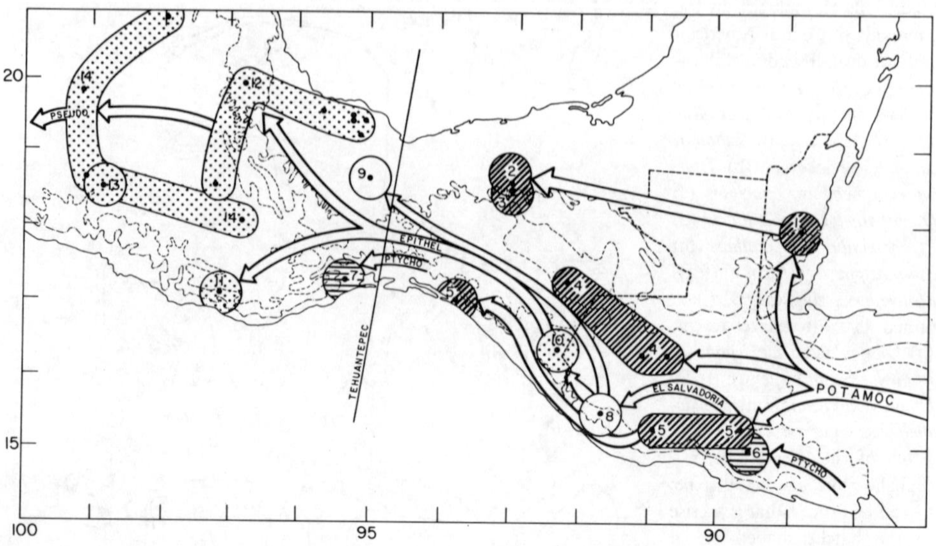

Figure 11. Distribution of main phyletic lines of Pseudothelphusidae around the Isthmus of Tehuantepec. Potamocarcinini, transversel hatching; *Ptychophallus*, horizontal hatching; *Elsalvadoria*, clear; Pseudothelphusini, dots. (1) *Potamocarcinus aspoekorum*; (2) *P.hartmanni*; (3) *Odonthothelphusa maxillipes*; (4) *Typhlopseudothelphusa* spp.; (5) *Potamocarcinus magnus*; (6) *Ptychophallus* spp.; (7) *Spirothelphusa verticalis*; (8) *Elsalvadoria zurstrasseni*; (9) *Lobithelphusa mexicana*; (10) *Epithelphusa chiapensis*; (11) *E.mixtepensis*; (12) *Tehuana* spp.; (13) *Pseudothelphusa guerreroensis*; (14) *P.americana*.

of these phyletic lines can be followed in the map of Figure 11. Numbers in parentheses refer to numbers in Figure 11.

1 — The Potamocarcinus phyletic line crosses north to Belize *Potamocarcinus aspoekorum*, (1) and Tabasco *Potamocarcinus hartmanni*, (2); *Odonthothelphusa maxillipes*, (3); to the caves in the highlands of Chiapas *Typhlopseudodthelphusa mocinoi* and other species, (4) and through the Suth, to the western limits of the stock in Chiapas *Potamocarcinus magnus*, (5). So this phyletic line stops at Tehuantepec and does not cross the isthmus.

2 — *Ptychophallus* is abundant in Costa Rica (6). One of the branches of this line is found near Tehuantepec *Spirothelphusa verticalis* (7). The other branch gives rise to the genus Elsalvadoria.

3 — The third Central American line is that of the genus *Elsalvadoria* (8) with affinities to *Ptychophallus*. One of the branches from this line is a dead end, terminating in Veracruz *Lobithelphusa mexicana* (9), the other branch crosses the isthmus and through *Epithelphusa* (10) gives rise to the Mexican Pseudothelphusinae.

4.3 *The Mexican morphoclines*

In many groups of Pseudothelphusidae it is possible to observe series of species in which the structure of the gonopod progressively changes over a geographical area. One such example is the progressive series (*Tehuana veracruzana, Pseudothelphusa guerereoensis, P.belliana, P.americana* and *P.peyotensis*) found in Southern Mexico (Fig.12). The lateral expansion of the gonopod is progressively reduced proceeding westward to a sharp spine; at the same time the superior margin of the front disappears from the surface of the carapace. This morphocline follows a general westward direction, with the exception of *P.americana*, suggesting an allopatric speciation of primitive demes, encompassing a progressive migration along the same geographical axis.

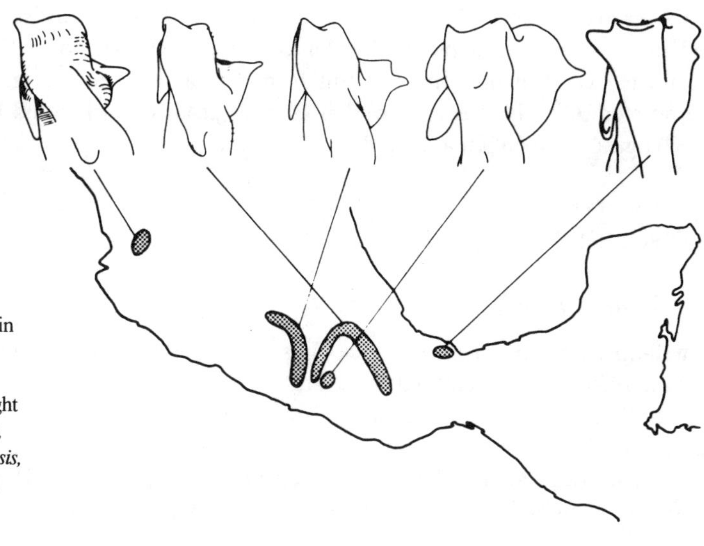

Figure 12. A morphocline in the gonopods of Pseudothelphusidae from southern Mexico. From right to left *Tehuana veracruzana, Pseudothelphusa guerreroensis, P.belliana, P.americana* and *P.peyotensis.*

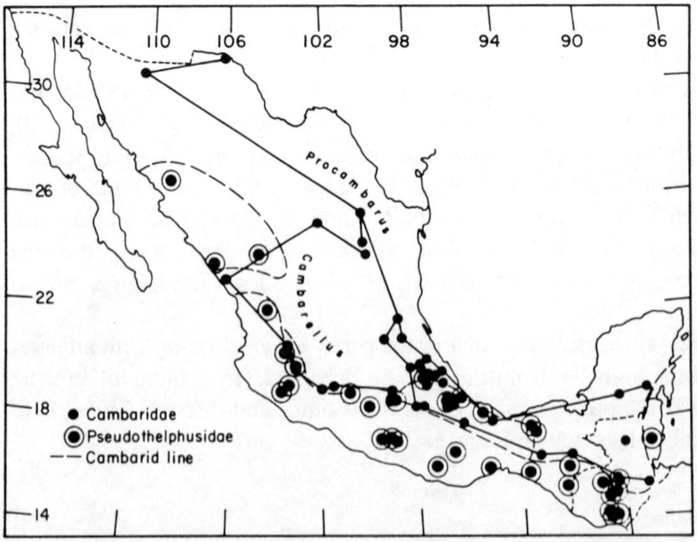

Figure 13. Distribution of the Cambaridae and Pseudothelphusidae in Mexico.

5 THE BIOGEOGRAPHICAL EXCLUSION OF PSEUDOTHELPHUSIDAE AND CAMBARIDAE

In the northernmost limit of their geographical range, the Pseudothelphusdae face the cambarid crayfishes along a front of 3000 km in Mexico and Guatemala (Fig.13). Overlap of the two groups occurs only in two areas: in Lake Chapala where the species of Pseudothelphusidae have lapped ranges with *Cambarellus* and in the Veracruz lowlands where the range of the crabs is nested within the ranges of *Procambarus.* This slight overlap suggests that one group excluded the other and it has been assumed that it was the Pseudothelphusidae that displaced the crayfish from a former more extended range. However, one of the few pieces of evidence supporting this hypothesis is the occurrence of the epizoic commensal branchiodellid are restricted to the crayfish in Northern latitudes and their presence in a pseudothelphusid is explained by Holt (1964) by passing from the former crayfish host (now absent from Costa Rica) to the crabs. Otherwise, there are no studies in the areas of overlap of both groups which could be used as evidences of ecological competition among them.

6 DISCUSSION

6.1 *The neotropical-holotropical disjunction*

A well-documented generalized track links tropical Africa and northeastern South America (Rosen 1976). This generalized track finds support in the connections between both land masses that existed up to the breakup of Pangaea in Mid-Cretaceous 110-100 m.a. BP (Pindall & Dewey 1982). India, which was attached to Eastern Africa through Madagascar, also broke away at this time and drifted north towards Asia. The present distribution of species with reduced exognath in America, Africa and India is thus evidence of a former Gondwanan distribution and shows at the same time that the transformation of buccal appendages had already begun in Mid-Cretaceous in basic groups whose modern representatives are the American Pseudothelphusidae and the African Gecarcinucidae. The Trichodactylidae, in which this buccal reaccommodation does not appear, derive from a separate basic group and are related to the Eupotamonea of Western Africa. Some authors (Rathbun 1906, Bouvier 1917) have even singled out *Potamonautes (Erimetopus) brazzae* from the Congo Basin, as the African species with greater resemblance to the American Trichodactylidae.

6.2 *The Greater Antillean-South American disjunction*

The morphology of the exognath of third maxilliped of Pseudothelphusidae, shows a disjunct distribution in the Greater Antilles (*Epilobocera*) and the northern mountains of Colombia (Strengerianini) (Fig.9). The long exognath found in species from both areas is categorized as plesiomorphic on the basis of accommodation of the mouth parts for aerial respiration. Consequently, explanation of this disjunction also has a bearing on the hypothetical origin of this family.

The origin of the Antillean terrestrial biota is currently explained by two opposing hypotheses. The classical theory suggests that the Greater Antilles were colonized by dispersal from continental America. One common explanation has invoked the former presence of a continuous land bridge between Yucatan and Cuba (see e.g. Rivas 1958) and more recently the role of overwater dispersal, at least for terrestrial vertebrates, has been re-emphasized (Pregill 1981). On the other hand the vicariance model (Rosen 1976) proposes that the present Antillean biota is the descendant of an ancestral biota that occupied a Proto-Antillean archipelago situated between North and South America during the Mesozoic. The islands moved to their present location during the late Mesozoic and Cenozoic when a portion of Eastern Pacific seafloor wedged itself between North and South America carrying the archipelago with it. This proto-Antillean archipelago was based on the plate tectonic models of Malfait & Dinkelman (1972), but additional evidence has been appended by recent studies (Hedges 1981, Pindell & Dewey 1982, but for an opposing view see Pregill 1981). The Antillean species appear to be the most primitive of the two disjunct groups based on the shape of the gonopod and the occasional presence of a palp on the exognath. On the other hand the Strengerianini appear as a basic group from which other continental groups could be derived.[3] It is more reasonable to explain the present distribution by a) vicariance of an ancestral epilobocerine group in the Antilles and b) a secondary dispersion from Northern South America towards Central America by the Strengerianini. Using the explanation according to the classical hypothesis of dispersal would require at least the following steps: (1) dispersion from the Antilles to Central America via a hypothetical Yucatan bridge, (2) migration to Northern Suth America, (3) secondary dispersal from a Northern South America center of origin.

6.3 *The Lesser Antilles-South American disjunction*

A third disjunction is presented by *Guinotia dentata.* This species is found in the islands of St Vincent, Martinique, Dominique and Guadelupe (Fig.9), but is unrelated to the species from the Greater Antilles (*Epilobocera*). All modern taxonomic arrangements place both genera in different subfamilies. *Guinotia* belongs indeed to the tribe Kingsleyini from northern South America, particularly by its short exognath. But peculiarly, its closest relative is not the genus *Eudaniela* from the nearby coastal range of Venezuela and Trinidad, but *Freddius* from the Guianas, which it resembles in the presence of a subapical rim in the apex of the gonopod. Several vertebrate terrestrial organisms show a similar disjunct distribution in South America and the Lesser Antilles, among them several snakes (*Boa, Chironius, Bothrops,* iguanid lizards, many bird groups, a mouse opposum (*Marmosa mitis*), an agouti (*Dasyprocta aguti*) and a racoon (Rosen 1976).

The Lesser Antillean freshwater and terrestrial fauna is not older than Upper Miocene, because a general submergence of these islands took place in the lower Miocene and lasted up to the end of the Miocene, when the earth movements that produced the Andean orogeny also caused the general uplift and emergence of the arc ridge (Tomblin 1975). Between the southern end of the Antillean arc and the South American mainland ocean depths are less than 500 m, and it is possible that emerged lands were closer than at present during the Pleistocene owing to eustatic changes of sea level (approximately 100 m in Barbados, according to Matthews 1972) related to glacial events. Nevertheless, geological evidence indicates that there was no land continuity between either area. Consequently, distribution attributed to a vicariance event or dispersal through land bridges has to be ruled out, and the only plausible explanation left is chance dispersal and the further speciation after isolation. This is not altogether improbable because the occurrence of rafts for dispersal is common in the Caribbean and even the arrival of a living South American snake (*Constrictor constrictor*) in St Vincent island is well-documented (King 1962a). The distance between the Orinoco River, which drains the area of distribution of *Freddius*, and the nearest island, St Vincent, where *Guinotia* is found, is 400 km. Pseudothelphusidae, owing to their aerial breathing do not require permanent submergence in freshwater, and possibly could have bridged this distance if adequately protected against desiccation. Direct development and retention of the young — usually 25 to 100 in number — under the abdomen of the mother, insure sufficient propagules for colonization.

6.4 *Dispersal*

The areas of high taxa density found within the geographical range of the Pseudothelphusidae are not necessarily equated to radiation centers for dispersal. The one found in Tehuantepec cannot indeed be considered as a radiation center because most species found there are either intermediate links or terminal forms of chorological series. Marine transgressions occurred in this area during the Cenozoic. Although these were not complete and produced disruptions of the land surface (Durham et al. 1955), conditions of aridity may have acted as temporary barriers and promoted repeated allopatric speciations, thereby explaining the high concentration of taxa.

The second area of high taxa concentration, found in northern Colombia, is more relevant. Evidence for dispersal from this center comes from several sources: a) Sympatry between closely related groups is an indication that dispersal has occurred because primary allopatry has been transformed into secondary sympatry (Croizat et al. 1974, Brundin 1981). Sympatry is observed in several areas of the pseudothelphusid geographical range, but it is of particular interest in northern Colombia, where species with plesiomorphic and apomorphic states of the same character (relative length of the endognath) occur together. b) Through serialization of gonopod patterns it is possible to draw three lines that radiate to northern South America, to Central America and the southern Andes from a common center in northern Colombia, with apomorphies gradually appearing away from this center of radiation. c) Conclusive evidence of dispersl are those cases of preserved multi-area vicariance where several steps in morphoclines can be followed over the map (Brundin 1981). An example was presented for the Mexican species of Pseudothelphusidae (section 4.3, Fig.12), but many others can be found, for instance in *Hypolobocera* of the Amazonian watershed of Ecuador, or the species of *Microthelphusa* from the Coastal Range of Venezuela. d) Another valuable argument for the dispersal of Pseudothelphusidae could be the possible exclusion of the cambarid crayfish from former territories that they might have occupied in Central America. However, except for the indirect evidence provided by the presence of epizoic branchiobdellids of supposedly crayfish origin in crabs from Costa Rica (Holt 1964), there is no other evidence in favor of this hypothesis.

6.5 *Summary*

The evidence discussed above suggests that the present distribution of freshwater crabs cannot be understood simply in terms of either vicariance or dispersal. The Neotropical/ Paleotropical disjunct distribution of groups that share synapomorphies of the respiratory structures strongly suggests that all freshwater crabs are a monophyltic group, separated at present by a vicariant event caused by the break up of Pangaea in mid-Cretaceous times. The disjunct Greater Antilles/South America distribution of plesiomorphic characters found in the Pseudothelphusidae can be explained by a second vicariance event in the Caribbean. But the presence of chorological series and morphoclines that radiate from northern South America, is evidence of dispersal from a common distribution center. Even chance dispersal cannot be ruled out because this is the only hypothesis that could explain the peculiar disjunct distribution of sister groups in the Lesser Antilles and Guayana.

ACKNOWLEDGEMENTS

The author greatly profited from many discussions with Carlos Schubert on Caribbean geological history and with Miriam Arends on the work of the late Léon Croizat. Ernesto Medina, Robert H.Gore and Kenneth L.Heck contributed many valuable criticisms to the first draft of the manuscript.

FOOTNOTES

1. Colosi derived his ideas from the theory of hologenesis proposed by Rosa in his book *Ologenesi. Nuova teoria dell'origine e della distribuzione geografica dei viventi*, R.Bemporand & Figlio, Firenze 1918. This work was a source of inspiration also for Croizat (Nelson 1973).

2. An extreme condition in the isolation of the openings of the efferent channels is found in the genera *Strengeriana* and *Chaceus* from northern Colombia (see Rodríguez 1982: Figs.6 and 11, Rodríguez & Díaz 1981: Fig.5) in which the lower border of the opening is closed, not by the exopod of the first maxilliped, but by a projection of the lateral lobe of the epistome and a spine of the jugal angle.

3. The following homologies can be established in the gonopods of these groups. Arrow indicates the direction of the apomorphies. For illustrations of the gonopods, see Rodríguez 1982. Strengerianini → Kingsleyini. The marginal plate of the Strengerianini is found in the same position in the Kingsleyini, although in some species of this last tribe it is outgrown and bent laterad. The cephalic plate of Strengerianini is bent counterclockwise

to form the spinose mesial process of Kingleyini, while the mesial plate is reduced to the characteristic hood of Kingleyini. The field of spines, although preserving its position is reduced. Strengerianini → *Hypolobocera.* The genus *Martiana* is equally related to *Chaceus* (Strengerianini) and to *Hypolobocera.* The two spinose processes of *Chaceus*, which are the rudiments of the mesial and marginal plates of *Strengeriana*, are homologous to the spiniform processes near the spermatic channels of *Martiana.* These processes are further reduced and appear as a small papillae in *Hypolobocera. Neostrengeria* → *Hypolobocera.* Vestiges of the double lateral lobe of *Neostrengeria* appear in *Hypolobocera orientalis*, and other species. It is not possible, however, to determine which one is the plesiomorphic character. *Neostrengeria* → *Ptychophalhus.* The double lateral lobe is homologus in both genera. Derivation of other Central American an Mexican groups is presented in Footnote 4.

4. The following homologies can be established in the gonopods of the species in Figures 10 and 11. For illustrations of the gonopods, see Rodríguez (1982). 1→3. The strong mesoproximal spine of *Odonthothelphusa* could be considered as homologous with the mesial spine of *Potamocarcinus.* If this interpretation is correct, *Odonthothelphusa* could be derived from a form like *P.aspoekorum.* 1→4. *Typhlopseudothelphusa* could be derived in a similar manner as the previous genus which it closely resembles, notwithstanding its cavernicolous habits. →6. *Elsalvadoria* has affinities with both *Ptychophallus* and *Camptophallus* in the lateral lobe which is divided in two in the middle, and in the meso-cephalic lobe which is similar to the mesial lobe of *Camptophallus.* →7. Although the gonopod in *Spirothelphusa* is considerably distorted, the caudal ridge and the oval shaped apex can be homologized to the respective parts of *Ptychophallus.* 6,8→9. The gonopod of *Lobithelphusa* is also considerably distorted. The mesocephalic process, which is bifid and with a basal spine in this genus, is homologous to that of *Elsalvadoria zurstrasseni*, although in *Lobithelphusa* it is strongly bent mesially. 8→10. The hatchet-shaped lobe of *Elsalvadoria* is analogous to the mesial lobe of *Epithelphusa mixtepensis* 1969; the bifid cephalic process is conical; the cephalo-lateral spine is weak; the lateral lobe is rudimentary. 10→12. The well-developed distal part of the marginal process of *Epithelphusa mixtepensis* and *E.chiapensis* is reduced and fused with the mesial lobe in *Tehuana*; the cephalic process is also reduced. 12→14. The distal projection of the marginal process, which in *Tehuana* is partially fused, is completely fused in *Pseudothelphusa.* The species of *Tehuana* closest to *Pseudothelphusa* is *T.veracruzana*, in which the laterocephalic process, characteristic of *Pseudothelphusa*, is already present.

REFERENCES

Bott, R. 1965. Die Süsswassekrabben von Madagaskar. *Bull. Mus. Hist. Nat. Paris* (2)37: 335-350.

Bott, R. Die Süsswassekrabben von Europa, Asien, Australien und ihre Stammesgeschichte. *Abhand. Senkenberg Naturforsch. gessels.* 526: 1-338.

Bouvier, E.L. 1917. Sur la distribution des Crabes d'eau douce de la famille des Potamonidés. *Acad. Sci. Paris* 165: 751-755.

Brundin, L.Z. 1981. Croizat Panbiogeography versus phylogenetic biogeography. In G.Nelson & G.Rosen (eds.), *Vicariance Biogeography, a Critique.* New York: Columbia University Press.

Cameron, J.N. 1975. Aerial exchange in the terrestrial Brachyura *Gecarcinus lateralis* and *Cardisoma guanhumi. Comp. Biochem. Physiol.* 52A: 129-134.

Colosi, G. 1921. La distribuzione geografica dei Potamonidae. *Rev. Biol. Roma* 3: 294-301.

Cracraft, J. 1981. The use of functional and adaptive criteria in Phylogenetic Systematics. *Amer. Zool.* 21: 21-36.

Croizat, L., G.Nelson & D.E.Rosen 1974. Centers of origin and related concepts. *Syst. Zool.* 23: 431-464.

Díaz, H. & G.Rodríguez 1977. The branchial chamber in terrestrial crabs. A comparative study. *Biol. Bull.* 153: 485-504.

Durham, J.W., A.Arellano & J.H.Peck 1955. Evidence for no Cenozoic isthmus of Tehuantepec Seaways. *Bull. Geol. Soc. America* 66: 977-992.

Hodges, S.B. 1982. Caribbean bigeography: Implications of recent plate tectonic studies. *Syst. Zool.* 31: 518-522.

Holt, P.C. 1966. A new branchilbdellid (Annelida) from Costa Rica. *Tulane Stud. Zool.* 12: 1-4.

King, W. 1962. The occurrence of rafts for dispersal of land animals into the West Indies. *Quart. J. Fl. Acad. Sci.* 25: 45-52.

Malfait, B.T. & M.G.Dinkelman 1972. Circum-Caribbean tectonic and igneous activity and the evolution of the Caribbean Plate. *Bull. Geol. Soc. America* 83: 251-272.

Matthews, R.K. 1973. Relative elevation of late Pleistocene high sea level stands: Barbados Uplift rates and their implications. *Quat. Res.* 3: 147-153.

McCoy, E.D. & K.L.Heck 1976. Biogeography of corals, sea grasses and mangroves: An alternative to the center of origin concept. *Syst. Zool.* 25: 201-210.

Nelson, G. 1981. Summary. In G.Nelson & G.Rosen (eds.), *Vicariance, Biogeography, a Critique.* New York: Columbia University Press.

Nelson, G. & N.Platnick 1980. A vicariance approach to historical biogeography. *Bioscience* 30: 339-343.

Ortmann, A.E. 1902. The geographical distribution of freshwater decapods and its bearing upon ancient geography. *Proc. Am. Phil. Soc.* 41: 267-400.

Pindall, J. & J.F.Dewey 1982. Permo-triassic reconstruction of western Pangea and the evolution of the Gulf of Mexico/Caribbean Region. *Tectonics* 1: 179-212.

Pregill, G.K. 1981. An appraisal of the vicariance hypothesis of Caribbean biogeography and its application to West Indian terrestrial vertebrates. *Syst. Zool.* 30: 147-155.

Pretzmann, G. 1973. Grundlagen und Ergebnisse der Systematik der Pseudothelphusidae. *Z. zool. Syst. Evol.-Forsch.* 11: 196-218.

Rathbun, M.J. 1906. Les crabs d'eau douce (Potamonidae). *Nouv. Arch. Mus. Hist. nat. Paris* (4)7: 159-321.

Rivas, L.R. 1958. The origin, evolution, dispersal, and geographical distribution of the Cuban poeciliid fishes of the tribe Girardinini. *Proc. Am. Phil. Soc.* 102: 281-320.

Rodríguez, G. 1982. *Les Crabes d'eau douce d'Amérique, Famille Pseudothelphusidae.* Paris: ORSTOM.

Rodríguez, G. & G.Díaz 1977. Note sur quelques restes de crabes d'eau douce (Pseudothelphusidae) provenant d'un 'Kjokken-moedding' du Venezuela. *Crustaceana* 32: 107-108.

Rodríguez, G. & H.Díaz 1981. New species of freshwater crabs from the Andes (Crustacea, Decapoda, Pseudothelphusidae). *Senckenb. biol.* 61: 305-312.

Rosen, D.E. 1976. A vicariance model of Caribbean Biogeography. *Syst. Zool.* 24: 431-464.

Tomblin, J.F. 1975. The Lesser Antilles and Aves ridge. In A.E.M.Nairn & F.G.Stehli (eds.), *The Oceans basins and margins*, vol.3, *The Gulf of Mexico and Caribbean.* New York: Plenum.

Crustacean biogeography of the Late Middle Miocene Middle Eastern landbridge

F.D.POR
Department of Zoology, Hebrew University, Jerusalem, Israel

ABSTRACT

The Middle East is an area of complex tectonic history through time, and this complexity is reflected in the biogeography of the region. Essentially, six historico-biogeographical 'strata' can be recognized among the crustaceans which currently inhabit the region. These are: 1) subterranean and interstitial forms which appear to extend back to the Permo-Triassic when Pangaea was still intact; 2) subterranean and interstitial species with an origin in the Mesozoic when the Tethys Sea was present; 3) euryhaline species related to the Miocene Messinian salinity crisis; 4) marine forms which originated with the Pliocene recolonization of the Mediterranean and Red Seas; 5) wet-tropical and wet-temperate species which migrated through adjacent river systems during the Pleistocene climatic fluctuations; and 6) Lessepsian migrants between the Mediterranean and Red Seas after the opening of the Suez Canal.

1 INTRODUCTION

The Middle Eastern landbridge, consisting of the Levantine countries of Syria, Lebanon, Jordan, Israel and the Sinai Peninsula, was responsible for the dissection of the ancient circumtropical Tethys Ocean. As classically exposed by Ekman (1967), the Tethys has been seen by marine biogeographers as the cradle, the 'aureum aetas', of all the marine biota. Starting with Sylvester Bradley (1967) and culminating with the modern theory of plate tectonics, our views of the history of the Tethys and its breakup became more focused and correct. It is, for instance, evident that the Tethys came into being only after the Mesozoic fragmentation of the Triassic Pangaea. Paleontological data accumulated, showing that even at its circumglobal maximum extension, the Tethys was inhabited by at least three biogeographically distinct faunas (Adams 1967). More than anything else the recognition of the fact that the Mediterranean portion of the Tethys nearly completely dried out following the establishment of the Middle Eastern landbridge in the late Miocene put an end to the exagerrated use of the 'paradisiac' term of 'Tethys relics'.

About six million years ago, during the so-called Messinian Salinity Crisis, the Mediterranean for a few hundred thousand years became a basin occupied only by a series of hypersaline lakes. This relatively recent and no doubt the most dramatic happening left its deep mark on the biogeography of the Middle East. The drying out of this sea went hand in hand with the older emergence of the Tethyan lands of North Africa and of the Middle East. The beginning of the Pliocene was marked by a renewed oceanic transgression into the Mediterranean through the Gibraltar area. However the Mediterranean Sea never again became tropical and salinities remained frequently abnormal.

Further biotic instability to the aquatic biota of the area followed in the wake of the rapid and pulsating sequence of the Pleistocene cold spells (Por 1975a). At the same time, but not necessarily in the same sequence, the arid lands surrounding the Mediterranean in the South and the East, underwent cyclical pluvial and arid periods. The Rift Valley extending from North to South, from Anatolia to the Dead Sea had been formed in the Pliocene and offered a 'steeple-chase' pathway for biotic dispersal for inland water biota (Por 1975b).

The Red Sea, with a much less well-documented history than the Mediterranean, was initially a gulf of the Miocene Mediterranean (see e.g. Williams & Williams 1980). As such it too underwent a Messinian salinity crisis. At the beginning of the Pliocene, more or less at the same time as the Mediterranean, the Red Sea was invaded by the tropical waters of the Indian Ocean. Pleistocene climatic fluctuations had also considerable effect on the Red Sea biota; however, unlike those of the Mediterranean, they never lost their tropical affinities.

Finally, recent man-made canals, and especially the modern Suez Canal, brought the Red Sea again into contact with the Mediterranean with all the resultant biogeographic consequences (Por 1978).

2 SPECIAL PROBLEMS OF THE REGIONAL CRUSTACEAN BIOGEOGRAPHY

It is not easy to define common biogeographic denominators for a large and ecologically diversified class such as the Crustacea. Ranging from open-sea and stenohaline orders such as the Stomatopoda (mantis shrimps) to the truly terrestrial Oniscoidea (isopods), one is faced with a nearly complete cross-section of distributional types. Yet, dealing with the Middle East and its adjacent seas, one general feature is immediately obvious: the paucity of species in this region. The low diversity is reflected in the relatively few, very eurytopic species, and there is a further decrease in diversity as one advances centripetally towards the central area of the Middle Eastern lands and shores. This appears, for instance, very clearly in the zoogeographical pattern of the Bosminidae (Cladocera), presented by Lieder 1983). In America and Eastern Asia, the tropical cladoceran taxa of Gondwanan origin intermingle with the Laurasian temperate taxa. On the other hand, in the Mediterranean area, there is a distinct hiatus, which has been filled by the euryhaline species of the nearly worldwide genus *Bosmina.* This is a pattern to be found in all the freshwater crustacean taxa.

Along the shores of the open oceans there is likewise a gradual transition from the tropical fauna to the temperate one. However, in the Middle Eastern region, the tropical biota of the Red Sea were, until 1869, isolated by the Middle Eastern landbridge from the temperate Mediterranean biota. Moreover, well before reaching the physical land barrier, a marked depletion is seen in the faunal diversity of both areas occurring from South to North in the Red Sea, and from West to East in the Mediterranean.

3 HISTORICAL-GEOLOGICAL STRATA

In order to organize our discussion, it is necessary to divide the crustacean fauna of the Levantine area into historical-biogeographical 'strata', i.e. relatively arbitrary groups established in order to facilitate our analysis.

The strata proposed are as follows: 1) The Pangaean Stratum (Permo-Triassic) represented by subterranean freshwater or interstitial taxa. 2) The Tethyan Stratum of Mesozoic age terminating with the late Miocene, and consisting predominantly of subterranean and interstitial taxa. 3) The Messinian Stratum of the latest Miocene, consisting of extremely euryhaline polymorphic and epigeic species belonging to a small number of taxa. 4) The Pliocene Stratum, comprising a fauna associated with the marine recolonization of the Mediterranean and the Red Sea as well as to the lagoons of the transgressing sea. 5) The Pleistocene Stratum of arid, wet-tropical and wet-temperate species that alternated within the Middle Eastern landbridge. 6) The Historically Recent Stratum characterized by the Lessepsian migrants from the Red Sea to the Mediterranean, and other biogeographical changes that occurred during the last century as a consequence of the activities of man, nature, or both.

One stratum is missing: that of the cold water Pleistocene invaders of the Mediterranean. Although this invasion has been an important biogeographic event in the Western Mediterranean and Black Sea, it did not leave any marks in the Levant basin. Likewise, the 'Caspian' stratum, so important in the recent distributional patterns of European freshwaters, does not exist in the area. For example, *Atyaephyra desmaresti* (Decapoda, Atyidae), one of the species of the expansive oligohaline Caspian group, is probably an old inhabitant of the area, and not a new invader.

3.1 *The Pangaean stratum*

Evidently, the few existing representatives of this very old stratum are all subterranean. Probably the best example among the crustaceans that still maintain a Pangaean distributional pattern is that of the Parabathynellidae (Syncarida), represented in the Middle East by *Cteniobathynella calmani* (Por 1968b). The Parabathynellidae are present in the subterranean freshwaters of all the continents, except North America. Schminke (1973, 1975) attempted a reconstruction of the distributional history of this family, using the *Progressionsregel* of Hennig. He considered that *Cteniobathynella* served as a link between the Laurasian and the Gondwanian Parabathynellidae. The idea is supported by the fact that *C.calmani* of the Jordan Valley is the most plesiomorphic species of this genus, while *C.noodti* of Brazil, is the most apomorphic one. Recently, Dumont (1981), in describing *C.essameuri* from the Central Sahara, commented that this species is morphologically closer to *C.noodti* although geographically it is situated much closer to *C.calmani.* According to Schminke, the Parabathynellidae had to spread over Africa during the Trias, since Madagascar (which became isolated in the Jurassic) is inhabited by a genus derived from *Cteniobathynella.* The spread from Africa to South America thus had to take place before the late Cretaceous separation of these continents.

The harpacticoid copepod family Parastenocarididae, usually associated with the Parabathynellidae, has an even broader cosmopolitan distribution. Therefore it should also be considered as belonging to the Pangaean stratum. The worldwide revision of this large family, necessary for a biogeographical analysis, is still lacking (see Schminke

1976). The Middle Eastern representative of the family is *Parastenocaris nominae* (Por 1982).

The subterranean gammarid Amphipoda of the genus *Bogidiella*, considered by Stock a family: Bogidiellidae (1977, 1983) is represented in the Middle East by *Bogidiella hebraea* (Ruffo 1963) from the Jordan Rift valley. Karaman (1982) considers the *Bogidiella* 'group of genera' (in his view) to be a very old freshwater taxon, because it is found worldwide, and also on isolated oceanic islands. It occurs for instance on Kerguelen Island in the sub-antarctic Atlantic Ocean. Stock, in the papers mentioned above, believes the group to be part of the Pangaean freshwater fauna. However, he accepts that some species may have had a later origin. It is nevertheless interesting that no *Bogidiella* has yet been found in marine waters.

The situation is different with the amphipod suborder Ingolfiellidea, in whic a few species are still living in the deep sea. Stock (1977) held that the widespread species of subterranean and freshwater Ingolfiellidea are also Pangaean elements. *Ingolfiella putealis* on the other hand is a mixohaline species from Bonaire in the Caribbean, an island that was mostly covered by sea in the Pleistocene. Therefore Stock (1977) asserts that colonization of the subterranean waters is still a dynamic process. Interestingly, no Ingolfiellidea have been reported yet from the Middle East.

In the Mystacocarida, two species of the genus *Derocheilocaris*, represented by *D.tehiyae* and *D.remanei akhzivi* (Masry & Por 1970) from the Mediterranean interstitial of Israel are both Mesozoic faunal elements. As shown by Hessler (1972), the exclusively circumatlantic distribution of the Mystacocarida speaks for a pre-Cretaceous origin of this order, because the Atlantic Ocean opened during the Cretaceous.

Again among the Harpacticoida (Copepoda), many taxa of the marine interstitial habitats are characterized by a worldwide and often nearly cosmopolitan distribution. In conjunction with the centrifugal movements of the Pangaea, one may speak of a 'rafting' apart of their distributional areas.

3.2 *The Tethyan stratum*

This stratum is represented by a wealth of subterranean crustacean taxa. We speak of Tethyan relics when dealing with subterranean species which have no congenerics in the open sea, and which appear in areas that were covered by the Eocene-Miocene Tethys Sea. The classically known representatives of this fauna are Circummediterranean species (see Hubault 1938, Por 1962a,b, Fryer 1964). In recent years many taxa of Tethyan origin have been described from the Caribbean Islands by Stock and by Botosaneanu & Stock (1979, 1982, see below). There are very vew reports from India and Southeastern Asia. This situation requires a certain care when speaking of a Tethyan pattern in its circumtropical sense.

It is, of course, simplistic to see the Tethys as a uniform biogeographic realm. Several distributional patterns merely indicate old Tethyan provinciality. For instance the palaemonid genus *Typhlocaris* — represented in the Levantine area by *T.galilea* — has two other East Mediterranean species but none in the Western Mediterranean and in the tropical East Atlantic shoreland. *Typhlocaris* are relatively big prawns and could have hardly been overlooked. According to Holthuis (1956), these prawns represent a very old sidebranch of the Palaemonidae.

The amphipod genus *Metacrangonyx* often reported from high altitudes in the Sahara, the Balearic Islands and the Northeastern Sinai (*M.sinaicus*, Ruffo 1982) also has a peculiar, restricted distribution. The same is true for the subterranean cirolanide isopod *Typhlocircolana*, which has a remarkably similar distribution pattern to that of *Metacrangonyx* (i.e. deep in the Sahara, the Balearic Islands and Israel). Two species live in the Middle East: *T.steinitzi* in the northern part of Israel, and *T.reichi* south of the Dead Sea and in the Negev desert (Strouhal 1961, Por 1962b, Herbst 1982). The two species provide evidence for two separate marine transgressions in the North and in the South of Israel respectively. More recently, Botosaneanu & Stock (1979) reported a subterranean cirolanid, *Anabolana imula* from the Caribbean. Thus, the distribution of the subterranean cirolanids seems to be amphiatlantic.

The classical case of Miocenic amphiatlantic patterns are those of the Thermosbaenacea, the Microparasellidae, Stenasellidae, the hadziid amphipods, and the genus *Stygiomysis*. From this complex, only *Monodella relicta* of the thermosbaenaceans (Por 1962a), *Microcerberus remanei israelis* (Masry & Por 1970) and *Microcharon* sp. (Coineau 1970), both microparasellids, are known from the Middle East.

For the origin of all these taxa Stock (1976a, 1977) developed the concept of late Oligocenic to Miocenic 'stranding' because of a worldwide marine regression followed by successive geological 'uplifting'. In Stock's view this process of penetration into the subterranean habitats is still going on in the Caribbean area. Among the many subterranean crustaceans discovered there during the last decade, some (such as the Atyidae) are not known to have subterranean representatives in the Mediterranean area. An instructive case is that of the anthurid isopod genus *Cyathura* which, according to Botosaneanu & Stock (1982), has no less than eight troglobitic species in the Caribbean and three more in the Indo-Pacific. In the well-investigated Mediterranean there is only one single, epigeic species, the cosmopolitan *Cyathura carinata* (Negoescu 1981). Evidently, in the Mediterranean one can hardly speak of a continued process of colonization of the subterranean waters by marine crustaceans. Even though some of the species were indeed stranded during the Miocene marine regression, in the Messinian Upper Miocene the Mediterranean lost almost all of its original tropical marine fauna, including (apparently) those members able to colonize subterranean waters.

3.3 *The Messinian stratum*

The Messinian salinity crisis, that led to an almost complete drying-out of the Mediterranean during the last 700,000 years of the Miocene, was first brought to worldwide attention by Ruggieri (1967). As geological data and especially those supplied by the Deep Sea Drilling Project (DSDP) continue to accumulate (see e.g. Hsü et al. 1976) the findings have already been used for biogeographic reconsiderations by Fredj (1974) and Por (1978). A paper by Por & Dimentman (1985) deals especially with biogeographic analysis of the salinity crisis.

The very compressed picture of the Messinian event, starts with the closure of the Middle Eastern isthmus in the area of Mesopotamia and Northern Syria, around 8 million years ago (Gwirtzmann & Buchbinder 1977). The closure left two branches of the Tethys: 1) the Paratethys, a brackish sea extending north of the isthmus, and 2) the Mediterranean, or at least its Eastern basin of today. The contact of the old Mediterranean with the Atlantic Ocean was thus interrupted some 6.2 million years ago. Probably during the eight million years that preceded this event the old Mediterranean had already gone through a progressive increase in salinity. When contact with the world ocean had been lost, the surrounding areas had already undergone considerable aridization, so that the isolated Mediterranean proper, with its concomitant large scale water balance-deficit formed a series of Dead Sea-like hypersaline basins. The westernmost part, the Sea of Alboran, might have been in contact (at least temporarily) with the Atlantic Ocean and if so the conditions there were probably not so extreme. However most of the Eastern and Southern Mediterranean basins almost totally lost their original marine inhabitants. In adjacent areas, such as the Adriatic Sea, several events of riverine contact and even overflow from the brackish Paratethys may have taken place. This probably explains the improved Messinian environment in that area and the fact that Dalmatia and Apulia (Southern Italy) have a much richer subterranean relic fauna than other parts of the Mediterranean. This has been well exemplified for the subterranean Mysidacea by Ariani (1980).

The Eastern Mediterranean waterbodies were probably populated during the Messinian crisis by a very small diversity of extremely euryhaline species. Por & Dimentman (1985) speak of an association involving *Cyprideis torosa* (Ostracoda), *Diamysis bahirensis* (Mysidacea), and *Aphanius* (Pisces, Cyprinodontes). *C.torosa* is most probably a complex of physiological species widespread today in Mediterranean and inland waters and existing over a wide range of salinities from 0.9% to 70%. A similar range of salinities characterizes the Mediterranean waters inhabited by *Diamysis bahirensis* (Almeida Prado-Por 1982), although this species is known only from environments open to the sea. *Diamysis* has two other species, living in the Pontocaspian waters. The presence of *Cyprideis* is well dcumented from the Messinian deposits (Benson 1973). These crustaceans were able to survive in the residual lagoons of the Messinian Mediterranean, and develop, like today, populations that are genetically different and physiologically adapted to the different salinities. As shown by Por & Dimentman (1985), this is probably the reason for the morphological polymorphism exhibited by these taxa. Some other crustaceans, especially the harpacticoid genera *Nitocra* and *Onychocamptus* as well as the cyclopoid *Neocyclops salinarum* and several other ostracods can be added to the list of the supposed Messinian survivors: they exhibit today a remarkable euryhalinity in the waters surrounding the Mediterranean. Other taxa, like *Sphaeroma hookeri* and *S.serratum* as well as *Cyathura carinata* (all Isopoda), might have outlived the salinity crisis owing to their ability to take refuge in the diluted river estuaries. Interestingly, the extremely euryhaline calanoid copepod genus *Pseudodiaptomus*, with many species all over the world is not represented in the Mediterranean (Dussart & Defage 1983). Instead, one closely related monospecific genus, *Calanipeda aquaedulcis* is distributed in most of the Mediterranean, in the Black Sea and Caspain estuaries. It has not been reported yet from the Levant basin. According to Grindley (in press), *Calanipeda* is the ancestral form of the genus *Pseudodiaptomus.* If this is indeed so, it would be a very remarkable old survivor in the Messinian basins.

Probably none of the stenohaline marine Tethyan species survived the salinity crisis. As far as crustaceans are concerned, using Fredj's (1974) authoritative work, only *Synalpheus gambarelloides*, appears as a tethyan species of the Mediterranean: it has no Atlantic relatives and belongs to a widespread Indopacific genus-complex.

3.4 *The Pliocene stratum*

The Trubi transgression, 5.5 million years ago re-established the oceanic contact of the Mediterranean with the Atlantic. This event roughly coincides with the start of the Pliocene period. However, the Atlantic fauna that reinvaded the Mediterranean was not a tropical, but was at best a warm-temperate fauna. Coral reefs did not develop again in the Mediterranean.

At about the same time, the uplift of the Gulf of Suez and the Suez isthmus effectively separated the Red Sea basin from the Mediterranean (Ross & Schlee 1977). The opening of the Red Sea to the Indian Ocean occurred about five million yars ago. However, unlike the Mediterranean's, the Indian Ocean invaders were typical tropical biota.

The transgressive Pliocenic Mediterranean covered extensive areas of shoreland. According to Horowitz (1979), there were two such transgressions in Israel: one entering the Jordan Rift Valley, as far as the Northern Dead Sea; the other, penetrating in the south, but not reaching the Rift Valley. The brackish lagoons that became established by these transgressions are dated between 4.7-5.0 million years. By late Pliocene, all the lagoons and lakes finally became isolated from the sea and some of them partly filled-up with alluvium (Horowitz 1979). In general, one may assume that species typically found in the Jordan Rift Valley, represented by congeneric species in the Mediterranean, are relics of this Pliocenic transgression. This can also be the criterion by which to separate them from the Tethyan, pre-Messinian relics: i.e. these have no vicariant congenerics in the open Mediterranean waters. Typical Pliocenic relics are the amphipods *Echinogammarus veneris* of the Jordan valley and *Echinogammarus foxi* of the Mediterranean shores (Herbst & Dimentman 1983). A third not yet identified species of *Echinogammarus* is found south of the Dead Sea. *Echinogammarus palmyrensis* lives in the Syrian oasis of Palmyra (Tadmor), more than 200 km from the Mediterranean shores (Alouf 1975). Five other related species of *Echinogammarus* have been described from Lebanon (Alouf 1976).

Some crustacean species described from Lake Kinneret fall in the same category, namely the harpacticoid copepods *Pseudobradya barroisi* and *Nannopus palustris tiberiadis* (Por 1968a) and the ostracod *Loxoconcha galilea* (Lerner-Seggev 1968).

All these Pliocene relics are epigeic animals of marine origin, adapted to oligohaline or freshwaters. Even though these relics appear in some waters together with the very euryhaline Messinian taxa, like *Cyprideis* or *Onychocamptus*; in the latter there is no indication of morphological speciation in the Pliocenic transgressive area and they are present both in the open Mediterranean and the inland basins. This is probably due to their possible capacity to be passively dispersed from waterbody to waterbody.

In the Middle Eastern seas, the Pliocene marked the re-establishment of a marine fauna. Basically, this is a fauna which still forms the bulk of the Mediterranean and Red Sea faunas of today. In the Mediterranean, the relationship to the Atlantic fauna is evident: during the last decades, many species thought to be endemic to the Mediterranean were discovered to occur along the less investigated North African Atlantic coast. Among the 268 species of Decapoda listed from the Mediterranean, Fredj (1974) indicates 52 endemic species (18.2%). Of the Mediterranean endemics in general, Fredj considers that 86% have related, Atlantic congenerics, and the majority of these coexist with them. One could accept the view that in the short time since the Pliocene transgression some populations have evolved into new species under isolated conditions. The extremely fluctuating hydrographic conditions of the Pleistocene, undoubtedly provided such conditions. Again quoting Fredj, the differences in the fauna found below 200 m depth is instructive; for example, the temperature in the deep Mediterranean waters is about 10°C higher than in the respective depths in the Atlantic. For the bathyal Decapoda, the following vicariant pairs are given by this author: *Munidopsis marioni* (Galatheidae), *Plesionika gigliolii*

(Pandelidae), and Richardina fredericii (Stenopodidae) in the Mediterranean and *Munidopsis* sp., *Plessionika edwardsii*, and *Richardina spinicincta* in the Eastern Atlantic. Interestingly, the two species of *Plesionika* also coexist.

One would have to assume a very rapid speciation process in the Mediterranean in order to account for the numerous 'neoendemics'. In a previous publication (Por 1978) I have expressed doubts about the reality of the Mediterranean endemism. Some doubts still persist, especially for the shallow water species.

The situation in the Red Sea which also received a Pliocenic influx of marine species, has also been recently revaluated. Ekman (1967) considered that 30% of the Decapoda species are endemic to this sea. Lewinsohn (1969), dealing with the Anomura of the Red Sea disagreed with this view and found no endemic anomuran species in the Red Sea. Holthuis (1958a,b) writing about the Brachyura and the Natantia of the Northern Red Sea, also presents a picture of much less endemism than previously thought. Lewinsohn (1969) generalized from this and stated that there are no endemic littoral Decapoda in the Red Sea. It would appear that, unlike the Mediterranean, where under less extreme Pleistocene conditions species survival and speciation was possible, in the Red Sea the Pleistocene conditions exterminated most of the marine fauna. Postglacial repopulation of the Red Sea from the Gulf of Aden refugium is postulated by Klausewitz (1983) for the fishes, and might also have been the source for the Crustacea.

3.5 *The Pleistocene stratum*

The climatic fluctuations during the Glacial periods were extremely strong in the Middle Eastern continental waters as well as in the Mediterranean and the Red Sea. According to Gosse et al. (1980) the present arid conditions first started in North Africa about 1.6-2.1 million years ago. Dumont (1978) found that as late as 5500-3000 years ago there had been a stage of 'hyperaridity' in the Sahara area, responsible for many presently fragmented and isolated distributional patterns. According to Gat & Magaritz (1980) the same period was a humid maximum in the Middle East.

The eustatic sea levels also fluctuated considerably during this period. During the Calabrian, at the start of the Pleistocene, sea level in the Mediterranean was 100-125 m higher and after several fluctuations, during the last Pleniglacial, i.e. some 20,000 years ago, the level was about 130-150 m lower than today (Horowitz 1979). Extensive changes in the regime of the water circulation over the Gibraltar sill, and the concomitant fluctuations in the humidity and the temperature of the region, must have caused considerable environmental oscillations (Por 1975a). In the Red Sea the situation has been, if anything, even more extreme.

On land, the result has been the fragmentation of the old Paleotropic biogeographic realm, by an intervening desert belt, the Paleoeremic region (Por 1975b) forming the present Ethiopian and Oriental wet-tropical realms. In the Paleoeremic desert belt, comprising the Sahara, Sinai, the Negev, the Arabian desert and the Iranian desert, several crustacean taxa underwent a specific adaptation. Among the terrestrial Isopoda Oniscoidea, desert inhabiting species evolved, such as *Hemilepistus reaumuri, Porcellio olivieri* and to some degree *Metoponortus pruinosus.* These species listed by Pretzmann (1974) for the Israeli fauna, are probably only a small sample of the desert-adapted oniscoids. In the aquatic fauna, diaptomid copepods adapted to conditions of rapidly drying-out desert rain pools. Such is the case of the copepod *Mixodiaptomus chevreuxi* which lives in the Sahara and the Arabian desert (Dumont 1979). To some extent, the anostracan *Branchipus schaefferi* can also be considered to be adapted to the arid zone rainpools (Dimentman 1976). No doubt, there are more 'entomostracans' of desertic waters in addition to these two examples.

On the Middle Eastern landbridge, unlike the Sahara, the gradual formation of the Rift valley, established a steeple-chase' of north-south rivers that could allow an advance into, and a retreat from the desert area by species belonging to the wet Ethiopian and the wet Palearctic faunas. In a sense, the Nile is the first link of this steeple-chase, continued to the north by the river systems of the Jordan, the Litani and finally the Orontes. The headwaters of the Orontes have a steeple-chase connection with the Euphrates on the one hand and with the rivers of the Anatolian highland on the other (Por 1975b).

Along this pathway, wet-tropical species advanced northwards. Species of the African freshwater prawns *Caridina* (Atyidae) expanded to the lower Nile but did not reach the Jordan system.

Another species of the Nile, the Ethiopian copepod *Afrocyclops gibsoni* (Cyclopoida) lives along the Jordan River and reaches north to Lake Kinneret (Ortal & Por 1978). The African cyclopoid *Mesocyclops ogunnus* is a dominant species in the plankton of Lake Kinneret (Van de Velde 1984). A similar tropical element in Lake Kinneret was the cladoceran *Daphnia lumholtzi*, a species that disappeared from there in the last 30 years. Finally, *Bryocyclps absalomi* (Por 1982) represents an isolated report of this wet-tropical genus north of the Sahara.

Potamon potamios, the amphibious freshwater crab, widespread in the Middle Eastern landbridge, is the only representative of the rich African fauna of Potamidae. Its several subspecies in the area clearly indicate that it utilized the north-south riverine steeple-chase, mentioned above (Kinzelbach 1980).

In general, when it comes to the freshwater fauna (and also to the terrestrial one!) the Middle Eastern landbridge has to be considered a 'transition-subtraction area' in the sense of Darlington (see Por 1975b), viz. only a few species of the bordering areas penetrate into the area which basically functions as a break between two major biotic realms.

This is true also for Palearctic species, filtering into the subtraction area. Several taxa of crustaceans that advanced south along the Orontes-Litani systems stop either in the Lebanon, or in the area of the upper Jordan River ('the Nehring Line' of Por 1975b). There are several examples for this pattern: the copepod genus *Moraria* has its southernmost representative in Lebanon (Chappuis 1955), whereas *Elaphoidella* still appears with two species in the area of the headwaters of the Jordan (Por 1983). Very illuminating is the southward decrease in the diversity of the amphipod *Gammarus sensu stricto.* In Lebanon Karaman & Pinkster (1977) and Alouf (1979) recorded six species, whereas in the headwater region of the Jordan River only two of these species are still present: *G.syriacus* and *G.pseudosyriacus.* Only the last species reaches south to the 'Bodenheimer Line' (Por 1975b) at the Southern outlayers of the Judean hills (Herbst & Dimentman 1983). *G.pseudosyriacus* is, according to Alouf (1979), the most eurytopic of all the six Lebanese species.

Another freshwater amphipod, the subterranean genus *Niphargus*, has two Middle Eastern species, *N.nadarini* (Alouf 1972) still appears in the Jordan headwaters (Herbst, pers. comm.).

The southernmost representative of this distributional pattern and the only atyid shrimp in the area, is *Atyaephyra desmaresti.* At one time it was very frequent in Lake Huleh and Lake Kinneret. It disappeared from the Huleh area following the draining of this lake in 1954; in Lake Kinneret it became much less frequent, probably because of competition with the marine shrimp *Palaemon elegans*, inadvertently introduced there (see below).

Some Palearctic taxa spread south to the Bodenheimer line (see above). They form a monotonous assemblage of low species richness. The isopod *Proasellus coxalis*, the only representative in Israel of this species-rich Mediterranean genus, can still be found in isolated wells in the northernmost, now arid Negev. The diaptomid copepods, very species-rich in Anatolia, suffer a gradual restriction, until south of the Nehring line only *Arctodiaptomus similis* and *Hemidiaptomus gurneyi hebraicus* appear (Dimentman & Por 1985). In the Negev, they are replaced by the peculiar circummediterranean diaptomid of African origin *Lovenula alluaudi.* In the area of the Judean hills, the harpacticoid fauna is exclusively represented by an impoverished Palearctic stock, consisting of *Canthocamptus microstaphylinus, Bryocamptus minutus* and *Atheyella crassa* (Por 1984).

In the adjacent seas the picture is somewhat similar. In the Mediterranean, some of the more cold-stenothermic Atlantic species do not reach the warm and saline Levantine Basin. Among the well-investigated Decapoda the list contains such species as the lobsters Nephrops norveicus, Homarus gammarus, Palinurus elephas, Panulirus regius and the brachyuran *Carcinus maenas.* This is an expression of the 'cul de sac' situation of the Eastern Mediterranean (Por 1975b). The eastward impoverishment of the Atlantic taxa resulted in an ecological imbalance which has recently improved with the establishment of a tropical fauna in the area.

One extremely important fact results from the Pleistocene depauperation of the Mediterranean: there is no ongoing colonization of the subterranean near-shore waters by crustaceans. The Pleistocene Mediterranean biota, as in any other temperate biota, did not provide the supply of species for such colonization. By contrast, the Caribbean did not suffer such a strong Pleistocenic depauperation, and the process of subterranean colonization is going on.

A similar 'cul de sac' situation also developed in the Pleistocenic Red Sea: the presently existing north-south gradient of increase in salinity and decrease in temperature became extremely steep during the low eustatic sea levels of the glacial maxima. Referring to the Gulf of Elat (Aqaba), Reiss et al. (1980) and Halicz & Reiss (1981) speak of an increase of salinity values to 50‰ and a decrease of 4°C from the present 21°C, during the last glacial phase. Although at such metahaline salinities some marine biota are still able to survive it is highly probable that the present tropical fauna of the Red Sea is the result of a massive post-Pleistocenic colonization. The north-south gradient of depauperization is still very evident. Only two examples need be given here. Gorsky (1975) indicates for the Gulf of Elat four species of Euphausiacea, namely *Euphausia diomedae, E.sanzoi, Stylocheiron abreviatum* and *S.affine*, while the Red Sea 'in toto' contains six species, the Gulf of Aden ten species and the Arabian Sea 24 species. The brachyuran crab genus *Uca* has many species along the East African shores, but is only represented by two species, U.inversa and *U.tetragonon* in the Northern open Red Sea (Por et al. 1977). However, the colonization of the nearshore subterranean waters, the so-called anchialine waters (Holthuis 1973, Por 1985) in the Red Sea is most probably an active and dynamic process. The 'red' caridean shrimp *Antecaridina lauensis* found in the Southern and the Northern Red Sea, as well as *Calliasmata pholidota*, found till now only in the Northern Red Sea, have a very wide Indopacific distribution (Holthuis 1973). Another caridean inhabitant of the subterranean nearshore waters, *Periclimenes pholeter* has until now been described only from the Sinai Peninsula (Holthuis 1973). These red shrimp represent nothing else but a subterranean crustacean fauna 'in statu nascendi'. Although the post-Pleistocene origin of these shrimp in the Red Sea is possible, it is not less probable an assumption that they survived the metahaline events that befell the Red Sea during the Pleistocene, owing to their special niche, where the mitigating effect of a fresh subterranean watertable allowed survival.

3.6 *The Lesepsian and historical stratum*

The changes that the crustacean fauna of the Middle Eastern landbridge underwent during the last century are impressive. The Suez Canal was constructed across the Isthmus of Suez, in 1869, and opened to maritime traffic. About half a century later, Fox (1924), Calman (1927), Gruvel (1928), Steinitz (1929) and Monod (1930) reported on the penetration of Indopacific Decapoda into the Mediterranean. The closeness in time of these reports indicates that indeed something rather spectacular occurred around this time. The migration of the Red Sea biota into the Mediterranean, via the Suez Canal of de Lesseps, has been called 'Lessepsian migration' and summarized by Por (1978). Since that date, no new cases of Lessepsian migrant Crustacea have been brought to my knowledge. However, settlement of the Red Sea is going on in other taxa, including Mollusca (Barash & Danin 1982) and fishes (Ben Tuvia 1985). There is moreover no reason to believe that the Crustacean migration has come to an end.

There are only few cases in which the progressive spread of a species into the Mediterranean has been recorded. In other cases of putative migrants, the only circumstantial criterion that can be used is the actual presence of the species in the metahaline and winter-cold waterbodies of the Isthmus of Suez. It is usually accepted that these waterbodies and especially the Bitter Lakes, form the ecological filter for the Lessepsian migrants.

Using such criteria I determined (Por 1978) that there are 128 species which can be considered with high probability Lessepsian migrants. Of these, 35 species (27.37%) are Crustacea, 23 (18%) of which are species of Decapoda. Of the 76 species of low probability migrants, 16 are Crustacea and eight of them are Decapoda.

Crustacea are the second-most successful migrant phylum after fish. Even when ranked at lower taxonomic levels, Decapoda come after the Teleostei, and before the Gastropoda and Pelecypoda. Among the Decapoda of the Levant Basin, the migrant species already account for 21% of the total number of species.

Among the teleostean fishes that invaded the Mediterranean, Ben Tuvia (in press) mentions representatives of 13 families new to this sea. On the other hand, the Decapoda, although relatively successful migrants, supplied only a few additional genera to the Mediterranean fauna.

Only two crustaceans have been mentioned as eventual 'Anilessepsian Migrants', i.e. migrants from the Mediterranean to the Red Sea (Por 1978): the isopods *Cymodoce truncata* and *Sphaeroma serratum* (Glynn 1972). However both cases are problematic: the first probably a circumtropical species and the second, very euryhaline, could be an old inhabitant of the Suez Isthmus area.

Without doubt, only rugged and euryhaline species which had been previously selected during the adverse conditions of the Pleistocenic Red Sea, were successful in settling the Isthmus waters, including the metahaline Bitter Lakes. First to cross the Isthmus were the actively swimming peneid shrimp and portunid crabs. *Metapeneus stebbingi*, one of the early migrants is characterized by Miquel (1982) as very euryhaline and eurythermic; it lives at salinities up to 70‰ and actively migrates against increasing salinity gradients. The first migrant brachyuran decapod was the swimming crab *Portunus pelagicus*, reported by Fox (1924) in the Mediterranean. This species was already reported from the Bitter

Lakes by Krukenberg (1888) at a time when the lakes had a salinity of slightly over 60‰.

Another category of successful migrant Decapoda are the non-selective soft-bottom dwellers, such as the two leucosiid crabs *Myra fugax* and *Leucosia signata* as well as several species of pistol shrimp, Alpheidae.

Of the many reef and hard bottom species of the Red Sea only the xanthid crab *Atergatis roseus* was successful as a migrant. In this respect, the waters of the Suez Canal acted as a selective filter for all the stenohaline species of the reef community.

Among other crustacean orders only the colonization of the stomatopod *Oratosquilla massawensis* into the Mediterranean is well documented (Steuer 1936, Lewinsohn & Manning 1980). According to the later authors, *O.massawensis* is still limited in the Mediterranean to the Levant Basin.

Only a few cases can be considered Lessepsian migrants among the smaller-sized taxa, most of which were totally unknown in the Levant Basin before 1869. Such are two species of calanoid copepods, *Calanopia elliptica* and *C.media* (Berdugo 1968) and the harpacticoid copepod *Scottolana longipes* (Por 1964). There is, however, not the slightest doubt that there are many tens of species of migrants, to be discovered among the Peracarida and other diminutive taxa.

A clearly defined category are those Red Sea species that are still confined to the waters of the Suez Canal and did not yet settle into the open Mediterranean. Some of them are limited to the metahaline Sirbonian lagoon (Sabkhat Bardawil) on the Northern Sinai coast. Among these species are the isopod *Cyathura francispori* (Negoescu 1981), and the copepods *Pseudodiaptomus salinus* (Gurney 1927) and *Canuellina insignis* (Por 1969).

The geographical limits of the Lessepsian migrants in the Mediterranean are probably still similar to those given by Por (1978), i.e. to the north and northwest, the islands of Cyprus and Rhodes and the Gulf of Izmir on the Anatolian peninsula. A few decapod species (*Alpheus audouini, Metapeneus stebbingi* and *M.monoceros*, as well as *Portunus pelagicus* and *Pilumnopeus vaquelini*) seem to be successful also along the North African Coast as far as Tunis.

The Lessepsian migration is a man-made event that in its significance can easily serve as a replica for a natural process of barrier break-down and a newly established interaction between two major biogeogrphic realms. However, in addition to this, much less dramatic distributional changes have also occurred in the Middle East in recent years. For example, *Callinectes sapidus*, the edible blue crab, appeared during the 1930's on the Middle Eastern shores and lives now in the riverine estuaries there (Holthuis & Gottlieb 1955). *Palaemon elegans* has been accidentally transported to the Jordan system and expanded successfully there. This same species has also been inadvertently introduced, again with fishes for mariculture pond-stocking, to the shores of the Red Sea, where it competes with the local *P.debilis.* Mariculture also introduced to the shores of the Red Sea the Californian amphimyctic *Artemia* which competes now with the local parthenogenetic Artemia species.

4 CONCLUSIONS

From the data presented above, it is evident that the crustacean fauna of the Levantine Basin and also of the African Mediterranean shores, still bears the marks of the Messinian crisis. The freshwater fauna has been very much reduced by the desertification process that likewise started in the Neogene. The Mediterranean lost its tropical fauna and as a consequence, species richness decreased sharply, with only a small addition of cold-temperate species. In several crustacean groups a very few polymorphic species groups survived the Messinian crisis and still live in the various coastal habitats. The Red Sea on the other hand regained tropical fauna, or more precisely, underwent several population events during the Pleistocene, which were interrupted by only small-scale salinity crises.

The Mediterranean harbors some endemic species of Pliocene origin. These were able to survive the Pleistocene climatic fluctuations *in situ.* Conversely, in the Red Sea, where glacial conditions were much more extreme, there are probably very few (if any) endemic crustacean species.

The colonization of the subterranean near-shore waters by marine crustaceans, seems to have ceased in the Mediterranean, as a consequence of the Messinian crisis. In the Red Sea, the colonization might have started again, from a fresh stock of tropical marine crustaceans coming from the Indian Ocean.

The Middle Eastern landbridge is occupied today mainly by arid lands that separate the two wet tropical realms one from another, and both of them from the wet Palearctic realm. Only a few crustacean taxa became specifically adapted to desertic waters. Others, essentially eury-oligohaline species, survived in the riverine systems of the Rift valley as remnants of a Pliocenic transgression that reached the area.

The north-south riverine 'steeple-chase' system of the Rift valley facilitated the advance into the Middle Eastern landbridge, of some wet-tropical Ethiopian crustaceans and to a greater extent of a series of Palearctic crustaceans. The Palearctic crustaceans advanced southward to a biogeographic discontinuity line in the Upper Jordan Basin, the Nehring Line. To the south along the Cis-Jordanian hills a residual extremely eurytopic Palearctic crustacean fauna reached the limits of the Negev desert, the Bodenheimer Line.

The recent opening of the Suez Canal, as seen in a historical perspective, eventually began the only large-scale deviation from a ten million year old trend that turned the Levantine Basin and the Middle Eastern landbridge into a low-diversity area on the world map of crustacean distribution.

REFERENCES

Adams, C.G. 1967. Tertiary foraminifera in the Tethyan, American and Indo-Pacific provinces. In C.G.Adams & D.V.Ager (eds.), *Aspects of Tethyan Biogeography.* London: The Systematics Association 7.

Almeida Prado-Por, M.S. 1982. Two new subspecies of the *Diamysis bahirensis* Sars species group (Crustacea, Mysidacea) from extreme salinity environments on the Israel and Sinai coasts. *Israel J. Zool.* 30: 161-175.

Alouf, N.J. 1972. Description d'un nouvel Amphipode Gammaride des eaux souterraines du Liban, *Niphargus nadarini* n.sp. *Annales de Speleol.* 27/3: 547-558.

Alouf, N.J. 1975. Sur la presence d'*Echinogammarus* du groupe *pungens* au Proche Orient: *E.palmyrensis* n.sp., *E.canubiaensis* n.sp. et *E.tabu arcadiensis* n.ssp. *Bull. Zool. Mus. Univ. Amsterdam* 4/16: 125-129.

Alouf, N.J. 1976. Sur la présence d'*Echinogammarus* du groupe *pungens* au Liban: *E.berytensis* n.sp., *E.tripoliensis* n.sp. et *E.rashini* n.sp. (Crustacea, Amphipoda). *Bull. Zool. Mus. Univ Amsterdam* 5/20: 164-176.

Alouf, N.J. 1979. Sur la présence du genre *Gammarus* au Liban, avec description de deux nouveaux taxa (Crustacea, Amphipoda). *Bull. Zool. Mus. Univ. Amsterdam* 6/23: 177-185.

Ariani, A.P. 1980. Systématique du genre *Diamysis* et paléogeographie de la Mediterranée. *Journées Etud. System. et Biogeogr. Medit. Cagliari* CIESM: 121-130.

Barash, A. & Z.Danin 1982. Mediterranean Mollusca of Israel and Sinai: Composition and distribution. *Israel J. Zool.* 31: 86-118.

Benson, R.H. 1973. An ostracodal view of the Messinian salinity crisis. In C.W.Drooger (ed.), *Messinian Events in the Mediterranean.* Amsterdam: North Holland.

Ben-Tuvia, A. (1985). The impact of the Lessepsian (Suez Canal) migration on the Eastern Mediterranean ecosystems. In: M. Muraitou-Apostolopolou & Y. Kiortsis (eds.), *Mediterranean Marine Ecosystems:* 367-375. New York: Plenum.

Berdugo, V. 1968. Sur la présence en Mediterranée orientale de 2 espèces de *Calanopia* (Copepoda, Calanoida). *Rapp. Comm. Int. Explor. Mer. Medit.* 19/3: 447-448.

Botosaneanu, L. & J.H.Stock 1979. *Arubolana imula* n.gen., n.sp., the first hypogean cirolanid isopod crustacean in the Lesser Antilles. *Bijd. Dierk.* 49: 227-233.

Botosaneanu, L. & J.H.Stock 1982. Les *Cyathura* stygobies (Isopoda, Anthuridea) et surtout celles des Grandes et Petites Antilles. *Bijd. Dierk.* 52: 13-42.

Calman, W.T. 1927. Cambridge Expedition to the Suez Canal. XIII. Report on the Crustacea Decapoda. *Trans. Zool. Soc. London* 22: 211-219.

Chappuis, P.A. 1955. Mission Heinri Coiffait au Liban (1951). 9. Crustacés d'eau douce cavernicoles. *Arch. de Zool. Exp. et Gen.* 91: 533-536.

Coineau, N. 1970. Isopodes interstitiels de l'île d'Eubée (Grèce). *Biologia Gallo-Hellenica* 3: 99-104.

Dimentman, Ch. 1976. Distribution and Biology of Anostraca (Crustacea) in temporary pools in Israel and factors affecting it. Ph.D. Thesis, Hebrew University (English summary).

Dimentman, Ch. & F.D.Por 1985. Distribution and habitat characteristics of the Diaptomidae (Copepoda, Calanoida) in the inland waters of Israel and Northern Sinai. *Hydrobiologia.* 127: 89-95.

Dumont, H.J. 1978. Neolithic hyperarid period preceded the present climate of the Central Sahel. *Nature* 274: 356-358.

Dumont, H.J. 1979. Limnologie van Sahara en Sahel. Habilitation thesis. Rijksuniversiteit Gent (in Flemish).

Dumond, H.J. 1981. *Cteniobathynella essameuri* n.sp. the first representative of the Bathynellacea (Crustacea) in the Central Sahara. *Rev. Hydrobiol. Trop.* 14: 59-62.

Dussart, B. & D.Defaye 1983. *Repertoire mondiale des Crutacés Copepodes des eaux intérieures I. Calanoides.* Paris: Editions CNRS.

Ekman, S. 1967. *Zoogeography of the Sea.* London: Sidgwick & Jackson.

Fox, H.M. 1924. The migration of a Red Sea crab through the Suez Canal. *Nature* 113: 714-715.

Fredj, G. 1974. Stockage et exploitation des données en écologie marine. C — Considérations biogéographiques sur le peuplement benthique de la Mediterranée. *Mem. Inst. Océanograph. Monaco* 7: 1-88.

Fryer, G. 1964. Studies on the functional morphology and feeding mechanism of *Monodella argentarii* Stella (Crustacea:Thermosbaenacea). *Trans. Roy. Soc. Edinb.* 66: 49-90.

Gat, J.R. & M.Magaritz 1980. Climatic variations in the Eastern Mediterranean Sea area. *Naturwiss.* 67: 80-87.

Glynn, P.I. 1972. Isopoda of the Suez Canal. *Israel J. Zool.* 21: 275-300.

Goose, R., P.Rognon & F.A.Street 1980. Quaternary History of the Afar and Ethiopian Rift Lakes. In M.A.J.Williams & H.Faure (eds.), *The Sahara and the Nile.* Rotterdam: Balkema.

Gorsky, G. 1975. Distribution of the Euphausiacea (Crustacea) in the Gulf of Elat, Red Sea, Master of Sciences Thesis, Hebrew University, Jerusalem (in Hebrew with English summary).

Grindley, R.J. (in press). The Biogeography of the Pseudodiaptomidae. Crustaceana.

Gruvel, A. 1928. Repartition géographique de quelques crustacés comestibles sur les côtes d'Egypte et de Syrie. *Compt. Red. Soc. Biogeogr. Paris* 38: 45-46.

Gurney, R. 1927. Cambridge Expedition to the Suez Canal, 1924. VIII Report on the Crustacea: Copepoda and Cladocera of the plankton. *Trans. Zool. Soc. London* 22: 139-172.

Gwirtzman, G. & B.Buchbinder 1977. The desiccation events in the Eastern Mediterranean as compared with other desiccation events around the Mediterranean. In G.Bijou Duval & L.Montadert (eds.), *International Symposium on the Structural History of the Mediterranean*, Split.

Halicz, E. & Z.Reiss 1981. Paleoecological Relations of Foraminifera in a desert-enclosed sea: the Gulf of Aqaba (Elat), Red Sea. P.S.Z.N. *Mar. Ecol.* 2: 15-34.

Herbst, G.N. 1982. New records of *Typhiocirolana* species (Crutacea:Isopoda) from Israel: ecological and biogeographical significance. *Israel J. Zool.* 31: 47-53.

Herbst, G.N. & W.Dimentman 1983. Distributional patterns and habitat characteristics of Amphipoda (Crustacea) in the inland waters of Israel and Sinai. *Hydrobiologia* 98: 17-24.

Hessler, R.R. 1971. Biology of the Mystacocarida: A prospectus. *Smiths. Contr. Zool.* 76: 87-90.

Holthuis, L.B. 1956. An enumeration of the Crustacea Decapoda Natantia inhabiting subterranean waters. *Vie et Milieu* 7: 43-76.

Holthuis, L.B. 1958a. Crustacea Decapoda from the Northern Red Sea (Gulf of Aqaba and Sinai Peninsula). I. Macrura. *Bull. Sea Fisheries Res. St. Haifa* 17: 1-40.

Holthuis, L.B. 1958b. Crustacea Decapoda from the Northern Red Sea (Gulf of Aqaba and Sinai Peninsula). II. Hippidea and Brachyura (Dromiacea, Oxystomata and Grapsoidea). *Bull. Sea Fisheries St. Haifa* 17: 41-54.

Holthuis, L.B. 1973. Caridean shrimps found in land-locked saltwater pools at four Indo-West Pacific localities (Sinai Peninsula, Funafuti Atoll, Maui and Hawaii Islands) with the description of one new genus and four new species. *Zool. Verh. Leiden* 128: 48.

Holthuis, L.B. & E.Gottlieb 1955. The occurrence of the American Blue Crab, *Callinectes sapidus* Rathbun, in Israel waters. *Bull. Res. Council Israel* 5(B): 154-156.

Horowitz, A. 1979. *The Quaternary of Israel.* New York: Academic Press.

Hsu, K.J., L.Montadert, D.Bernoulli, M.B.Cita, A.Erickson, R.E.Garrison, R.B.Kidd, F.Melieres, C.Muller & R.Wright 1976. History of the Mediterranean Salinity Crisis. *Initial Reports DSDP* 42: 1053-1078.

Hubault, E. 1938. *Sphaeromicola sphaeromicola* nov.sp. commensal of *Sphaeromides virei* Valle, en Istrie et considerations sur l'origine de diverse éspèces cavernicoles périmediterranéennes. *Arch. Zool. Exp. Gen.* 180: 11-24.

Karaman, G.S. 1975. Revision of the *Echinogammarus* genus-complex (Fam. Gammaridae). *Arch. Sci. Biol. Beograd* 27: 69-93.

Karaman, G.S. 1981. Contribution to the Knowledge of the Amphipoda. 118. Revision of genus *Metacrangonyx* Chevreux 1909. *Glasnik. Republik. Zavoda. Zast. Priroda Titograd* 14: 31-48.

Karaman, G.S. 1982. Critical remarks to the recent revisions of *Bogidiella* group of genera with study of some taxa (Fam. Gammaridae). *Poljoprivreda I Sumarstvo. Titograd* 28: 31-57.

Karaman, G.S. & S.Pinkster 1977. Freshwater *Gammarus* species from Europe, North Africa and adjacent regions of Asia (Crustacea, Amphipoda). *Bijd. Dierk.* 47: 1-97.

Kiefer, F. 1981. Beitrag zur Kenntnis von Morphologie, Taxonomie und geographischer Verbreitung von *Mesocyclops leuckarti* auctorum. *Arch. Hydrobio./Suppl.* 61: 148-190.

Kinzelbach, R. 1980. Hydrobiologie am Orontes. *Nat. und Mus.* 110: 9-18.

Klausewitz, W. 1968. Remarks on the zoogeographical situation of the Mediterranean and the Red Sea. *Annali Mus. Civico Di Storia Naturale Genova* 77: 323-328.

Klausewitz, W. 1983. Die Entwicklung des Roten Meeres und seiner Küstenfische. II. Paläogeographie, Palökologie und Endemitenentwicklung. *Natur. u. Museum Frankfurt a. Main* 113: 349-368.

Lerner-Seggev, R. 1968. The fauna of Ostracoda in Lake Tiberias. *Israel J. Zool.* 17: 117-143.

Lewinsohn, Ch. 1969. Die Anomuren des Roten Meeres (Crustacea Decapoda: Paguridea, Galatheidea, Hippidea). *Zool. Verh.* 104: 1-213.

Lewinsohn, Ch. & R.B.Manning 1980. Stomatopod Crustacea from the Eastern Mediterranean. *Smiths. Cont. Zool.* 305: 1-22.

Lieder, U. 1983. Revision of the genus *Bosmina* Baird 1845 (Crustacea, Cladocera). *Int. Revue ges. Hydrobiol.* 68: 121-139.

Masry, D. 1970. *Microcerberus remanei israelis* new subspecies (Isopoda) from the Mediterranean shores of Israel. *Crustaceana* 19: 200-204.

Masry, D. & F.D.Por 1970. A new species and a new subspecies of Mystacocarida (Crustacea) from the Mediterranean shores of Israel. *Israel J. Zool.* 19: 95-108.

Miquel, J.C.E. 1982. Le genre *Metapenaeus* (Crustacea Penaeidae). Taxonomie, Biologie et Pêches Mondiales. *Zool. Verh.* 195.

Monod, Th. 1930. Uber einige Indo-Pazifische Decapoden der Meeresfauna Syriens. *Zool. Anzeig.* 92: 135-141.

Negoescu, I. 1981. New data about the fauna of Anthuridean Isopods (Isopoda, Anthuridea) from the Lybian coast and the Suez Canal. *Travaux Mus. Hist. nat. Grigore Antipa Bucarest* 23: 109-122.

Ortal, R. & F.D.Por 1978. Effect of hydrological changes on aquatic communities in the Lower Jordan River. *Verh. intern. Verein. Limnol.* 20: 1543-1551.

Por, F.D. 1962a. Un nouveau Thermosbaenacea, *Monodella relicta* n.sp. dans la depression de la Mer Morte. *Crustaceana* 4: 247-252.

Por, F.D. 1962b. *Typhlocirolana reichi* n.sp. un nouvel isopode cirolanide de la depression de la Mer Morte. *Crustaceana* 4: 247-252.

Por, F.D. 1968a. The benthic Copepoda of Lake Tiberias and of some inflowing springs. *Israel J. Zool.* 17: 31-50.

Por, F.D. 1968b. *Parabathynella calmani* n.sp. (Syncarida, Bathynellacea) from Israel. *Crustaceana* 14: 151-154.

Por, F.D. 1969. The Canuellidae (Copepoda, Harpacticoida) in the waters around the Sinai Peninsula and the problem of 'Lessepsian' migration of this family. *Israel J. Zool.* 18: 169-178.

Por, F.D. 1973. The nature of Lessepsian migration through the Suez Canal. *Rapp. Comm. int. Explor. Mer Medit.* 21: 679-682.

Por, F.D. 1975a. Pleistocene pulsation and preadaptation of biota in mediterranean seas; consequences for Lessepsian migration. *Syst. Zool.* 24: 72-78.

Por, F.D. 1975b. An outline of the zoogeography of the Levant. *Zool. Scripta* 4: 5-20.

Por, F.D. 1978. *Lessepsian Migration. The Influx of Red Sea Biota into the Mediterranean by Way of the Suez Canal.* Springer Verlag.

Por, F.D. 1982. A new species of *Bryocyclops* (Copepoda: Cyclopoida) and of *Parastenocaris* (Copepoda: Harpacticoida) from a cave in Israel and some comments on the origin of the cavernicolous copepods. *Israel J. Zool.* 30: 35-46.

Por, F.D. (in press)a. The freshwater Canthocamptidae (Copepoda: Harpacticoida) of Israel and Sinai. *Israel J. Zool.*

Por, F.D. (in press)b. An outline of the distribution patterns of the freshwater Copepoda of Israel and surroundings. *Hydrobiologia.*

Por, F.D. & Ch.Dimentman (in press). The continuity of the Messinian biota in the Mediterranean Basin. In D.J.Stanley & F.C.Wezel (eds.), *Geological Evolution of the Mediterranean Basin.* NATO Advanced Research Institute.

Por, F.D., I.Dor & A.Amir 1977. The mangal of Sinai: Limits of an ecosystem. In O.Kinne & H.P.Bulnheim (eds.), *Ecosystem Research. Helgoländer wiss. Meeresunters.* 30: 295-314.

Pretzman, G. 1974. Isopoden aus Israel (det. Strouhal). *Ann. Naturhistor. Mus. Wien* 78: 445-462.

Reiss, Z., B.Luz, A.Almogi-Labin, E.Halicz, A.Winter, M.Wolf & D.A.Rose 1980. Late Quaternary paleoceanography of the Gulf of Aqaba (Elat), Red Sea. *Quat. Res.* 14: 294-308.

Ruffo, S. 1963. Studi sui Crostacei Amfipodi. 57. Una nuova specie di *Bogidiella* (Crustacea Amphipoda) della depressione del Mar Morto. *Bull. Res. Council Israel* 11B: 188-194.

Ruffo, S. 1973. Studi sui Crostacei Amfipodi. 74. Contributo alla revisione del genere *Bogidiella* Hertzog (Crustacea Amphipoda, Gammaridae). *Boll. Soc. ent. Univ. Bologna* 5: 49-77.

Ruffo, S. 1982. Une nouvelle espèce de *Metacrangonyx* Chevreux (Amphipoda: Gammaridae) du desert du Sinaï. *Israel J. Zool.* 31: 151-156.

Ruggieri, G. 1967. The Miocene and later evolution of the Mediterranean Sea. In *Aspects of Tethyan Biogeography: 283-290.* London: System. Ass.

Schminke, H.K. 1973. Evolution, System und Verbreitungsgeschichte der Familie Parabathynellidae (Bathynellacea, Malacostraca). Mikrofauna des Meeresbodens 24. *Akad. Wiss. u. Lit. Mainz, Math. Naturwis. Klasse.*

Schminke, H.K. 1975. Phylogenie und Verbreitungsgeschichte der Syncarida (Crustacea, Malacostraca). *Verh. Dtsch. Zool. Ges.* 1974: 384-388.

Schminke, H.K. 1976. Systematische Untersuchungen an Grundwasserkrebsen-Eine Bestandsaufnahme (mit der Beschreibung zweier neuer Gattungen der Familie Parabathynellidae (Bathynellacea). *Int. J. Speleol.* 8: 195-216.

Steinitz, W. 1929. Die Wanderung indopazifischer Arten ins Mittelmeer seit Beginn der Quartärperiode. *Intern. Rev. Hydrobiol.* 22: 1-90.

Steuer, A. 1936. Cumacea und Stomatopoda von Alexandrien in Agypten. *Notizen de Deutsch-Italienischen Institutes für Meeresbiologie in Rovigno* 21: 1-19.

Stock, J.H. 1976a. A new genus and two new species of the Crustacean order Thermosbaenacea from West Indies. *Beijd. Dierk.* 46: 47-70.

Stock, J.H. 1976b. A new member of the Crustacean suborder Ingolfiellidea from Bonaire. *Studies on the Fauna of Curaçao and other Caribbean Islands* 50: 56-75.

Stock, J.H. 1977. Microparasellidae (Isopoda, Asellota) from Bonaire. *Studies on the Fauna of Curaçao and other Caribbean Islands* 55: 69-91.

Stock, J.H. 1983. Discovery of a Bogidiellid amphipod crustacean in inland waters of the East Indian Archipelago: *Bogidiella (Medigidiella) sarawacensis* n.sp. *Crustaceana* 44: 198-204.

Strouhal, H. 1961. Eine neue ostmediterrane *Typhlocirolana* Spezies (Isopoda, Cirolanidae). *Ann. Naturhist. Mus. Wien* 64: 245-256.

Sylvester-Bradley, P.C. 1967. The Concept of Tethys. In C.G.Adams & D.V.Ager (eds.), *Aspects of Tethyan Biogeography* 7. London: Systematics Association.

Van der Velde, I. 1983. Revision of the African species of the genus *Mesocyclops* Sars, 1914 (Copepoda: Cyclopidae). *Hydrobiologia* 109:3-66.

Williams, M.A.J. & F.M.Williams 1980. Evolution of the Nile Basin. In M.A.J.Williams & H.Faure (eds.), *The Sahara and the Nile.* Rotterdam: Balkema.

Zoogeographic patterns of North American Subterranean amphipod crustaceans

JOHN R.HOLSINGER
Department of Biological Sciences, Old Dominion University, Norfolk, Virginia, USA

ABSTRACT

The subterranean amphipod fauna of North America is highly diverse and presently includes 169 species representing 28 genera in 10 families or family groups. The vast majority of these species are stygobionts (i.e. obligate dwellers in aquatic subterranean habitats). Three major zoogeographic patterns are recognized among subterranean amphipods in North America: (1) ancient freshwater distribution; (2) marine relict distribution; and (3) marine-brackish water distribution. Taxa with pattern 1 distribution are exclusively freshwater forms restricted to the Holarctic region and have no known marine relatives. The large, exclusively subterranean genus *Stygobromus* (Crangonyctidae) is a principal example. Taxa with pattern 2 distribution are exclusively freshwater forms but have close relatives in marine/brackish water and are believed to have been derived from marine ancestors by stranding during marine transgressions in Late Cretaceous and Tertiary times. The weckeliid group of Hadziidae is a principal example. Taxa with pattern 3 distribution live at present in habitats ranging from fully marine to marginally fresh waters on and around islands in the West Indies, where they are believed to have become isolated during emergence of island groups in middle to late Cenozoic times. The genus *Metaniphargus* (Hadziidae) is a principal example.

1 INTRODUCTION

The subterranean amphipod fauna of North America (including Mexico, Central America and the West Indies) encompasses a highly diverse assemblage of species representing many genera and families (Table 1). The majority of species live in freshwater, but some are found in brackish and even fully marine water. Subterranean amphipods are defined herein as species that occur in subsurface habitats, including: (1) groundwaters of cave streams and pools; (2) phreatic waters exposed in wells and deep fissures; (3) interstitial media of small gravels or coarse sand either beneath (hyporheic) or beside (parafluvial) streams, or in coral rubble, coarse sand, etc. in littoral and sublittoral zones; (4) outflow of small springs and seeps through fine sediments or leaf litter above the water table

Table 1. Taxonomic representation of subterranean amphipods in North America. * Includes Mexico, Central America and the West Indies; ** includes all described species, both epigean and hypogean, from North America; *** probable family group; **** a single species recently discovered just outside of N.A. in a cave near the northern coast of Venezuela (Stock & Botosaneanu 1983).

Taxon	Total described species **	Number of stygobiont species	Number of stygophile species	Species reported outside of North America
GAMMARIDEA				
Allocrangonyctids***				
Allocrangonyx	2	2	–	No
Artesiidae				
Artesia	1	1	–	No
Bogidiellidae				
Actogidiella	1	1	–	No
Bogidiella	16	16	–	Yes
Marigidiella	1	1	–	Yes
Parabogidiella	1	1	–	No
Crangonyctidae				
Bactrurus	3	3	–	No
Crangonyx	19	5	2	Yes
Stygobromus	97	97	–	Yes
Synurella	4	–	1(?)	Yes
Gammaridae				
Gammarus	18	–	3 or 4	Yes
Hadziidae				
Allotexiweckelia	1	1	–	No
Alloweckelia	1	1	–	No
Holsingerius	1	1	–	No
Mayaweckelia	2	2	–	No
Metaniphargus	11	11	–	Yes****
Mexiweckelia	2	2	–	No
Paramexiweckelia	1	1	–	No
Saliweckelia	2	2	–	No
Texiweckelia	1	1	–	No
Texiweckeliopsis	1	1	–	No
Weckelia	1	1	–	No
Melitidae				
Paraweckelia	1	1	–	No
Psammogammarus	2	1	–	Yes
Quadrivisio	1	–	1	Yes
Pardaliscidae				
Spelaeonicippe	1	1	–	Yes
Sebidae				
Seborgia	1	1	–	Yes
INGOLFIELLIDEA				
Ingolfiellidae				
Ingolfiella	6	6	–	Yes

(hypotelminorheic); and (5) landlocked pools or cave lakes in limestone, coral or basalt with subterranean connections to the sea (anchialine).

The great majority of subterranean amphipods are stygobionts. Stygobionts (sometimes called phreatobites, or troglobites if they are associated with caves per se) are obligate subterranean species characterized by loss of eyes and pigment and often have attenuated bodies and/or appendages. A few species, however, occur in both epigean and hypogean habitats, show little appreciable morphological modification for subterranean existence, and are called stygophiles.

Table 1 includes a list of all genera associated with subterranean waters in North America. Excluded from the list are a few coastal marine species of Hadziidae (viz. *Dulzura* and *Protohadzia*), which, although eyeless or nearly so, have been collected from 'open' water where they apparently were not associated with subterranean habitats as defined above. A total of 28 genera representing 10 families or family groups and two suborders contain subterranean species in North America. These genera are in turn represented by 199 described species, of which 161 are stygobionts and seven or eight are stygophiles. Approximately 81% of the total number of species are stygobionts, undoubtedly a reflection of the strong tendency for these taxa to successfully colonize subterranean habitats.

With few exceptions, the classification system utilized in this paper follows that recently advocated by Bousfield (1978, 1982a, 1983). Parts of this system remain controversial, and a few workers still hold to the older concept of a much broader family Gammaridae which, by earlier definition, included all or most of the families listed in Table 1 except Bogidiellidae, Pardaliscidae, Sebidae and Ingolfiellidae. Many taxonomic problems remain unresolved among gammaridean amphipods, especially at the family and superfamily levels, but progress is steadily being made.

The principal objective of the present treatment is to examine the zoogeographic patterns of North American subterranean amphipods, with interpretations based on data from current distributions, hypothesized phyogenies, and postulated vicariant and/or dispersal events that are believed to have influenced the evolutionary history of these organisms. Much of the emphasis, however, will be on Crangonyctidae and the weckeliid group of Hadziidae, both of which have been the focal point of my research for a number of years (e.g. Holsinger 1966, 1967, 1969, 1971, 1973, 1974, 1977a,b, 1978, 1982, 1986a, Holsinger & Longley 1980, Holsinger & Minckley 1971, Holsinger & Peck 1968). Some of my earlier views on the zoogeography of subterranean crangonyctid amphipods (especially *Stygobromus*) have changed considerably since the 1960s and early 1970s (viz. Holsinger 1966, 1967, 1971, 1974) as more species have come to light and the concepts of plate tectonics and vicariance biogeography have become firmly established. More recent papers (Holsinger 1977a, 1978, 1986a), including the present one, reflect these changes in perspective. The zoogeography of Bogidiellidae and other hadziids (viz. *Metaniphargus*) has been analyzed in detail in recent papers by Ruffo (1973), Ruffo & Vigna Taglianti (1977) and Stock (1977a,b, 1980a, 1981, 1983) and will be touched on only briefly in my treatment. Subterranean amphipods would appear to be excellent candidates for zoogeographic studies. They are rapidly becoming well-known taxonomically, are restricted geographically by limited dispersal abilities, belong to lineages that apparently are comparatively old phylogenetically (see Bousfield 1982b, 1983), and are adapted to relatively old, stable habitats which, in contrast to surface (epigean) habitats, tend to be better protected against environmental perturbation.

2 MAJOR ZOOGEOGRAPHIC PATTERNS

The distribution of the North American subterranean amphipod fauna reflects three major zoogeographic patterns that I will designate: (1) ancient freshwater distribution; (2) marine relict distribution; and (3) marine-brackish water distribution. The first pattern is illustrated by taxa that occur in continental North America north of Mexico, that apparently have occupied freshwater habitats for long periods of geological time, perhaps in some instances since the early Mesozoic, and have no known closely related marine relatives at present. Included here are *Allocrangonyx* and the crangonyctids.

The second pattern includes those taxa that occur at present in freshwater habitats but live either near marine coastal areas or in areas that were exposed to shallow marine embayments in Late Cretaceous and/or Tertiary times. This pattern is exemplified by taxa found in Texas, Mexico, Central America and some islands in the West Indies and includes *Artesia*, many species of Bogidiellidae, the weckeliid group of Hadziidae, and a few species of *Metaniphargus, Paraweckelia* and *Seborgia.*

The third pattern is less well defined than the first two, because it is illustrated by taxa that live primarily in marine to brackish waters of littoral/sublittoral interstitial and anchialine habitats, but which are sometimes found in wells, caves and similar places near the coast with water that is marginally fresh (ca.400-1000 mg/l dissolved salts). This pattern is exemplified by taxa endemic to islands in the West Indies and is represented by a few species of Bogidiellidae, most species of *Metaniphargus, Saliweckelia, Psammogammarus, Spelaeonicippe, Ingolfiella* and probably *Quadrivisio.*

Possibly even a fourth pattern is suggested by the distribution of *Gammarus* (Gammaridae *sensu stricto*), a large genus represented in both marine and freshwaters over much of the Holarctic region (see Barnard & Barnard 1983). However, because this genus is predominantly epigean and contains only a few stygophiles that have apparently invaded cave stream habitats in recent times, I have not designated a major zoogeographic pattern for its distribution in this paper. Of the 10 or so strictly freshwater species described from North America, only four are recorded from subterranean habitats. Although three of these species (viz. *G.acherondytes, G.minus* and *G.troglophilus*) are represented by populations that sometimes exhibit degenerate eyes and weak pigmentation when found in cave waters (see Bousfield 1958, Holsinger & Culver 1970, Holsinger 1972), none is probably a stygobiont by strict definition. Subterranean populations probably have invaded cave streams since the Pleistocene by migrating upstream from springs. In many areas, distribution along an underground stream course may be more or less continuous from a spring outlet into an adjoining cave.

The first two patterns will be examined in some detail below, especially with respect to the crangonyctids and weckeliids. The taxa in pattern 3 have, for the most part, been analyzed taxonomically and zoogeographically in recent papers by Stock (1976, 1977a,c, 1979, 1980b, 1981, 1983) and Stock & Vermeulen (1982) and will not be discussed further. However, the zoogeographic significance of pattern 3 distribution should not be overlooked, because the ecology of these species may very well represent an important transitional stage between marine and fresh waters, similar no doubt to that experienced by ancestors of pattern 2 taxa now living in areas formerly exposed to marine or brackish waters. Pattern 3 taxa are frequently closely allied morphologically with forms now fully isolated in freshwater habitats, especially among the bogidiellioids and hadziids.

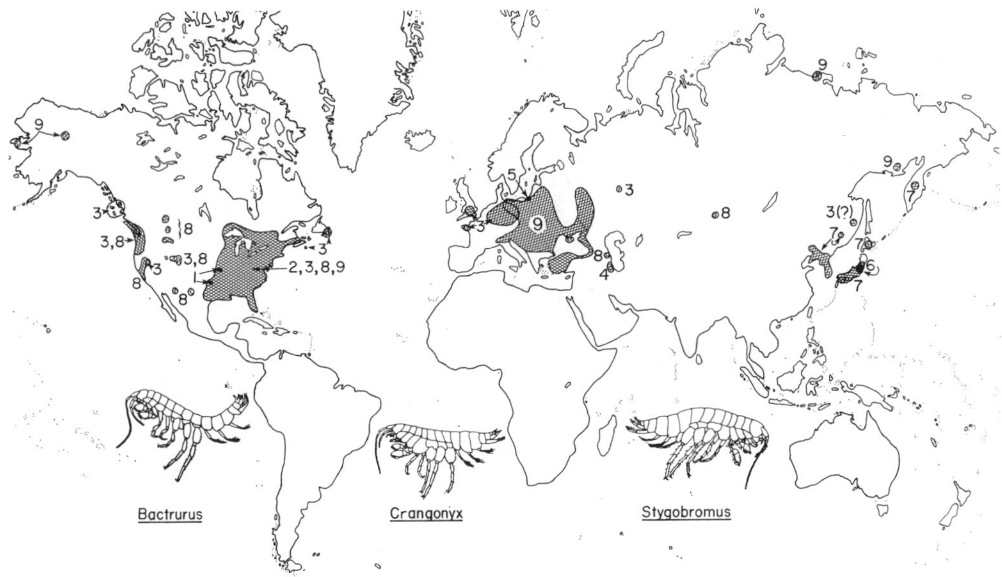

Figure 1. Generalized geographic distribution of Crangonyctidae *sensu stricto* and related genera (*Allocrangonyx, Procrangonyx* and *Pseudocrangonyx*) in the Holarctic region: (1) *Allocrangonyx*; (2) *Bactrurus*; (3) *Crangonyx*; (4) *Lyurella*; (5) *Palaeogammarus* (fossil); (6) *Procrangonyx*; (7) *Pseudocrangonyx*; (8) *Stygobromus*; (9) *Synurella*. (*Bactrurus* and *Crangonyx* redrawn from Holsinger 1986a).

3 ZOOGEOGRAPHIC ANALYSIS AND INTERPRETATION

3.1 *Ancient freshwater distribution*

This distributional pattern is clearly exemplified by both Crangonyctidae and *Allocrangonyx* which, along with the allied genera *Pseudocrangonyx* and *Procrangonyx* from eastern Asia, are restricted to continental freshwaters of the Holarctic region (see Fig.1) and contain numerous subterranean species. Although *Allocrangonyx* and the east Asian genera are sometimes included in the family Crangonyctidae in the broadest sense (see Bousfield 1982a, 1983, Holsinger 1986a), adherence to strict monophyly precludes their assignment to this family. The fully transformed state of character 10 described in Table 2 is apomorphic for Crangonyctidae *sensu stricto* when compared with gammaridean amphipod out-groups, and is most parsimoniously interpreted as having evolved only once in the crangonyctid clade, as indicated in Figure 2. Thus, whereas Crangonyctidae *sensu stricto* and its satellite genera probably share an early evolutionary history in Holarctic freshwaters and have been placed together in the superfamily Crangonyctoidea because of overall similarity (see Bousfield 1982a, 1983), these genera, in my opinion, represent independent clades that split off long ago from the lineage leading to modern crangonyctids and merit full family-level recognition (Holsinger in prep.).

Ecologically, crangonyctids and their allies are cold-stenothermal, photonegative amphipods, known almost exclusively from small, relatively secluded biotopes closely associated with groundwater (Holsinger 1977a, 1986a). Approximately 126, or 82% of the 154 decribed extant species, are stygobionts (Holsinger 1986a). The family Crangonyctidae *sensu stricto* is composed of six genera, of which three — *Crangonyx, Stygobromus* and *Synurella*

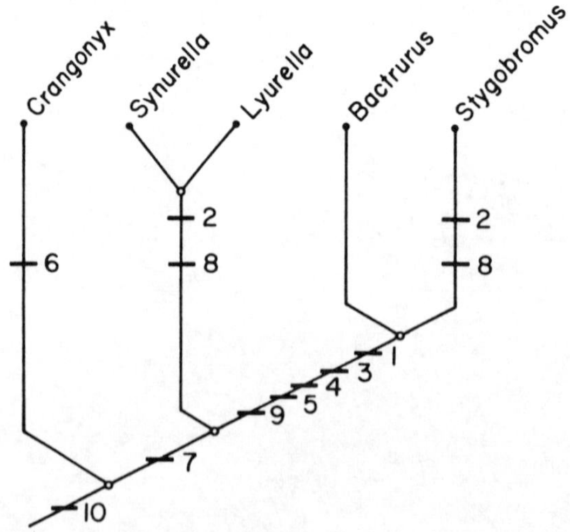

Figure 2. Hypothesized phylogenetic relationships among the extant genera of Crangonyctidae based on the character states given in Table 2. Black bars are inferred apomorphies. Plesiomorphies not shown.

Table 2. Characters used to analyze the phylogenetic relationships among the extant genera of Crangonyctidae sensu stricto illustrated by the cladogram in Figure 2. * Absent in a few species (possible reversal).

Character	Plesiomorphic state	Apomorphic state
1 Calceoli on antenna 2 of mature male	Present	Absent
2 Inner margin of outer plate of maxilliped	With row of spines	Spines reduced to 1 or 2, or absent
3 Coxal plates (especially 1-4)	Broad, deep	Reduced in overall size
4 Posterior corner of pleonal plates (especially 1 and 2)	Produced and/or acuminate	Not produced; rounded or indistinct
5 Peduncular process of uropod 1 of male	Absent	Present*
6 Outer ramus of uropod 2 of mature male	Unmodified	Modified
7 Outer ramus of uropod 3	2 times length of peduncle	Subequal to or shorter than peduncle
8 Inner ramus of uropod 3	Scale-like/vestigial	Absent
9 Eyes	Present	Absent
10 Palmar spines of propods of gnathopods	Simple	Distally bifid; densely packed in double row

— occur in both North America and Eurasia. *Bactrurus* is endemic to eastern North America, whereas *Lyurella*, a monotypic genus very closely allied with *Synurella* (see Holsinger 1977a), is endemic to a small area near the Caspian Sea. The sixth genus, *Palaeogammarus*, is based on three fossil species of Eocene to Oligocene age described from Baltic amber.

Crangonyctid amphipods and the satellite genera mentioned above appear to represent a very old group of freshwater organisms that were well established on the Laurasian landmass prior to the separation of North America and Eurasia in the late Mesozoic.

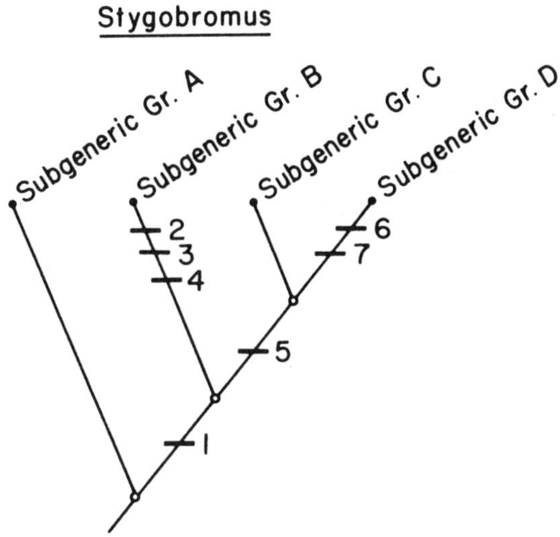

Figure 3. Hypothesized phylogenetic relationships among the subgeneric groups of *Stygobromus* based on the character states given in Table 3. Black bars are inferred apomorphies. Plesiomorphies not shown.

Table 3. Characters used to analyze the phylogenetic relationships among the subgeneric groups of *Stygobromus* illustrated by the cladogram in Figure 3.

Character	Plesiomorphic state	Apomorphic state
1 Sexual dimorphism	Weak; male smaller than female	Usually strong; male larger than female
2 Propod of gnathopod 2 of mature male	Subequal to or little larger than that of first	2-3 times larger than that of first
3 Palmar margin of gnathopod propod 2 of mature male	Even; without excavation	With excavation
4 Dactyl of gnathopod 2 of mature male	Normal; not thickened	Thickened
5 Propod of gnathopod 1 in both sexes	Subequal to or little smaller than that of second	Typically larger than that of second
6 Palmar margin of gnathopod propod 1 in mature male	Even; without excavation	With excavation
7 Anterodistal lobe of basis of pereopod 7 of mature male	Indistinct	Well developed

This hypothesis is corroborated by four lines of evidence: (1) crangonyctids and their allies are exclusively freshwater amphipods without close taxonomic relationship to any marine forms; (2) the three largest genera (i.e. *Crangonyx, Stygobromus* and *Synurella*) have a Holarctic distribution pattern, with representative species in both North America and Eurasia; (3) the large number of species, diversity of genera, and colonization of subterranean waters on a grand scale over an immense geographic area may suggest a long period of adaptive radiation and speciation in the freshwater environment of the Northern Hemisphere; and (4) the fossil genus *Palaeogammarus*, dated as being of Eocene or Oligocene age and apparently closely allied morphologically with modern *Crangonyx* (see Just 1974, Holsinger 1977a), implies that the 'crangonyctid facies' had already evolved in freshwater forms by the early Tertiary.

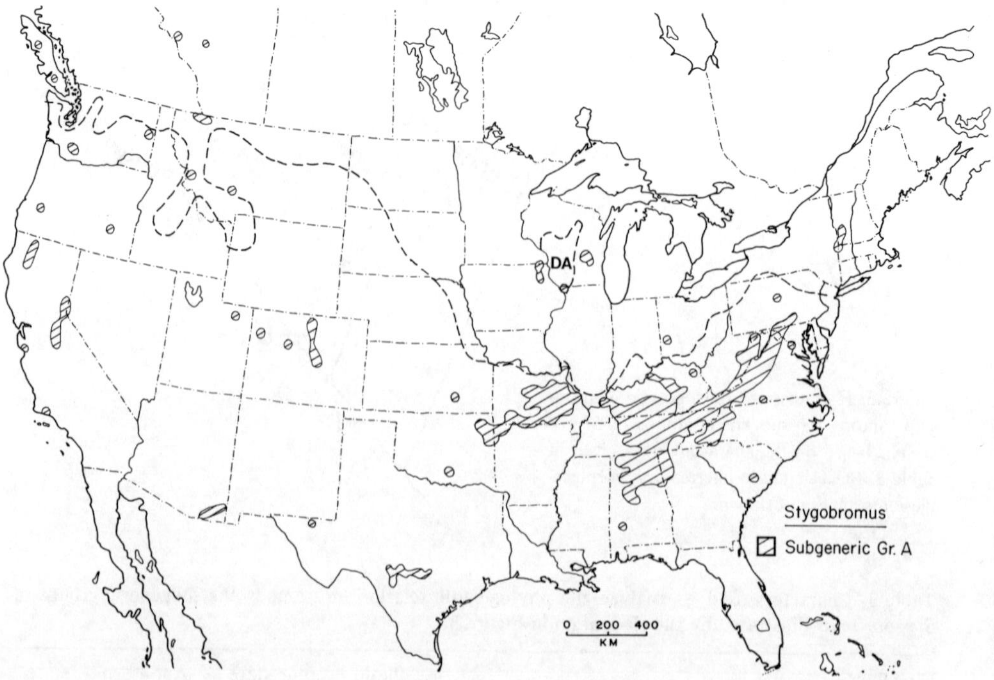

Figure 4. Geographic distribution of species in subgeneric group A of *Stygobromus* in North America. Continuous shading indicates generally contiguous distribution. Distribution data from both described and undescribed species. Heavy dashed line indicates approximate maximum extent of continental glaciation during the Pleistocene. DA – Driftless Area of Wisconsin.

Let us now turn our attention to a brief zoogeographic analysis of individual genera. Chief among these in importance is the exclusively subterranean genus *Stygobromus*, which contains 97 described species in North America and two in Eurasia. The Eurasian species occur in the Soviet Union, one from Teletzkoye Lake in south-central Siberia (see Holsinger 1977a, 1986a), and the other from a spring just west of the Caspian Sea. The latter species, *S.apscheronia*, has until now been assigned to *Synurella* (Holsinger 1986a), but a careful re-examination of its description suggests that it should be placed in *Stygobromus* (Holsinger in prep.).

Taxonomically, *Stygobromus* can be divided into at least four subgeneric groups (formal recognition of subgenera is in preparation), the suggested phylogenetic relationships of which are illustrated by the cladogram in Figure 3. For purposes of discussion here, these subgeneric groups are designated A, B, C and D. Their geographic distributions are indicated in Figures 4 and 5. Clearly, the largest and most widely distributed subgeneric group is A (Fig.4) which, in addition to containing the majority of species, also has fewer apomorphic characters (see Fig.3). This subgeneric group contains 79 described species (plus approximately 62 provisionally recognized, undescribed species) of which 63 have been assigned to 14 species groups by Holsinger (1967, 1978, 1980). Subgeneric group B is composed of four described and one undescribed species, representing three species groups, only one of which has been formally recognized to date (Holsinger 1978). Subgeneric group C is based solely on the *tenuis* group, previously recognized by Holsinger

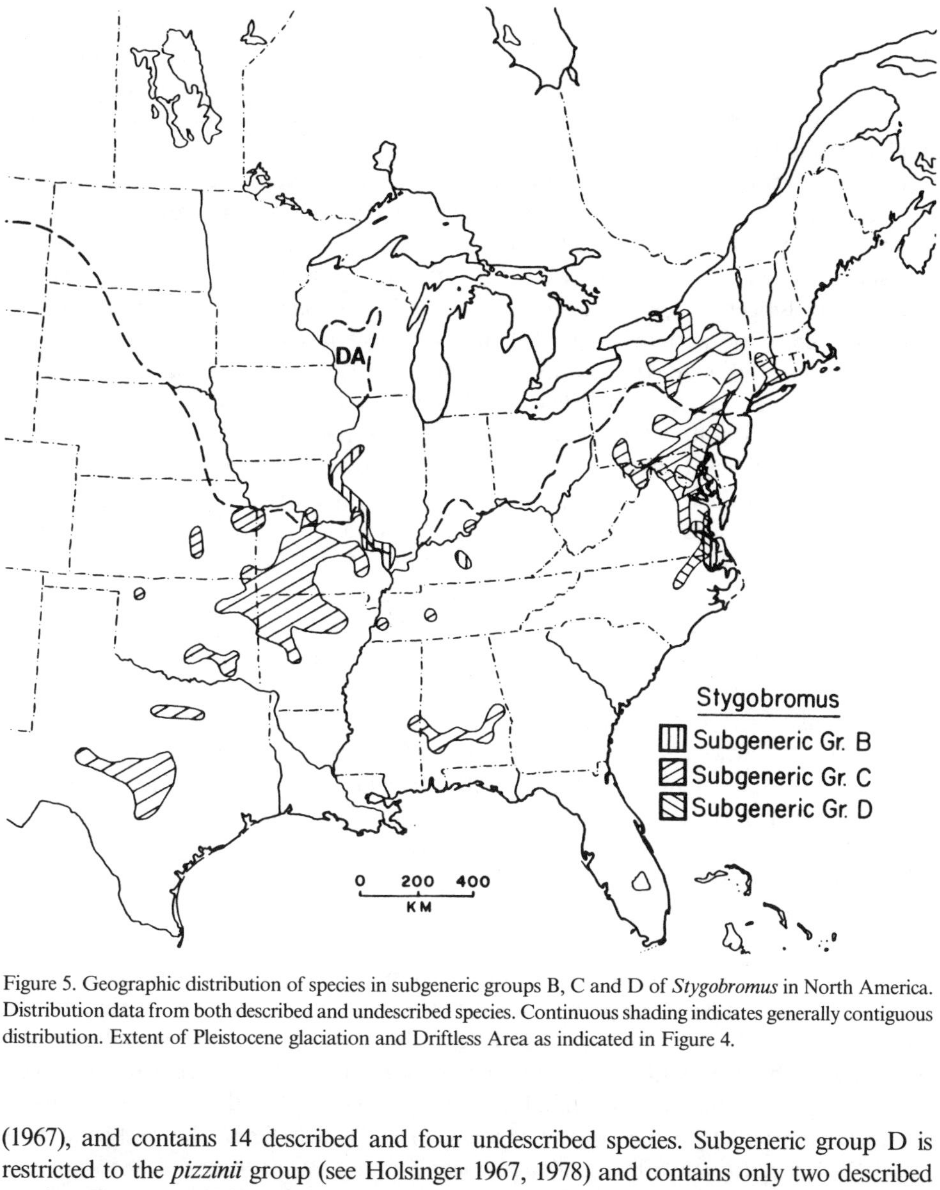

Figure 5. Geographic distribution of species in subgeneric groups B, C and D of *Stygobromus* in North America. Distribution data from both described and undescribed species. Continuous shading indicates generally contiguous distribution. Extent of Pleistocene glaciation and Driftless Area as indicated in Figure 4.

(1967), and contains 14 described and four undescribed species. Subgeneric group D is restricted to the *pizzinii* group (see Holsinger 1967, 1978) and contains only two described species. The geographic distribution of these subgeneric groups corresponds rather closely to the branches of the biological cladogram in Figure 3. For instance, the oldest and presumably most 'primitive' subgeneric group (A) is also the largest and most widespread, with significant concentrations of cavernicolous species in major karst regions of the eastern and central United States (e.g. Appalachians, Interior Low Plateaus, Ozarks, and Edwards Plateau of south-central Texas). Subgeneric group A also occurs in the Cordillera of western North America and is represented by species as far north as southwestern Canada. The two Eurasian species belong to this subgeneric group as well.

In contrast to A, the ranges of subgeneric groups B, C and D are much more restricted (Fig.5). With the exception of some significant overlap of their ranges with that of A in the Ozarks and Edwards Plateau and a small amount of overlap in central Kentucky and the central Appalachians, groups B, C and D occur on the periphery of the range of A. Except in a few areas where species of group C have invaded caves (viz. in Texas, New York and Missouri), more than half of the species in the latter three groups inhabit groundwater habitats outside of karst areas,

It is clear from Figures 4 and 5 and the data reviewed above that species richness in *Stygobromus* is far greater in the eastern half of North America. West of the Great Plains, *Stygobromus* is represented by only 23 described species (plus approximately 10 provisional recognized but undescribed species) of which 18 are assigned to the *hubbsi* group (see Holsinger 1974, Ward 1977) and one to the *canadensis* group (Holsinger 1980). Moreover, all of the western species belong to subgeneric group A. Probably the best explanation for this marked disparity in diversity is based on regional differences in geomorphology and to a lesser extent on climate. Many of the physiographic regions in the eastern and central United States having large numbers of species, such as the Appalachians, Interior Low Plateaus, Ozarks, and Edwards Plateau, are characterized locally by karst terranes with numerous caves and extensive groundwater aquifers. These conditions have provided many opportunities for the invasion and colonization of aquatic subterranean habitats over broad areas (Holsinger 1967, 1978). In western North America, on the other hand, cave density is much lower and karst areas with extensive groundwater aquifers are greatly limited. Moreover, tectonic activity since the middle Cenozoic has been very pronounced in the Cordilleran region of western North America, resulting in extensive faulting, uplifting and vulcanism (King 1958, Dott & Batten 1976). Consequently, many potential habitats for stygobromids have probably developed more recently there than in the east, where, for the same period of time, geological conditions have been relatively stable (see Ward & Holsinger 1981). In addition, the humid climatic regime of the eastern half of North America would continually enhance the development and recharge of groundwater aquifers, whereas the drier climate that prevails in many parts of western North America would not be generally favorable to the development or maintenance of these conditions.

Although it is not yet possible to date the arrival of stygobromid amphipods in western North America, a few clues are available which suggest an approximate time. Either progressive aridity in the Great Plains since the Miocene or the destruction of groundwater habitats by glaciation in southern Canada and the northern United States during the Pleistocene, or both, might have served as major vicariant events, splitting what was once almost certainly a continuous distribution of *Stygobromus* across the Great Plains from the central United States to the Cordilleran region in late Tertiary times. The most persuasive piece of evidence for an earlier continuous distribution and subsequent vicariance is the presence in southeastern Wisconsin of an unequivocal member of the *hubbsi* group (Holsinger 1974), presently separated from its nearest closely related species in Colorado by a distance of approximately 1400 km (see Fig.4).

Pleistocene glaciation has undoubtedly affected the distribution of *Stygobromus* in North America. Whereas most of a pre-Pleistocene groundwater amphipod fauna would have been extirpated by glaciation, recent discoveries of species endemic to areas far north of the southern limits of continental glaciation (see Figs.4, 5) suggest that some species of *Stygobromus* were able to survive the potentially inimical effects of glaciation in deep groundwater refugia beneath the ice. A strong case for subglacial refugia has been made by me elsewhere (see Holsinger 1980, 1981), but the most compelling evidence assessed

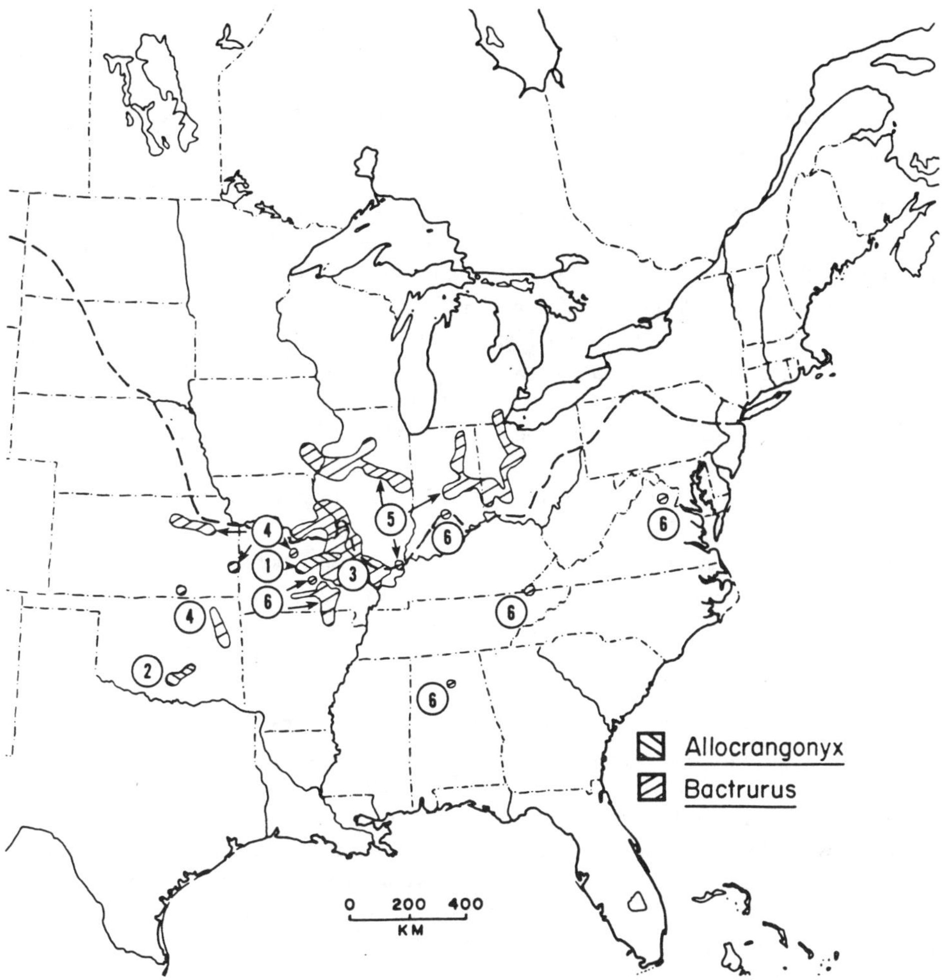

Figure 6. Geographic distribution of species of *Allocrangonyx* and *Bactrurus* in North America. *Allocrangonyx*: (1) *A.hubrichti*; (2) *A.pellucidus*. *Bactrurus*: (3) *B.brachycaudus*; (4) *B.hubrichti*; (5) *B.mucronatus*; (6) *Bactrurus* spp. (5 or 6 undescribed species). Extent of Pleistocene glaciation as indicated in Figure 4.

to date is based on the occurrence of *S.canadensis* (Holsinger 1980), which inhabits a cave in the Canadian Rockies that is still covered in part by an active glacier.

Bactrurus and the non-crangonyctid genus *Allocrangonyx* are, as is *Stygobromus*, exclusively subterranean genera containing only stygobiont species. Unlike *Stygobromus*, however, the distributions of these genera are much more narrowly circumscribed and restricted to small areas in the central and eastern United States (see Fig.6). Among the crangonyctids, *Bactrurus* appears to be more closely related cladistically to *Stygobromus* than to other genera in the family (Fig.2).

The highly disjunct nature of the ranges of species of *Bactrurus* is difficult to interpret but may reflect a more continuous distribution in the past, now long since fragmented by vicariant events which still remain to be elucidated. One species of this genus,

B.mucronatus, occurs almost entirely in glaciated areas of the east-central United States (Fig.6), where it inhabits the interstices of loosely consolidated drift near major stream valleys and is usually sampled from the outlets of shallow drain tiles on the edge of cultivated fields in glacial drift plains (Holsinger 1981). Unlike a number of species of *Stygobromus* recorded from glaciated areas, *B.mucronatus* is not believed to have survived glaciation in groundwater refugia, but instead has probably colonized glaciated areas in post-Pleistocene times. This hypothesis is based on the fact that the species lives in biotopes that are composed directly from glacial deposits and is represented by a single cave population in southern Illinois just south of the glacial limit (Holsinger 1981).

Allocrangonyx shares a number of important characters with the crangonyctids but also shows some rather remarkable morphological convergence with the Palearctic subterranean genera *Niphargus* and *Pseudoniphargus* (Holsinger 1971, 1986a). Nevertheless, the origin of this genus remains obscure, and its present distribution, which is limited to small areas in the Ozarks of central Missouri and the Arbuckle Mountains of south-central Oklahoma, suggests that it is a very old relict.

Crangonyx is relatively widely distributed in North America (Fig.1) but only a small remnant persists today in Europe (three stygobionts, plus one epigean species apparently introduced into Great Britain from North America) and possibly eastern Asia (viz. *C.arsenjevi*, questionably a congener — see Holsinger 1986a). Of the 22 or 23 described species, less than one-half are stygobionts. Five species from North America are stygobionts and primarily inhabit caves in the eastern half of the United States. Two other species, *C.floridanus* and *C.forbesi*, are stygophiles (see Fig.7). In addition, there are a number of undescribed species in the *packardi* complex which, when described, may raise the total number of stygobionts in this genus in North America to about 10.

The extensive distribution of *Crangonyx* (Fig.1) suggests that it is a very old genus in Holarctic freshwaters. However, despite a number of stygobionts, the genus has probably been represented in subterranean waters, in North America at least, for a much shorter time than the more highly specialized (stygomorphic) stygobiont genera *Allocrangonyx, Bactrurus* and *Stygobromus.* Not only do the stygobionts in *Crangonyx* appear to be less specialized morphologically for a subterranean existence, but they are, with the exception of *C.hobbsi*, closely allied with epigean congeners, suggesting a recent evolutionary nexus between surface and subterranean species. In contrast, the exclusively stygobiont genera have no known epigean congeners.

The stygobiont species of *Crangonyx* are found primarily in the Appalachians, Interior Low Plateaus, and karst regions of Florida. In the latter, *Crangonyx* is the only subterranean amphipod, a fact that may appear rather curious at first glance, because the Floridan peninsula is the only part of unglaciated North America north of Mexico with extensive groundwater aquifers lacking species of *Stygobromus.* There may be two explanations for the presence of *Crangonyx* and the absence of *Stygobromus* in the caves of Florida. First, the limestone terranes of this region are much younger (i.e. Cenozoic) than the Paleozoic limestones of the Appalachians, Interior Low Plateaus, and Ozarks. According to a study on the distribution of troglobitic crayfishes in Florida by Franz & Lee (1982), the colonization of caves by aquatic freshwater crustaceans in this region would be no older than late Miocene or Pliocene, prior to which time freshwater habitats were probably nonexistent. Second, the distribution of epigean *Crangonyx* extends along the Coastal plain to north Florida and suggests the possibility that potential ancestors of subterranean species were in the vicinity of the Florida peninsula during late Tertiary or Quaternary

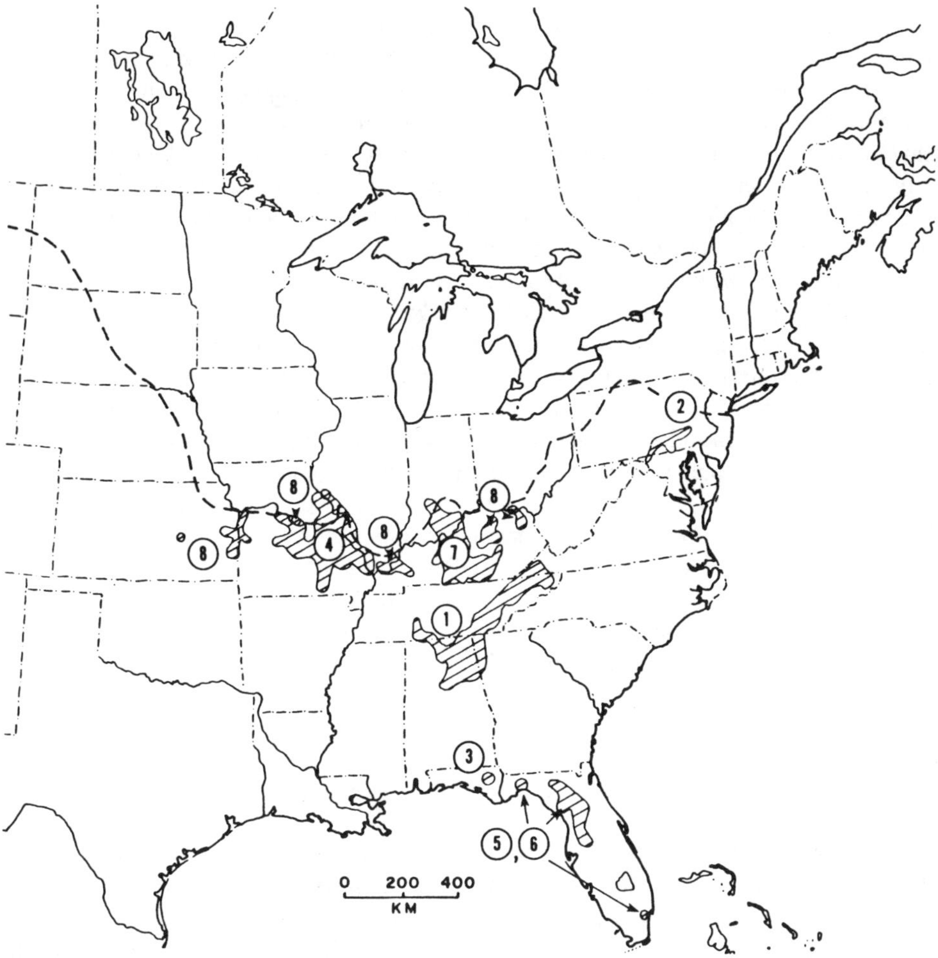

Figure 7. Geographic distribution of subterranean species of *Crangonyx* in North America: (1) *C.antennatus*; (2) *C.dearolfi*; (3) *C.floridanus* (cave localities only); (4) *C.forbesi* (including epigean localities); (5) *C.grandimanus*; (6) *C.hobbsi*; (7) *C.packardi*; (8) *Crangonyx* spp. (undescribed species in the *packardi* complex). Continuous shading indicates generally contiguous distribution. A few scattered cave localities for stygophile species in the *gracilis* group from southern West Virginia, middle Tennessee and elsewhere not shown. Extent of Pleistocene glaciation as indicated in Figure 4.

times. *Crangonyx grandimanus*, one of the two Florida troglobites, is closely allied morphologically with members of the *obliquus-richmondensis* complex which are recorded from southern Georgia and northern Florida (Holsinger 1972). In contrast, the species of *Stygobromus* nearest to Florida are found in southern Alabama (see Figs.4, 5).

Synurella, like *Crangonyx*, contains more epigean than hypogean species. In North America only one of four species (viz. *S.dentata*) has been recorded from subterranean waters (Holsinger 1986a). This species has been collected only rarely from small streams just inside of cave entrances and is at best a marginal stygophile (see Table 1). In eastern Europe *Synurella* is represented by a few stygobionts, but they are generally closely allied with epigean forms and several have vestigial eyes (Holsinger 1986a). On the basis of overall similarity, Barnard & Barnard (1983) synonymized *Synurella* with *Stygobromus*, but as I have pointed out previously (Holsinger 1977a), and as indicated by the cladogram in Figure 2, these two taxa should be considered separate genera. Two of the similarities between them probably represent homoplasous characters, as indicated in Figure 2.

In summary, there appear to be two distinct levels of adaptation to subterranean waters by amphipods exhibiting what I have termed a pattern 1 distribution. The first level includes those taxa most firmly entrenched in subterranean waters — the stygobiont genera *Allocrangnonyx, Bactrurus* and *Stygobromus.* Collectively, these genera are widespread over the North American continent and are represented by species that inhabit the entire spectrum of freshwater subterranean biotopes. These taxa or their putative ancestors have presumably occupied subterranean groundwaters for a long period of geological time, possibly since the early Tertiary or even before, and there are no epigean congeners known to date.

The second level includes *Crangonyx* (and the crangonyctid genus *Synurella* in eastern Europe) which is represented in subterranean waters by less than 50% of its species, none of which, except perhaps *C.hobbsi*, is either far removed taxonomically from surface congeners or has evolved the highly specialized stygomorphisms so characteristic of stygobionts on the first level. Vestigial eyes and even occasionally very weak pigmentation have been retained by many of these stygobionts, and the eye structure, when absent, has probably been lost secondarily as an ad hoc adaptation. Chronologically, these stygobionts are probably no older than the late Tertiary. They live primarily in caves of the eastern United States and are rarely found in interstitial or hypotelminorheic habitats outside of karst areas.

3.2 *Marine relict distribution*

Although a number of diverse taxa fit this pattern, one of the most indicative is the subset of nine, largely monotyic genera in the family Hadziidae, referred to as the weckeliid group (Holsinger & Longley 1980, Barnard & Barnard 1983). The taxonomic limits of Hadziidae are still unresolved and currently undergoing revision (see Stock 1977a, 1980b, 1983, Ledoyer 1982), but the distributions of most of the genera assigned to this family by recent workers are indicated in Figure 8. Many of the hadziids, similar to their relatives in the closely allied family Melitidae, occur in shallow marine and brackish waters in warmer parts of the world. The distribution of Hadziidae *sensu stricto* is circumtropical, with a significant concentration of species in the old Tethyan remnants of the peri-Mediterranean and greater Caribbean-Gulf of Mexico regions (Fig.8). In both of these regions a number of species have colonized subterranean freshwater habitats, although many others have remained in hypogean mixohaline environments on or around small islands. Stock (1977a) explained the origin of *Metaniphargus* stygobionts by a regression model which holds that putative marine ancestors became stranded and isolated on and around many Antillean islands in the Caribbean during emergence of these small landmasses during the middle Tertiary.

The regression model is a vicariance explanation because it involves the isolation of faunas through geological events which broke up more continuous distributions hypothesized to have existed earlier. Stock (1977a) also attributes the remarkable taxonomic affinity between hadziids in southern Europe and southern North America (especially *Metaniphargus* and *Hadzia sensu lato*) to a continuous distribution of their ancestral stock during the

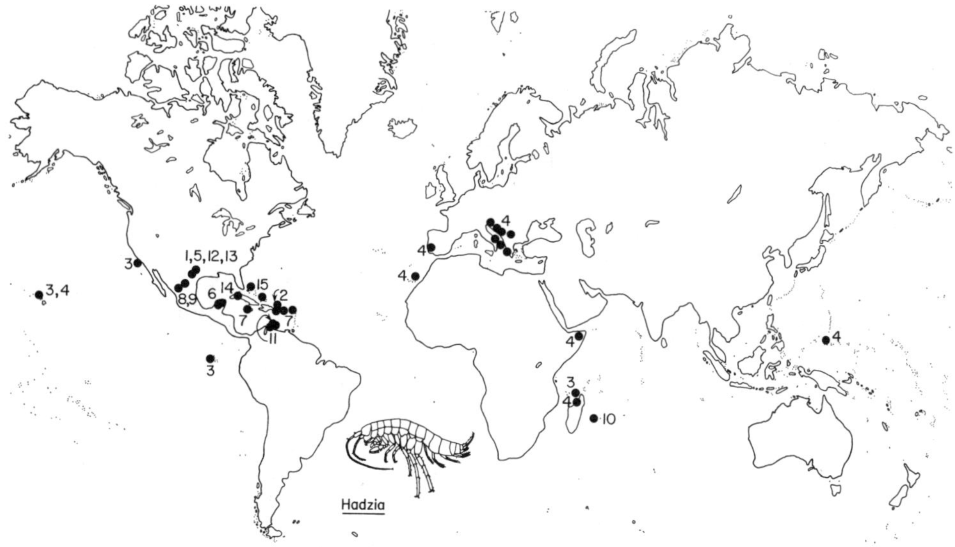

Figure 8. Geographic distribution of Hadziidae *sensu stricto*: (1) *Allotexiweckelia*; (2) *Alloweckelia*; (3) *Dulzura*; (4) *Hadzia sensu lato* (including *Liagoceradocus* and *Metahadzia*); (5) *Holsingerius*; (6) *Mayaweckelia* (7) *Metaniphargus*; (8) *Mexiweckelia*; (9) *Paramexiweckelia*; (10) *Psammoniphargus*; (11) *Saliweckelia*; (12) *Texiweckelia* (13) *Texiweckeliopsis*; (14) *Weckelia*; (15) new genus from the Bahamas and Turks & Caicos (ms. in prep.). *Protohadzia*, a fully marine genus from Corona Reef, Puerto Rico, Bimini Lagoon, Bahamas, and the Great Reef of Tuléar, Madagascar, not shown. A closed circle may represent more than one locality or taxon if two or more localities or taxa are in close geographic proximity. (*Hadzia* redrawn from Karaman 1943).

Jurassic, when the Tethys seaway is believed to have been a shallow marine connection extending westward from the present Mediterranean area to the region of the northern Gulf of Mexico and the Caribbean (see also Dott & Batten 1976). However, the occurrence of hadziids in such non-Tethyan regions as the Hawaiian Islands remains a zoogeographic enigma and has so far not been satisfactorily explained.

The hypothesized phylogeny of the weckeliids is shown by the cladogram in Figure 9 and their geographic distribution in Figure 10. Weckeliids are diagnosed cladistically from other hadziids by the transformed character state of uropod 3, which has lost the second segment of the outer ramus. The congruence between the branches of the cladogram and geographic distribution is striking (see Figs.9, 10), inasmuch as each derived taxon or nested subset of taxa is found in a discrete, geographically separate area. For example, *Alloweckelia* and *Weckelia*, which are endemic to Puerto Rico and Cuba, respectively, are separated both cladistically and geographically from the remainder of the weckeliid genera that live on mainland North America. Moreover, *Mayaweckelia* from the Yucatán Peninsula in southern Mexico is separated geographically from *Mexiweckelia* and *Paramexiweckelia* in northern Mexico, and the latter two are in turn separated from a subset of four genera endemic to the Edwards Aquifer in south-central Texas.

To gain a better understanding of when the putative ancestors of these taxa first became isolated from each other through colonization of subterranean freshwater habitats, I have plotted approximate shorelines of former marine embayments on the continental region of the map in Figure 10. Assuming, as Stock's regression model does, and as I and others have also hypothesized in earlier papers (Holsinger 1973, 1977b, Holsinger & Peck

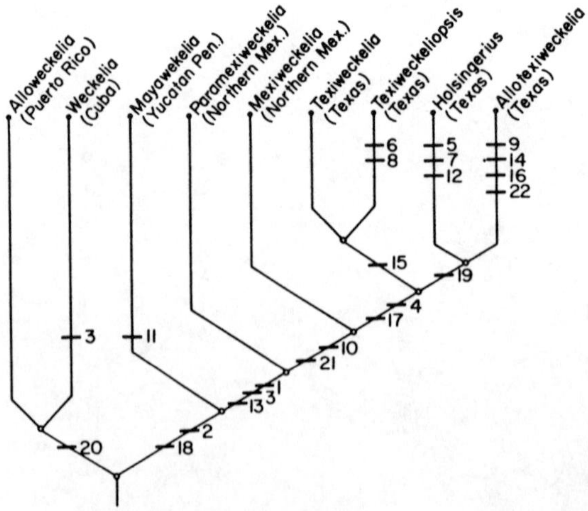

Figure 9. Hypothesized phylogenetic relationships among the genera of the weckeliid group of Hadziidae based on the character states given in Table 4. Black bars are inferred apomorphies. Plesiomorphies not shown.

Table 4. Characters used to analyze the phylogenetic relationships among the genera of the weckeliid group (family Hadziidae) illustrated by the cladogram in Figure 9.

Character	Plesiomorphic state	Apomorphic state
1 Accessory flagellum of antenna 1	2-4 segments	1 segment or vestigial
2 Mandibular palp	Present	Absent
3 Mandibular molar seta	On right and left	On right only
4 Lacinia mobilis of right mandible	Present	Absent
5 Number of apical setae on inner plate of maxilla 1	Few to 25	40
6 Number of spines on outer plate of maxilla 1	7-11	15
7 Number of facial setae in oblique row on inner plate of maxilla 2	10-50	100
8 Inner plate of maxilliped	Normal	Greatly expanded
9 Palm of propod of gnathopod 1	Short and transverse	Relatively long, oblique
10 Posterior margin of segment 5 of gnathopod 1	Without distinct lobe	Not enlarged, without lobe
11 Segment 4 of gnathopod 1	Enlarged, with lobe	Not enlarged, without lobe
12 Dimorphism of gnathopod 2 of male	Weak to moderate	Strong
13 Coxal plate of pereopod 4	Expanded, with posterior excavation	Reduced, excavation shallow or obsolete
14 Bases of pereopods 3 and 4	Narrow, not expanded	Broadly expanded
15 Pereopods 5, 6 and 7	Not attenuate	Attentuate, especially bases
16 Setae on segment 6 of pereopod 6	Short	Elongate
17 Number of setae on anterior margin of dactyls of pereopods 5, 6 and 7	1	3-8
18 Number of basofacial spines on peduncle of uropod 1	1	2-4
19 Size of basofacial spine(s)	Small	Enlarged
20 Rami of uropod 3	With spines and setae	With spines only
21 Lobes of telson	Separated or nearly so	Partly fused
22 Telson (overall)	Lightly spinose	Heavily spinose

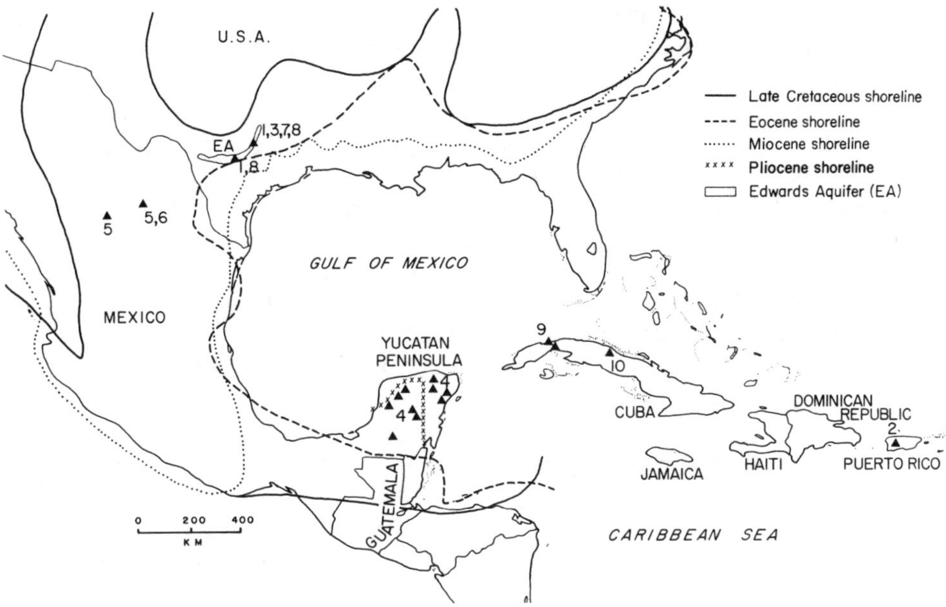

Figure 10. Geographic distribution of genera of the weckeliid group of Hadziidae and *Paraweckelia* (Melitidae) in southern North America: (1) *Allotexiweckelia*; (2) *Alloweckelia*; (3) *Holsingerius*; (4) *Mayaweckelia*; (5) *Mexiweckelia*; (6) *Paramexiweckelia*; (7) *Texiweckelia*; (8) *Texiweckeliopsis*; (9) *Weckelia* (10) *Paraweckelia*. Recorded localities indicated by closed triangles. Approximate shorelines of old marine embayments and the subterranean Edwards Aquifer of south-central Texas indicated on map (adapted from Dunbar 1960, Dott & Batten 1976, Holsinger & Longley 1980, Wilkins 1982).

1968, Holsinger & Minckley 1971, Holsinger & Longley 1980), that these groups evolved directly from marine/brackish water ancestors by gradually adapting to freshwater as marine waters receded from the continent and as Cuba and Puerto Rico emerged by uplifting, the geological changes that have transpired in this region since the Cretaceous provide a good insight into when ancestors of these taxa might have become isolated.

The Mesozoic and early to middle Cenozoic geological history of the greater Caribbean-Gulf of Mexico region is complex, and it is further complicated by many aspects of plate tectonics and island arc movements not yet fully understood or adequately explained (e.g. Malfait & Dinkelman 1972, Rosen 1975, Pregill 1981). Presumably, however, a protoweckeliid fauna existed in shallow marine waters of this region in the Cretaceous and was represented in areas that were later decoupled or became emergent during uplifting and/or marine transgressions.

Stranding in incipient freshwater habitats by putative ancestral weckeliids in northern Mexico and south-central Texas probably occurred sometime between Late Cretaceous and Eocene times, as shallow marine waters receded from these areas (see Fig.10). The high intergeneric diversity of the Edwards Aquifer weckeliid fauna might imply that diversification of lineage leading to present taxa had already begun during or even before marine regressions and has probably continued to some extent since that time (Holsinger & Longley 1980). Holsinger & Peck (1968) suggested that subterranean freshwater weckeliids probably originated in western Cuba and Puerto Rico at a time roughly coincident with emergence of these islands, beginning perhaps in the late Oligocene for western Cuba and continuing into the middle or late Miocene for Puerto Rico. The genus *Mayaweckelia* of Yucatán probably colonized subterranean freshwater much more recently than the other weckeliids, inasmuch as this area was not fully freed of marine water until the late Tertiary (Holsinger 1977b, Wilkens 1982). Possibly further evidence that *Mayaweckelia* is a very recent freshwater genus is the presence of one population of *M.cenoticola* in a cave with slightly brackish water (Holsinger 1977b).

The origin of *Paraweckelia*, a monotypic genus recorded from a single cave in central Cuba (Fig.10), can also be explained by the vicariant regression hypothesis. Although this genus was formerly assigned to the weckeliid group, it has been removed and tentatively assigned to Melitidae because of its assumed close affinity with the marine ceradocids (see Holsinger & Longley 1980, Barnard & Barnard 1983).

Bogidielloid amphipods (families Artesiidae and Bogidiellidae), like the hadziids, are widely distributed, but unlike the latter they show a much stronger freshwater orientation and are better represented in the Southern Hemisphere (especially South America) (see Fig.11). In addition, bogidielloids are exclusively of stygobiont facies. Stock (1981) postulated that bogidiellids (bogidielloids as used herein, because Stock does not recognize them as a superfamily group composed of two families) constitute a very old group of amphipods that originated and dispersed prior to the breakup of Pangaea in the early Mesozoic. He proposed two alternative hypotheses to account for their present distribution and ecology (which encompasses a broad spectrum of subterranean biotopes including mesopsammic marine, littoral interstitial, freshwater hyporheic and freshwater caves): (1) they originated in a marine environment and subsequently invaded freshwater habitats over a long period of time; or (2) they originated in a continental (Pangaean) freshwater environment and subsequently invaded brackish/marine waters. Ruffo (1973), in a revision of the genus *Bogidiella*, favored the first hypothesis; whereas Karaman (1982), though a bit vague, apparently favored the second.

The presence of two bogidielloid genera (*Artesia* and *Parabogidiella*) in south-central Texas and a third (*Bogidiella*) in southern Mexico and Guatemala (Fig.11), suggests that these taxa were derived from marine/brackish water ancestors by stranding in accordance with the regression model (see also Holsinger & Longley 1980, Reddell 1981). Both of these areas were subjected to marine embayments in the Late Cretaceous, and the Mexico-Guatemala area was again inundated during the Miocene (see Fig.10). Stock (1981) also suggested that the four species of *Bogidiella* living in caves and wells with fresh to slightly brackish waters on several islands in the Lesser Antilles (Fig.11) were derived from marine ancestors through stranding when these islands emerged above sea level in the middle Tertiary.

Undoubtedly, the continued discovery of bogidielloids in remote parts of the world (many new discoveries have been made in recent years), combined with further taxonomic refinements, will shed new light on the zoogeography of this large and significant group of subterranean amphipods.

One of the most interesting presumptive marine relicts in North America subterranean freshwater is *Seborgia relicta*, a member of the marine family Sebidae known only from the Edwards Aquifer in south-central Texas (Holsinger & Longley 1980). The Sebidae include two genera (*Seba* and *Seborgia*) and 14 described species. Outside of Texas, only one other species of *Seborgia* is recorded, *S.minima*, an eyeless form collected from an oligohaline-brackish water lake on Rennell Island in the Indo-West Pacific (see Fig.12). The available evidence strongly indicates that *S.relicta* represents the relict of a marine

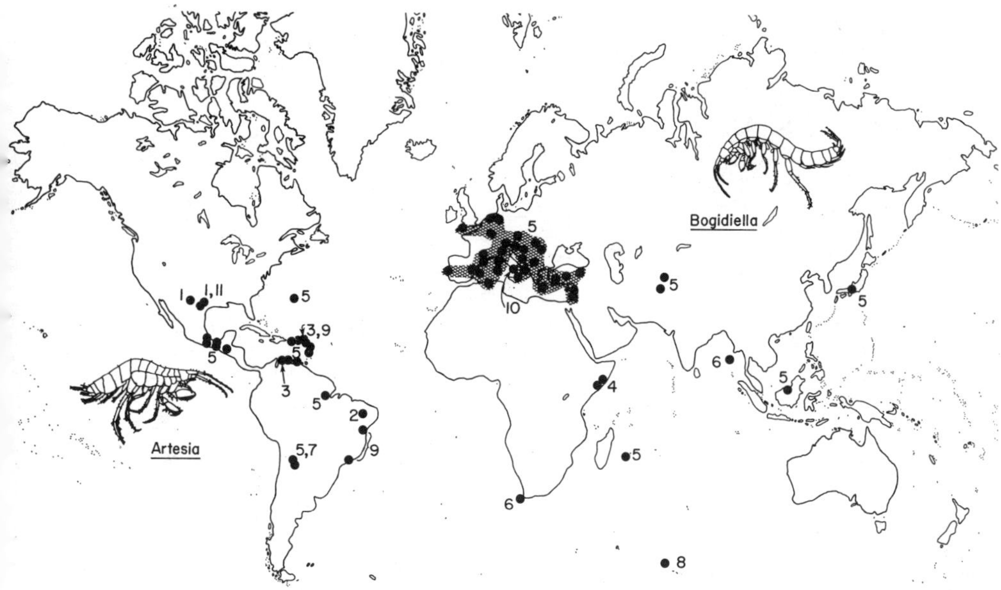

Figure 11. Geographic distribution of bogidielloid amphipods. Artesiidae: (1) *Artesia*; (2) *Spelaeogammarus*. Bogidiellidae *sensu stricto* (excluding *Dussartiella, Paracrangonyx* and *Pseudingofiella* – see Bousfield 1982a, 1983); (3) *Actogidiella*; (4) *Afridiella*; (5) *Bogidiella*; (6) *Bolligidia*; (7) *Eobogidiella*; (8) *Kergueleniola*; (9) *Marigidiella* (10) *Marinobogidiella*; (11) *Parabogidiella*. A closed circle may represent more than one locality or taxon if two or more localities or taxa are in close geographic proximity. (*Artesia* redrawn from Holsinger & Longley 1980). (*Bogidiella* redrawn from Ruffo 1973).

embayment in southern North America during the Late Cretaceous or early Tertiary. All other sebids are small, weakly pigmented, mostly eyeless species reported from benthic marine habitats. Given these characteristics, sebids would appear to be good candidates for the colonization of interstitial habitats during marine regressions (Holsinger 1986b). Moreover, the occurrence of the closely allied *S.minima* in a land-locked, oligohaline-brackish water lake, presumably isolated there for a long period of time, may well indicate the manner in which the putative ancestor of *S.relicta* colonized the transitional aquatic environment that may have existed in south-central Texas in late Mesozoic-early Cenozoic times.

4 SUMMARY

The North American subterranean amphipod fauna is highly diverse taxonomically. At present it includes 10 families or family groups, 28 genera and 169 species. Most of these species are obligate subterranean forms (i.e. stygobionts) that are restricted to aquatic, hypogean biotopes.

Three major zoogeographic patterns are reflected in the distribution of the subterranean amphipod fauna. The first pattern, termed Ancient Freshwater Distribution, includes taxa which are restricted to continental freshwaters of the Holarctic region, have no known marine relatives, and are hypothesized to have been derived from ancestors living in a

Figure 12. Geographic distribution of *Seborgia* (Sebidae): (1) *S.minima*; (2) *S.relicta.* Closed circles indicate single localities. (*Seborgia* redrawn from Holsinger 1986b).

freshwater environment prior to the separation of North America and Eurasia in the late Mesozoic. Members of the family Crangonyctidae and the allocrangonyctid family group show this pattern. Among the genera which fit this pattern, several (e.g. *Allocrangonyx, Bactrurus*, and *Stygobromus*) are exclusively subterranean, whereas others (e.g. *Crangonyx* and *Synurella*) contain both epigean and hypogean species.

The second pattern, termed Marine Relict Distribution, includes freshwater taxa which have close relatives living in marine/brackish water and are hypothesized to have been derived from marine ancestors in southern North America by stranding during marine transgressions in Late Cretaceous and Tertiary times. Members of several families, including Artesiidae, Bogidiellidae, Hadziidae and Sebidae, show this pattern. Probably the most interesting example of this pattern is indicated by the weckeliid group of Hadziidae, which is composed of nine mostly monotypic genera recorded from south-central Texas, northern Mexico, the Yucatán Peninsula, Cuba and Puerto Rico.

The third pattern, termed Marine-Brackish Water Distribution, includes taxa which are endemic to islands in the West Indies and live at present in habitats ranging from fully marine to marginally fresh waters. Some members of the family Bogidiellidae, many members of the family Hadziidae and a few members of several other families show this pattern.

ACKNOWLEDGEMENTS

Stimulating conversations and exchange of unpublished data with J.L.Barnard, E.L.Bousfield and J.H.Stock have benefited me in the preparation of this paper. This is not to imply that I agree with them on all aspects of amphipod systematics, or that they concur with all my concepts or conclusions. L.Botosaneanu and J.R.Reddell also supplied useful information. Preparation of maps and graphs assisted by D.M.Carson, O.D.U. Center for Instructional Development and K.D.Payne, and final typing of the manuscript by K.Mukherjee of the O.D.U. Research Foundation, are gratefully acknowledged. E.L.Bousfield, A.H.Savitzky and J.H.Stock read earlier drafts and offered valuable comments. This study was supported by NSF Grant DEB 82-06716.

REFERENCES

Barnard, J.L. & C.M.Barnard 1983. *Freshwater Amphipoda of the World.* Pt. I. Evolutionary patterns. Pt. II. Handbook and bibliography. Mt Vernon: Hayfield Associates.

Bousfield, E.L. 1958. Fresh-water amphipod crustaceans of glaciated North America. *Can. Fld.-Nat.* 72: 55-113.

Bousfield, E.L. 1978. A revised classification and phylogeny of amphipod crustaceans. *Trans. Roy. Soc. Can.* (4)16: 343-390.

Bousfield, E.L. 1982a. Amphipoda, Gammaridea, and Ingolfiellidea. In *Synopsis and Classification of Living Organisms.* Vol.II. New York: McGraw-Hill.

Bousfield, E.L. 1982b. Amphipoda. In *McGraw-Hill Yearbook of Science and Technology for 1982-1983.* New York: McGraw-Hill.

Bousfield, E.L. 1983. An updated phyletic classification and palaeohistory of the Amphipoda. In F.R.Schram (ed.), *Crustacean Phylogeny.* Rotterdam: Balkema.

Dott, R.H. Jr & R.L.Batten 1976. *Evolution of the Earth.* New York: Wiley.

Dunbar, C.O. 1960. *Historical Geology.* New York: Wiley.

Franz, R. & D.S.Lee 1982. Distribution and evolution of Florida's troglobitic crayfishes. *Bull. Fla. St. Mus. Biol. Sci.* 28: 53-78.

Holsinger, J.R. 1966. Subterranean amphipods of the genus *Stygonectes* (Gammaridae) from Texas. *Amer. Midl. Nat.* 76: 100-124.

Holsinger, J.R. 1967. Systematics, speciation and distribution of the subterranean amphipod genus *Stygonectes* (Gammaridae). *US Nat. Mus. Bull.* 259: 1-176.

Holsinger, J.R. 1969. The systematics of the North American subterranean amphipod genus *Apocrangonyx* (Gammaridae), with remarks on ecology and zoogeography. *Amer. Midl. Nat.* 81: 1-28.

Holsinger, J.R. 1971. A new species of the subterranean amphipod genus *Allocrangonyx* (Gammaridae), with a redescription of the genus and remarks on its zoogeography. *Int. J. Speleol.* 3: 317-331.

Holsinger, J.R. 1972. The freshwater amphipod crustaceans (Gammaridae) of North America. *Biota of Freshwater Ecosystems, Identification Manual US Environmental Protection Agency* 5: 1-89.

Holsinger, J.R. 1973. Two new species of the subterranean amphipod genus *Mexiweckelia* (Gammaridae) from Mexico and Texas, with notes on the origin and distribution of the genus. *Assoc. Mex. Cave Stud. Bull.* 5: 1-12.

Holsinger, J.R. 1974. Systematics of the subterranean amphipod genus Stygobromus (Gammaridae), Pt. 1: Species of the western United States. *Smith. Contr. Zool.* 160: 1-63.

Holsinger, J.R. 1977a. A review of the systematics of the Holarctic amphipod family Crangonyctidae. *Crustaceana* Suppl.4: 244-281.

Holsinger, J.R. 1977b. A new genus and two new species of subterranean amphipod crustaceans (Gammaridae s.lat.) from the Yucatán Peninsula in Mexico. *Assoc. Mex. Cave Stud. Bull.* 6: 15-25.

Holsinger, J.R. 1978. Systematics of the subterranean amphipod genus *Stygobromus* (Crangonyctidae), Part III: Species of the eastern United States. *Smiths. Contr. Zool.* 266: 1-144.

Holsinger, J.R. 1980. *Stygobromus canadensis*, a new subterranean amphipod crustacean (Crangonyctidae) from Canada, with remarks on Wisconsin refugia. *Can. J. Zool.* 58: 290-297.

Holsinger, J.R. 1981. *Stygobromus canadensis*, a troglobitic amphipod crustacean from Castleguard Cave, with remarks on the concept of cave glacial refugia. *Proc. 8th Int. Cong. Speleol.* 1: 93-95.

Holsinger, J.R. 1982. *Paramexiweckelia*, a new genus of subterranean amphipod crustacean (Hadziidae) from northern Mexico. *Int. J. Speleol.* 12: 37-44.

Holsinger, J.R. 1986a. Amphipoda: Holarctic Crangonyctidae. In L.Botosaneanu (ed.), *Stygofauna Mundi.* Leiden: Brill.

Holsinger, J.R. 1986b. Amphipoda: Sebidae. In L.Botosaneanu (ed.), *Stygofauna Mundi.* Leiden: Brill.

Holsinger, J.R. & D.C.Culver 1970. Morphological variation in *Gammarus minus* Say (Amphipoda, Gammaridae), with emphasis on subterranean forms. *Postilla* 146: 1-24.

Holsinger, J.R. & G.Longley 1980. The subterranean amphipod crustacean fauna of an artesian well in Texas. *Smiths. Contr. Zool.* 308: 1-62.

Holsinger, J.R. & W.L.Minckley 1971. A new genus and two new species of subterranean amphipod crustaceans (Gammaridae) from northern Mexico. *Proc. Biol. Soc. Wash.* 83: 425-444.

Holsinger, J.R. & S.B.Peck 1968. A new genus and species of subterranean amphipod (Gammaridae) from Puerto Rico, with notes on its ecology, evolution and relationship to other Caribbean amphipods. *Crustaceana* 15: 249-262.

Just, J. 1974. On *Palaeogammarus* Zaddach, 1864, with a description of a new species from western Baltic amber (Crustacea, Amphipoda, Crangonyctidae). *Steenstrupia* 3: 93-99.

Karaman, G.S. 1982. Critical remarks to the recent revisions of Bogidiella-Group of genera with study of some taxa (fam. Gammaridae). *Poljoprivreda I Sŭmarstvo* 28: 31-57.

Karaman, S. 1943. Die unterirdischen Amphipoden Südserbiens. *Sr. Akad. Nauka Pos. Izd. 135* 34: 161-312.

King, P.B. 1958. Evolution of modern surface features of western North America. In C.L.Hubbs (ed), *Zoogeography* 51. Washington D.C.: American Association for the Advancement of Science.

Ledoyer, M. 1982. Crustacés Amphipodes Gammariens, Familles des Acanthozomatidae à Gammaridae. *Faune de Madagascar* 59: 1-598.

Malfait, B.T. & M.G.Dinkelman 1972. Circum-Caribbean tectonic and igneous activity and the evolution of the Caribbean plate. *Geol. Soc. Amer. Bull.* 83: 251-272.

Pregill, G.K. 1981. An appraisal of the vicariance hypothesis of Caribbean biogeography and its application to West Indian terrestrial vertebrates. *Syst. Zool.* 30: 147-155.

Reddell, J.R. 1981. A review of the cavernicole fauna of Mexico, Guatemala, and Belize. *Tex. Mem. Mus. Bull.* 27: 1-327.

Rosen, D.E. 1975. A vicariance model of Caribbean biogeography. *Syst. Zool.* 24: 431-464.

Ruffo, S. 1973. Studi sui Crostacei Amfipodi, LXXIV: Contributo alla revisione del genere *Bogidiella* Hertzog (Crustacea Amphipoda, Gammaridae). *Boll. Ist. Entomol. Univ. Bologna* 31: 49-77.

Ruffo, S. & A.Vigna Taglianti 1977. Secondo contributo alla conoscenza del genera *Bogidiella* in Messico e Guatemala (Crustacea, Amphipoda, Gammaridae). *Quad. Accad. Nazion. Lincei* 171: 125-172.

Stock, J.H. 1976. A new member of the crustacean suborder Ingolfiellidea from Bonaire, with a review of the entire suborder. *Stud. Fauna Curaçao* 50: 56-75.

Stock, J.H. 1977a. The taxonomy and zoogeography of the hadziid Amphipoda with emphasis on the West Indian taxa. *Stud. Fauna Curaçao* 55: 1-130.

Stock, J.H. 1977b. *Bogidiella martini*, un nouvel Amphipode souterrain de l'Ile Saint-Martin (Antilles) et al zoogéographie des Bogidellidae. *Int. J. Speleol.* 9: 103-113.

Stock, J.H. 1977c. The zoogeography of the crustacean suborder Ingofiellidea with descriptions of new West Indian taxa. *Stud. Fauna Curaçao* 55: 131-146.

Stock, J.H. 1979. New data on taxonomy and zoogeography of ingolfiellid Crustacea. *Bijdr. Dierk.* 49: 81-96.

Stock, J.H. 1980a. Un *Metaniphargus* (Amphipoda) nouveau de l'Ile Marie-Galante (Antilles). *Bull. Mus. natn. Hist. Nat. Paris* (4)2: 413-419.

Stock, J.H. 1980b. A new cave amphipod (Crustacea) from Curaçao: *Psammogammarus caesicolus* n.sp. *Bijdr. Dierk.* 50: 375-386.

Stock J.H. 1981. The taxonomy and zoogeography of the family Bogidiellidae (Crustacea, Amphipoda), with emphasis on the West Indian taxa. *Bijdr. Dierk.* 51: 345-374.

Stock, J.H. 1983. The stygobiont Amphipoda of Jamaica. *Bijdr. Dierk.* 53: 267-286.

Stock, J.H. & L.Botosaneanu 1983. Première découverte d'amphipodes Gammaridae du groupement des hadziides dans des eaux souterraines de l'Amérique du Sud: Description de *Metaniphargus venezolanus* sp.n. *Bijdr. Dierk.* 53: 158-164.

Stock, J.H. & J.J.Vermeulen 1982. A representative of the mainly abyssal family Pardaliscidae (Crustacea, Amphipoda) in cave waters of the Caicos Islands. *Bijdr. Dierk.* 52: 3-12.

Ward, J.V. 1977. First records of subterranean amphipods from Colorado with descriptions of three new species of *Stygobromus* (Crangonyctidae). *Trans. Amer. Microsc. Soc.* 96: 452-466.

Ward, J.V. & J.R.Holsinger 1981. Distribution and habitat diversity of subterranean amphipods in the Rocky Mountains of Colorado, USA. *Int. J. Speleol.* 11: 63-70.

Wilkens, H. 1982. Regressive evolution and phylogenetic age: The history of colonization of freshwaters of Yucatán by fish and Crustacea. *Assoc. Mex. Cave Stud. Bull.* 8/*Tex. Mem. Mus. Bull.* 28: 237-243.

2 Dispersal

Prologue

Part 2 consists of four authors whose specialities lie within the peracaridan and malacostracan Crustacea. Again, while not intending to label the authors, the general thrust of their subject material and associated concepts tends toward dispersal as the important biogeographical concept to be considered in explaining the distributions of their respective groups. Although not totally ignored, vicariance mechanisms are not as heavily emphasized as in Part 1.

Gale Bishop provides evidence for a generalized dispersal track of the Upper Cretaceous brachyuran decapod *Dakoticancer* complex, using an extensive data set taken from examinations of all brachyuran fossils of this group in North America, as well as other genera. The *Dakoticancer* generic complex is examined in both its stratigraphic and geographic distributions, and important zoogeographic conclusions are derived, including some postulates on evolution within this group.

John Markham examines for the first time the general distribution of the epicaridean family Bopyridae, which are parasitic on decapod crustaceans in the branchial or abdominal areas. Thus, the distribution of the bopyrids is, in part, a reflection of that of their hosts. Markham hypothesizes that the most primitive subfamily, the branchially-parasitic Pseudioninae, probably first infested anomuran Galatheidae, a decapod family known from the Middle Jurassic to the present. All other subfamilies of branchial parasites are derivable from the Pseudioninae, with a series of genera infesting thalassinideans in what has been termed the 'Thalassinidean Transition'. The two major subfamilies that are abdominal parasites apparently arose separately. Markham points out the evidence of abdominal infestation would be poor or non-existent in the fossil record, compared to that of branchial parasitiziation, owing to the more 'permanent' preservation of cephalothoracic features. He suggests that the first appearance of the bopyrids may be more than a fortuitous correspondence with the first appearance of nearly all extant decapod Crustacea in the Jurassic. Markham's paper offers an interesting way to examine evolutionary data by employing zoogeographic and stratigraphic information shared among hosts and parasites, both in the present and by implication in the past.

Jurgen Sieg examines the world distribution of the Tanaidacea. Endemism within regions and bipolarity are discussed, as are previously unconsidered data on the correlation of eye development and depth distribution in the various taxa. The Tanaidae are postulated to have a southern 'center of evolution', and the order Tanaidacea can be characterized as exhibiting a relatively constant reduction in size as evolution progressed, with all large deep-sea or cold water species probably plesiomorphic compared to smaller members in the same region. Sieg's paper provides a wealth of information on this small but diverse group and should become the pivotal study for further investigation on both shallow water and deep-sea taxa.

Jan Stock provides a most intriguing paper on stygobiont Malacostraca in the Caribbean, and in the process questions geological data on the age of Caribbean Islands in relation to vicariant hypotheses recently proposed by several authors. By sampling the groundwater biota of 43 West Indian islands, Stock was able to obtain a data base allowing a comparison of vicariance and dispersal hypotheses within the Caribbean area. He suggests that species-area (S/A) curves might provide a better 'biological clock' marking an island's age than hypotheses relying on vicariance biogeography. But Stock also holds that stygobiont Crustacea evolved from marine ancestors via a vicariant process which he terms the Regression Model; i.e. stranding and gradual adaptation by marine populations (primarily meiofauna) to mixohaline and eventually fresh groundwater on the various islands. Thus, the 'center of origin' for these stygobionts was the pelagic areas interdicted by island emergence. Long distance dispersal over barriers was unnecessary, because as each island arose, the stranded populations that could survive and adapt, diverged from their marine ancestors. Stock's paper provides a simple hypothesis for the origin of island stygobionts, by invoking a vicariant event which more or less interrupted a continuing dispersal event. This concept might profitably be considered with other allegedly isolated groups both within the Crustacea, and in the other invertebrate and invertebrate groups as well. The majority of attempts begin with geology to date the origins of taxonomic distributions; this paper suggests that taxonomic distributions (viz. S/A curves) may provide an alternative data base for the ages of oceanic islands.

Occurrence, preservation, and biogeography of the Cretaceous crabs of North America

GALE A.BISHOP
Department of Geology and Geography, Georgia Southern College, Statesboro, Georgia, USA

ABSTRACT

The preserved record of the Cretaceous crabs of North America is affected by the preservation potentials of their habitats, their mode of life, their necessity for molting, and diagenesis. These factors lead to a limited, fragmented record which is very strongly biased. The known crabs from the Cretaceous of North America are assigned to 47 genera and 95 species. Their stratigraphic and geographic distribution are disjunct and tied directly to a few favorable lithosomes which occur throughout the Cretaceous. Taxa illustrated from several localities demonstrate the appearance and preservational range encountered in the field. The record of North American Cretaceous crabs is far from complete but at this stage it appears that the Western Interior, the Gulf Coastal Plain (and the Mississippi embayment), and the northern Atlantic Coastal Plain developed crab faunas which exhibit distinct provincialism.

1 INTRODUCTION

This paper represents the first attempt to synthesize current knowledge of the Cretaceous crabs of North America. Crab faunas of the northern Atlantic Coastal Plain, Gulf Coastal Plain, US Western Interior and Pacific Slope are reviewed in order to: (1) demonstrate regional (i.e. provincial) differences and similarities in crab faunas, (2) present a current summary of known crabs, (3) postulate possible reasons for observed differences, and (4) accumulate pertinent citations in one bibliography.

In this study the term 'crab' is used in a broad sense to include not only the brachyuran decapod crustaceans (i.e. true crabs) but also the crab-like anomuran (Williams 1965) decapod crustaceans (i.e. the hermit crabs and the galatheids). This review joins Feldmann's (1981) review of the paleobiogeography of the North American lobsters and shrimp.

1.1 *Scarcity of crab fossils*

Crabs are among the scarcest of fossils. Reasons for this scarcity are myriad but may include paleoecological, diagenetic and collecting conditions, all of which introduce severe biases for or against certain taxa which further fragment an already limited record.

Crabs inhabit nearly all environments closely associated with the sea, as well as some which are associated with fresh water. Because the habitats have different potentials for preservation, the crabs inhabiting them will leave records which are biased accordingly. Fresh water crabs (e.g. *Aegla* and *Potamon*) are rare and the sediments in which they could be preserved are generally destroyed by erosion. *Aegla* is poorly represented as a fossil and *Potamon* is known only in Late Miocene to Pliocene strata of Europe and India in spite of its relatively widespread recent distribution in southern Europe, Africa, southern Asia, eastern Asia, and the East Indies (Glaessner 1969). Supratidal crabs are usually tied rather closely to salt water, using it in respiration and in propagation. These crabs are among the most visible and therefore best known of Recent crabs (*Ocypode, Cardisoma, Gecarcinus, Sesarma*). In spite of their great numbers and fair diversity, their fossil record is severely limited because their habitat has limited potential for preservation. Intertidal crabs also inhabit environments which have relatively low preservational potential. In order to protect themselves from predators, desiccation and rapid temperature changes, intertidal crabs must be adapted to move with the strandline (*Emerita, Callinectes, Hippa, Pagurus*), burrow (e.g. *Uca, Panopeus*) or live in other protected environments (i.e. commensally within worm burrows (*Pinnixa*) or within oysters or clams (*Pinnotheres*)). Many crabs inhabit the subtidal zone. Schäfer (1954, 1972) has differentiated the free-living subtidal crabs into four locomotion groups: (1) crawlers and runners, (2) diggers, (3) climbers, and (4) swimmers. Each group has characteristically shaped claws, cephalothorax, and pereopods (Schäfer 1972). The size of the crab and its parts, and the degree of mineralization also vary from group to group. The burrowing crabs (e.g. *Raninella, Notopocorystes, Ranina, Nostosceles*) are well represented as fossils compared to remains of the other three main locomotory groups. Of the three remaining groups, fossils of the crawlers and swimmers are more abundant than those of the climbers (e.g. *Maja, Leptomithrax, Micromithrax*) probably because the climbing crabs are commonly found in rocky or shelly environments or as epibionts on kelp (Schmitt 1921, Rathbun 1925, Williams 1965). The crawling crabs (*Carcinus, Cancer, Geryon, Xantho*) and the swimming crabs (e.g. *Portunus, Callinectes*) often seek refuge in the bottom, which enhances their preservational potential (Schäfer 1972).

An overall pattern of habitat correlation with preservation potential and the subsequent fossil record is thus suggested. In the habitats which are more likely to be preserved the chances for preservation of some types of crabs are enhanced. Other types will be found rarely as fossils because their habitat has little chance for preservation (*Uca, Sesarma*).

Another factor which must be reflected in the fossil record is the wide range of relative abundance among different crab species, or within a single crab species, depending on environmental differences. Crab communities described from the Holocene (Thorson 1957) of Ise Wan, Japan (*Pinnixa rathbuni* community) and the Persian Gulf (*Xenophthalamus pinnotheroides* community) are characterized by 3000 and 1500 mature individuals per square meter, respectively, which must be close to the extreme limit of possible number of individuals per square meter. Burial of such crab concentrations would certainly give us a very different picture than that presented by burial of a community having much lower density.

An additional ecological factor which biases the record is the fact that crabs molt periodically. Being relatively long lived (Schäfer 1972), a single specimen has the potential to produce numerous fossils. In some cases molts can be differentiated from corpses (Schäfer 1951, Bishop 1972b, Richards 1975) but this is unusual.

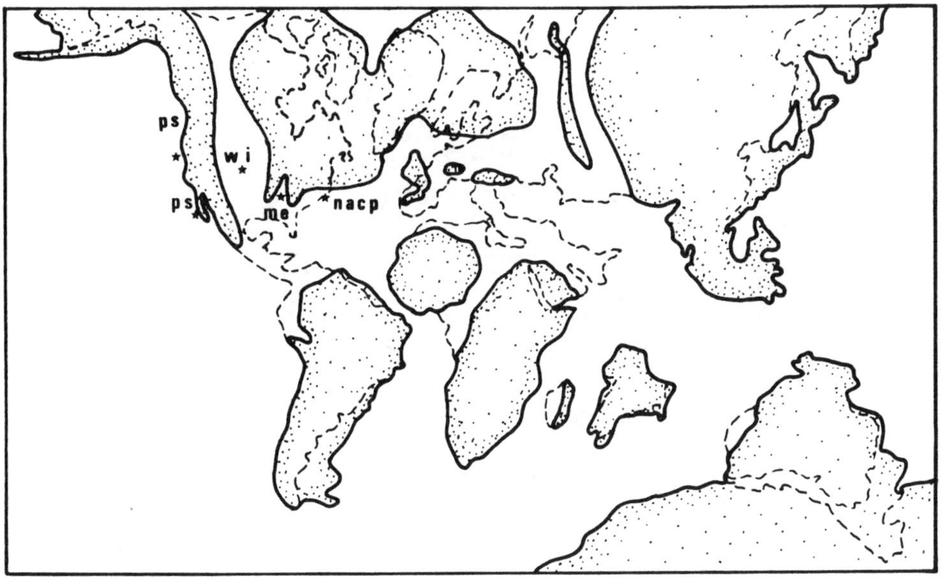

Figure 1. Cretaceous paleogeography (after Charig 1979).

The processes of selective preservation may act on corpses or molts during or after burial. Some of the biases entering the record at this stage (Glaessner 1969) are: (1) weakly calcified forms are selected against except when buried in a reducing and acid environment; (2) more heavily calcified body parts are preserved selectively (claws, spines, carapaces); (3) crabs living in quiet environments are more apt to be preserved; (4) crabs living in areas of rapid sedimentation are more often preserved; (5) crabs living in the intertidal zone, where sediment is often reworked (pagurids and mole crabs), are selected against; and (6) crabs which are prey animals for many predators are selected against. Once buried, crab remains decompose rapidly, being destroyed by chemical decomposition in mud deposits in as little as four weeks (Schäfer 1972: 140) and probably in shorter times by physical or biological disintegration.

Thus the preserved record is biased and fragmentary in nature in terms of taxa *and* individual specimens. Collecting bias adds further to imperfection of the record. During routine measurement and collection of sections, only the astute non-specialist is likely to notice and collect the rare, small, nonspectacular Cretaceous fossil crabs. It is a tribute to the objectivity of many Cretaceous workers that so many fossil crabs have been found. However, the author knows of and has collected fossil crabs from localities crossed by measured sections of other geologists who overlooked these little fossils. One objective of this paper is to furnish a better knowledge of North American Cretaceous crabs so that future workers will have a better chance of discovering these fossils.

1.2 *The Cretaceous marine setting*

The generally accepted concept of Cretaceous paleogeography (Fig. 1) and paleoclimate has been ably presented by Kauffman (1975) (see also Charig 1979). The Cretaceous world was dominated by continental plates which were drifting apart after the breakup

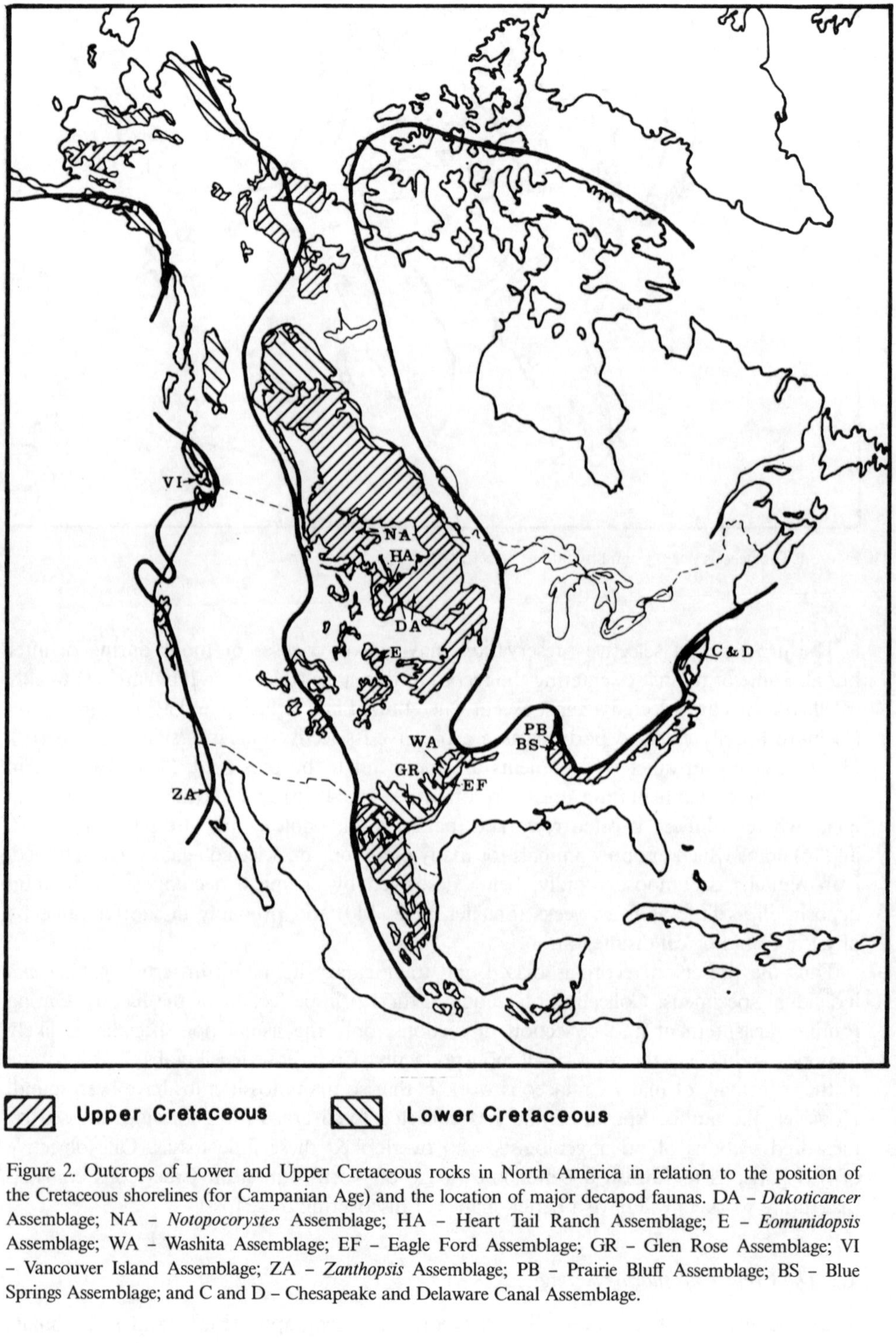

Figure 2. Outcrops of Lower and Upper Cretaceous rocks in North America in relation to the position of the Cretaceous shorelines (for Campanian Age) and the location of major decapod faunas. DA – *Dakoticancer* Assemblage; NA – *Notopocorystes* Assemblage; HA – Heart Tail Ranch Assemblage; E – *Eomunidopsis* Assemblage; WA – Washita Assemblage; EF – Eagle Ford Assemblage; GR – Glen Rose Assemblage; VI – Vancouver Island Assemblage; ZA – *Zanthopsis* Assemblage; PB – Prairie Bluff Assemblage; BS – Blue Springs Assemblage; and C and D – Chesapeake and Delaware Canal Assemblage.

of Pangaea. During this break-up, a broad equatorial ocean, the Tethys Seaway, dominated the paleogeographic setting. Branching off the Tethys Seaway were north-south oriented seaways, three of which are of interest here: the Western Interior Seaway, the Proto-Atlantic Seaway, and the Pacific Slope.

The distribution of water masses and current patterns controls weather and climate. Within these paleo-oceans currents were present just as they are today. By using the present oceanic current distribution as a model, the Cretaceous paleocurrent (and climate and weather) pattern can be reconstructed as has been done by Kauffman (1975).

Using available data (oxygen-isotope analysis, floral and faunal analogy) the Cretaceous climate is interpreted as having been significantly warmer and more uniform than Recent climate. Apparently there was less seasonal change and a smaller temperature differential in the Cretaceous oceans. This resulted in broad, uniform, stable marine climatic zones. Kauffman (1975) recognizes the tropical equatorial Tethyan Realm, a narrow subtropical zone, and broad warm-temperate, mid-temperate, and cool-temperate zones. This model can be used as a reasonable approximation of Cretaceous paleogeography and paleoclimatology.

Superimposed on this basic pattern of ocean distribution are a series of transgressive-regressive cycles (Hattin 1962, 1964; Weimer 1959, 1960; Kauffman 1962, 1967) which were probably caused by changes in base level as the continental or ocean plates pulled apart. In the Western Interior the pattern of transgression is well documented. The seas transgressed across the Western Interior southward from the Arctic Ocean and northward from the Tethys Seaway (Gulf of Mexico), finally joining in early late Albian time (refer to Fig.3 for approximate absolute ages of relative geologic time units). The seas then immediately began regressing and the connection was broken by late late Albian time. [Rates of transgression have been calculated to have been 75 to 95 miles/million years for Late Cretaceous rocks in the Western Interior (Gill & Cobban 1973).] During this time the Gulf of Mexico region was fringed by rudist reefs behind which lagoonal sediments accumulated (Hendricks & Wilson 1967). The reefs formed a barrier to migration of marine organisms as illustrated by ammonites so that endemic forms developed in the lagoons. Occasionally the reefs were breached and at such times new faunal elements entered the shelf areas. The 'cyclic' pattern gave rise to ammonite faunas which are alternately endemic and cosmopolitan (Young 1974). The crustacean faunas are still poorly known but are expected to yield a pattern similar to the ammonites including endemic faunas from the lagoons and reefs.

The late late Albian regression was followed rapidly by transgression which re-established the Western Interior Seaway by Cenomanian time (Williams & Stelck 1975). The seaway then remained open until final withdrawal of the Western Interior Sea in the Maastrichtian. Thus during the Early Cretaceous, conditions suitable for decapods were present on the Gulf Coast region and perhaps to the northern end of the Western Interior Seaway although Early Cretaceous decapod faunas have not yet been discovered in the Western Interior Seaway.

Transgression of the Late Cretaceous sea into the Western Interior was marked by numerous small sea level fluctuations which gave rise to repetitive facies and, consequently, faunas (Kauffman 1967). As the strandline fluctuated, size and shape of the Western Interior Seaway and the configuration of the coastal margins also changed. Such conditions were amenable for fossilization of shallow water taxa.

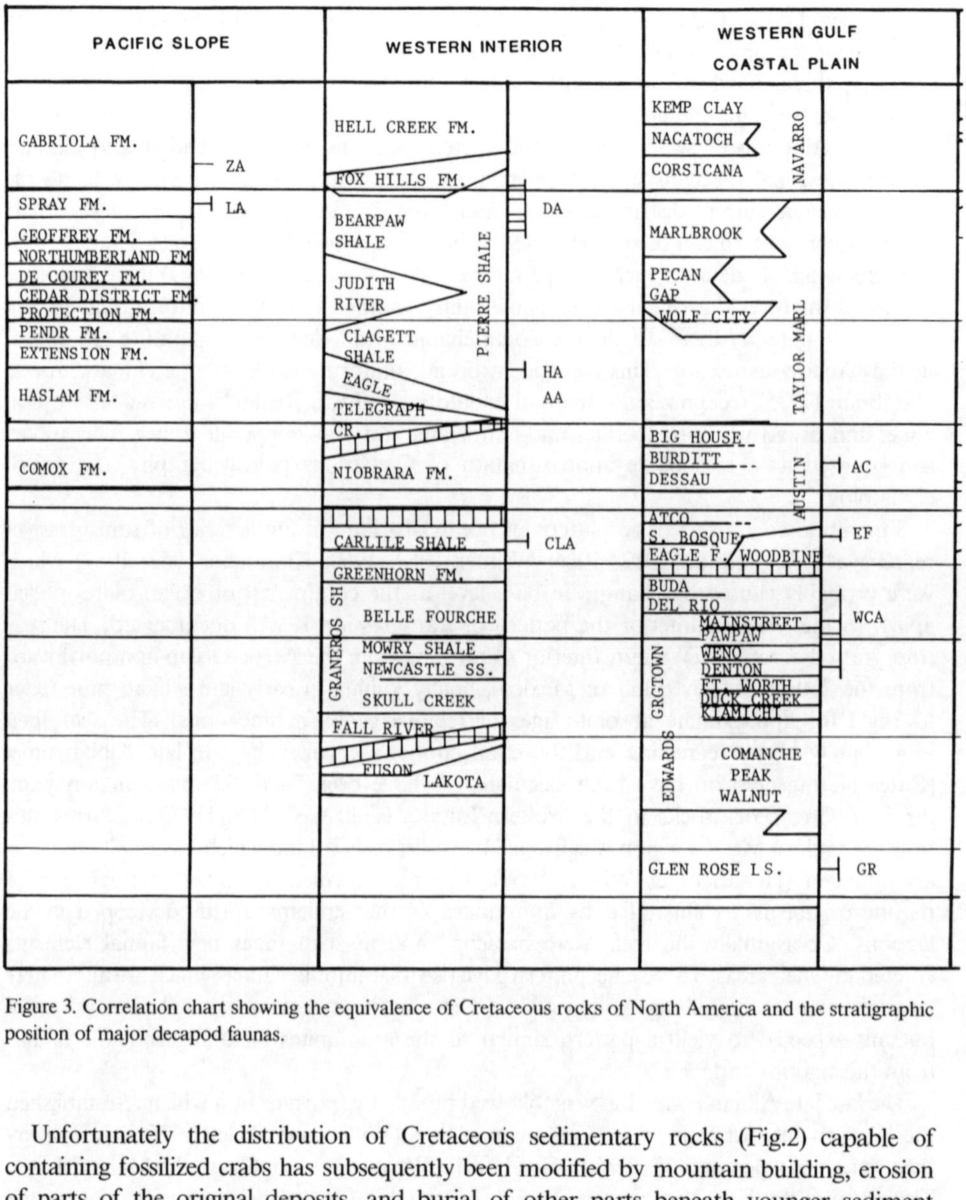

Figure 3. Correlation chart showing the equivalence of Cretaceous rocks of North America and the stratigraphic position of major decapod faunas.

Unfortunately the distribution of Cretaceous sedimentary rocks (Fig.2) capable of containing fossilized crabs has subsequently been modified by mountain building, erosion of parts of the original deposits, and burial of other parts beneath younger sediment limiting the available preserved record.

1.3 *Stratigraphy*

Fossil crabs from the North American Cretaceous furnish a spotty record of the group (Fig.3). However, some stratigraphic units contain significant numbers of taxa and specimens. In Figure 2 the stratigraphic position is indicated for crab-rich beds in the Atlantic and Gulf Coastal Plains and in the Western Interior. The stratigraphic position of crabs from the Western Interior is known accurately because of good biostratigraphic control in that region. Crabs from the Gulf Coastal Plain were collected by many workers and were

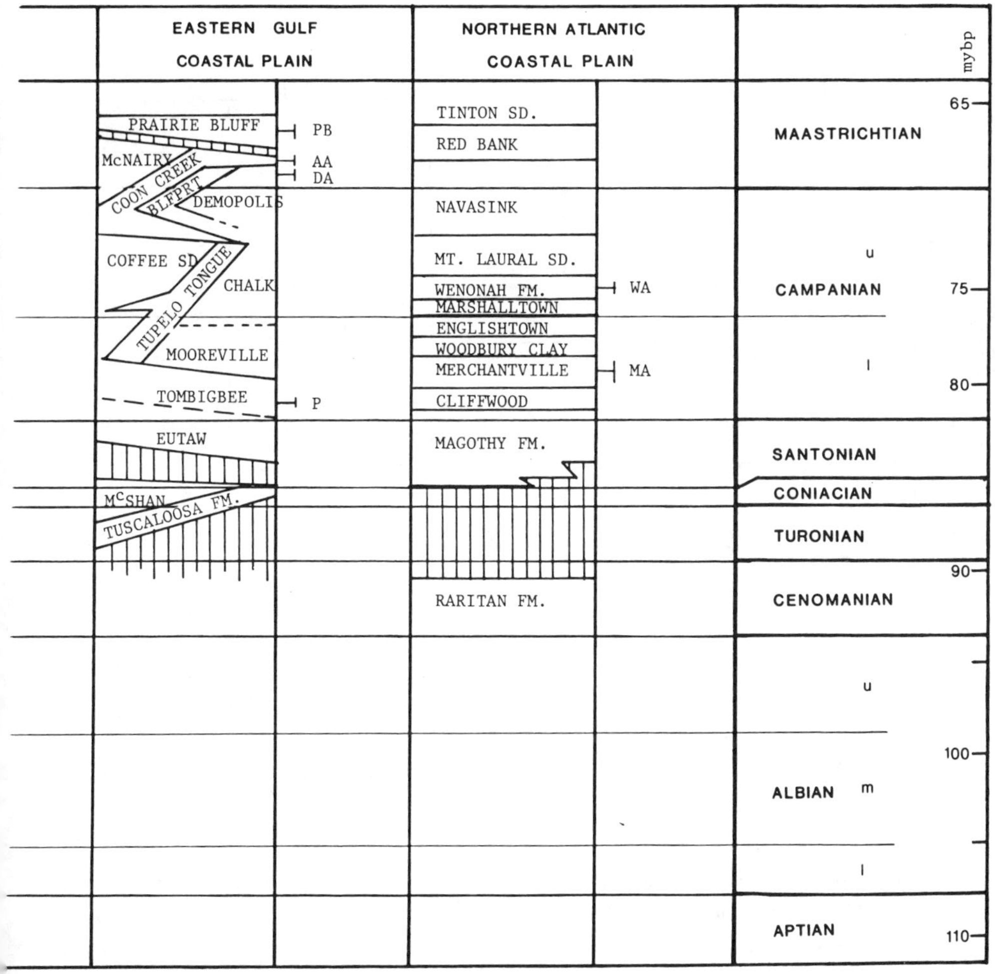

described subsequently by Rathbun (1935b). Stratigraphic position of these specimens is less certain. Crab specimens from the Atlantic Coastal Plain come from formations which have been assigned to various time-rock units over the years (Owens et al. 1970: Fig.24). This had led to great confusion in the ages of these sediments. Many collections from the Chesapeake and Delaware Canal and other localities are float from spoil banks. Their stratigraphic position is undeterminable because the rock containing the fossils has been jumbled by construction.

2 TAXONOMIC AND GEOGRAPHIC DISTRIBUTION

Crab taxa described from the Cretaceous of North America are listed in Appendix 1

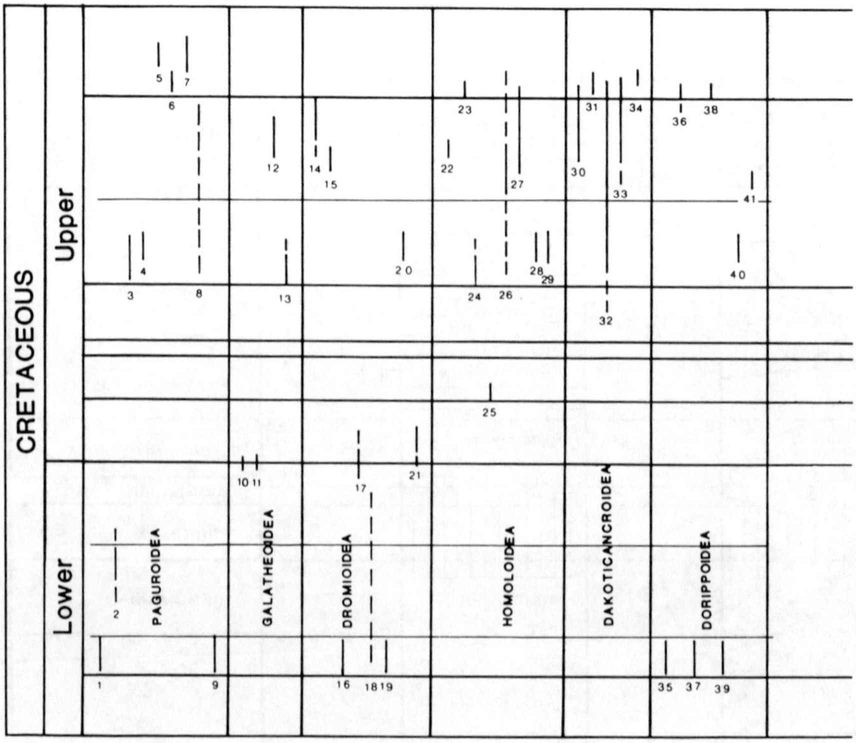

Figure 4. Range chart of decapod species in North America. Numbers correspond to taxa in the Appendix.

according to the classification used by Williams (1965), modified somewhat after Wright & Collins (1972) and Glaessner (1969). Each species-level taxon includes a notation of its original publication. Subsequent parts of the synonomy are presented if the taxon has been reassigned or if a subsequent reference illustrates the taxon more clearly. Known geographic locations for each taxon are presented in general terms as is the stratigraphic range, and an estimate of the number of known specimens.

Although this list is a limited synonomy, it will situate the taxa within the taxonomic heirarchy and furnish a reference to published figures (both original and subsequent) and descriptions.

I have included a series of plates which illustrate some taxa so as to furnish a general guide to readers who wish to identify selected specimens in their collections. Time and space do not permit the illustration of all taxa but those illustrated are from important faunas or represent better illustrations of older taxa. Each taxon has the same number in the hierarchy, on the range chart, and on the maps.

2.1 *Stratigraphic distribution*

The pattern of stratigraphic distribution of the taxa in the Appendix (Fig.4) reflects the fragmentary nature of the record. In rocks which yield crab remains the record is respectable (i.e. lower Albian, upper Campanian-lower Maastrichtian). It will be interesting in the

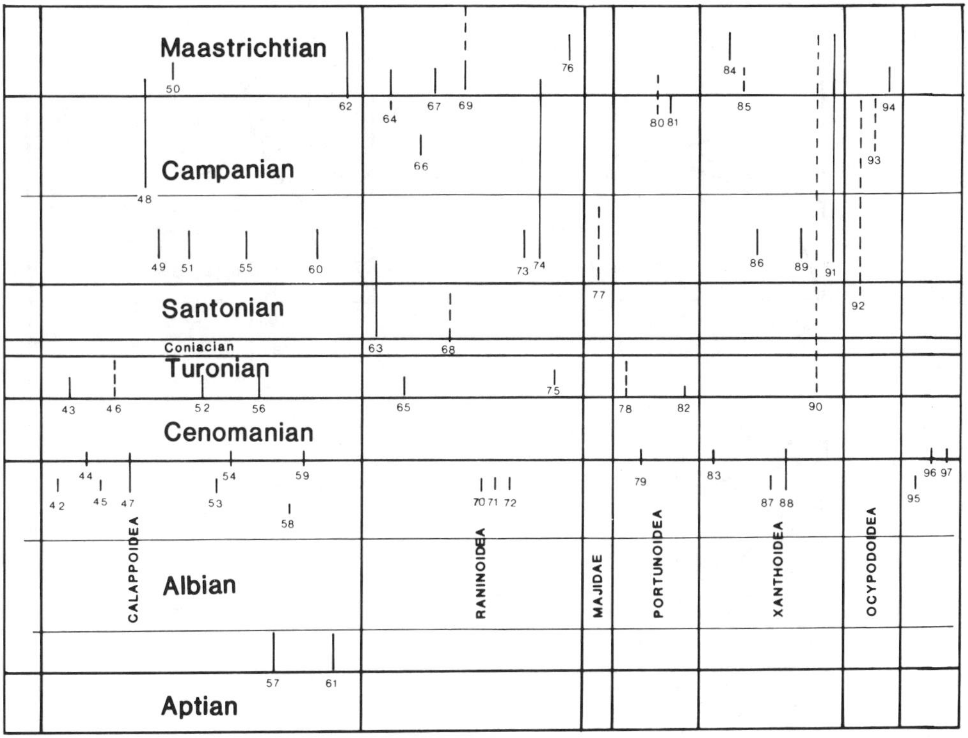

future to observe the intervals currently barren of crabs to see if the gaps are due to lack of fossils or to lack of collecting.

Certain biases are built into the range chart. Many taxa are shown as having very long ranges because data on their actual biostratigraphic range is limited (i.e. Gulf Coast *Dakoticancer australis* (taxon 30) is from the 'Ripley Formation' which means nearly anywhere in the upper Campanian or lower Maastrichtian). Placement of stage boundaries also fluctuates [i.e. the Campanian-Maastrichtian boundary was placed at the base of the *B.aculites baculus* zone (Gill & Cobban 1966) but is now placed at the base of the *B.aculites reesidei* zone (Obradovich & Cobban 1975) by W.A.Cobban].

Wherever possible, and especially in collections from the Western Interior, I have attempted to plot ranges as precisely as possible. Ranges known less precisely are indicated by dashed lines.

2.2 *Cretaceous decapod occurrences*

The Cretaceous decapods of North America occur as 1) uncommon remains in molluscan fossil assemblages, 2) as common remains in molluscan dominated fossil assemblages, or 3) as dominant elements in decapod assemblages (e.g. the *Dakoticancer* Assemblage described by Bishop 1981). The occurrence of fossil decapods as rare or occasionally common elements in molluscan assemblages has led to the logical conclusion that the

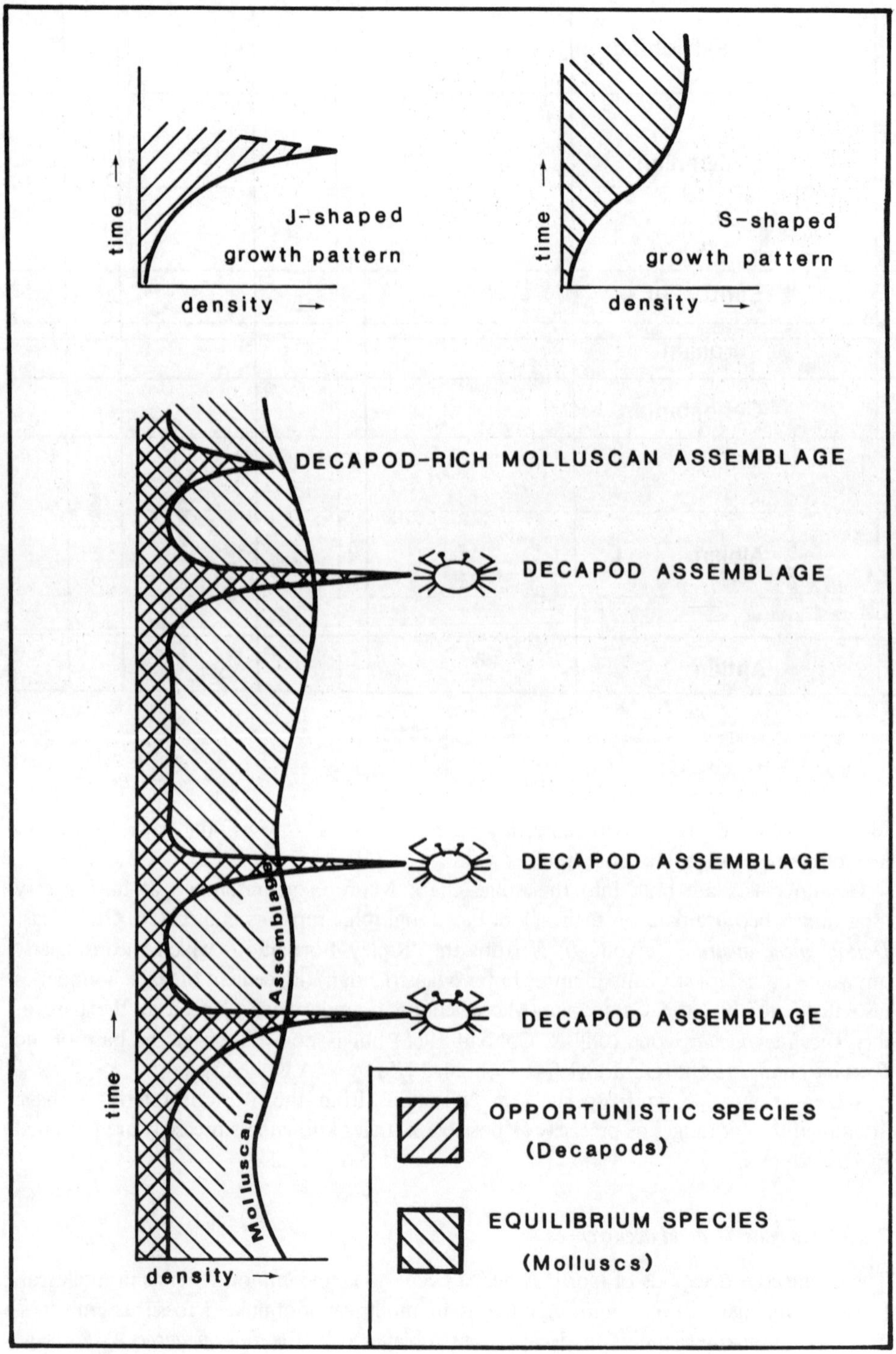

Figure 5. Diagram illustrating modes of occurrence of Cretaceous North American decapod faunas.

Figure 6. Early Cretaceous decapods from the Glen Rose Limestone of Central Texas (see Bishop 1983a). (A) Carapace of *Palaeodromites naglei* (San Diego Society of Natural History (SDSNH) 23644, holotype); (B) Carapace of *Pseudonecrocarcinus stenzeli*, reconstructed about mirror plane of symmetry (SDSNH 23641, holotype); (C) Carapace steinkern of *Hillius youngi* (SDSNH 23643, holotype); (D) Right propodus of *Protocallianassa klofi* (SDSNH 23665, holotype); (E) Center of dorsal shield of *Dioratiopus scotti* (SDSNH 23642, holotype); (F) Carapace of *Diaulax roddai* (SDSNH 23640, holotype); (G) Right propodus of *Torynomma? densus* (SDSNH 23664, holotype); (H) Right propodus (less fixed finger) of *Prehepatus hodgesi* (SDSNH 23644, holotype); (I) Right chela of *Palaeopagurus banderensis* (Univ. Texas 45473); (J) Left chela of *Roemerus robustus* (UT 45704, holotype). (A-C) (E-H) and (J) × 2.0 (Scale 1); (D) × 1.5 (Scale 2); and (I) × 1.0 (Scale 3).

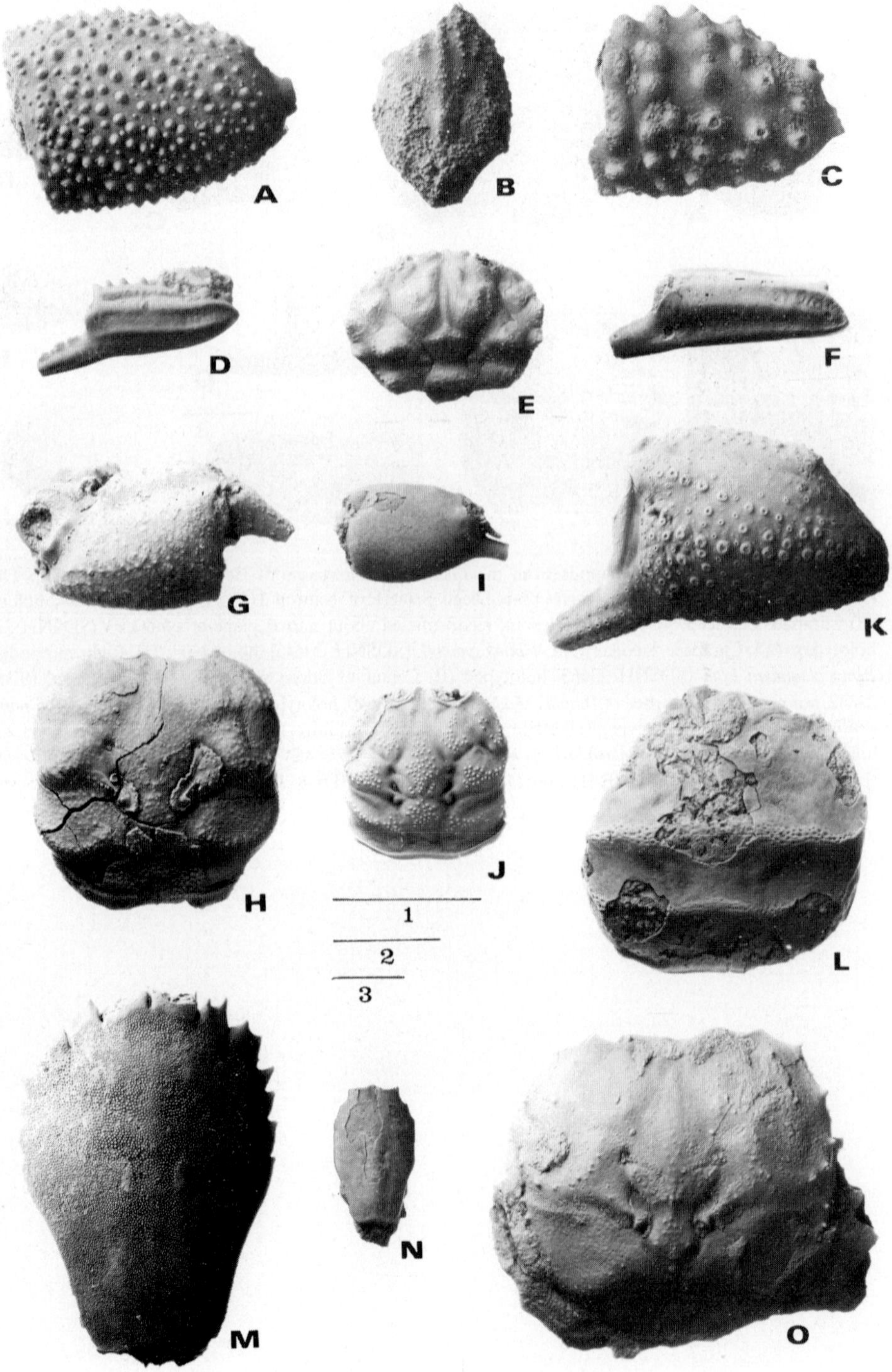
A
B
C
D
E
F
G
I
K
H
J
L
1
2
3
M
N
O

geological record of decapods is very poor. This conclusion is supported by the disjunct, fragmentary record of fossil decapods. Of the 97 known Cretaceous decapods from North America 42 species are known from a single specimen, 8 species each are represented by 2, 3 and 4 specimens, and 3 species each are represented by 5, 6 and 7 specimens. Only 18 species are represented by 9 or more specimens and of these only 4 are represented by 100 or more specimens. These data suggest that fossil decapods are scarce and our knowledge of them is very incomplete. If this is true, then our knowledge of their stratigraphic and geographic distribution must be thought of as very fortuitous and random (in the sense of being useful for analysis). With this limitation in mind it must be emphasized that however limited the geologic record, it must be considered in evolutionary and paleoecologic studies.

Decapod dominated assemblages have been described (Bishop 1981, 1983c) in terms of associated taxa, distribution, and mode of preservation. Bishop (1981) described 5000 fossil decapods collected from six localities at five stratigraphic levels in the Upper Cretaceous Pierre Shale of South Dakota. The decapod dominated assemblages are preserved in apatite concretions distributed through 3 to 7 m of shale which are laterally continuous over areas of 15 to 1500 km^2. The decapods were buried as molts and/or corpses. The thin intersegmental membranes rapidly decomposed allowing the decaying bodies to be filled by intrusion of mud. The organic-rich sediment was ingested by deposit-feeding organisms as evidenced by filled and open burrows and ubiquitous masses and strings of fecal pellets (Bishop 1977). Compression due to weight of overburden flattened, broke, and rearranged parts of some decapods. Concentration of phosphates in the fecal pellets combined with the chemical micro-environments of the enclosed spaces within the crabs caused the precipitation of the apatite concretions leading to their preservation.

The uniformity of distribution, faunal composition, and mode of preservation of these faunas lead to the conclusion that they are the preserved parts of a recurrent decapod community. This community was dominated by the crab *Dakoticancer overanus* (75.0%), baculitid cephalopods and inoceramid bivalves. Less common preserved taxa are the decapods *Homolopsis punctata* (~0.8%); *Necrocarcinus pierrensis* (~7.6%); *Raninella oaheensis* (~0.3%); *Sodakus tatankayotankaensis* (~0.1%); *Zygastrocarcinus mendryki* (~0.05%); *Palaeo-*

Figure 7. Late Cretaceous decapods from the Mississippi Embayment. (A-E) Prairie Bluff Formation: (A) Left propodus (less fixed finger) of a new pagurid species (Georgia Southern College Museum) (GAB 52-1, holotype); (B) *Raninella* n.sp. (GAB 52-13); (C) Left propodus of *Prehepatus harrisi* (less fixed finger) (GSCM 1688, paratype); (D) Left propodus of *Eryma* n.sp. (GAB 52-14); (E) Anterior of carapace of a new xanthid species (GAB 52-3, holotype). (F-N) *Dakoticancer* Assemblage, Coon Creek Formation, Union Co. MS.: (F) Left propodus of *Eryma* cf. *flecta* (GAB 37-909); (G-H) Right propodus (G) and carapace (H) of *Dakoticancer australis* (GAB 37-1094, 37-515); (I-J) Right propodus (I) and carapace (J) of *Tetracarcinus subquadratus* (GAB 37-882, 37-1113); (K) Left propodus of *Paguristes? whitteni* (GSCM 1683, holotype); (L) Carapace of *Cristipluma mississippiensis* (GAB 37-1142, topotype); (M) Carapace of *Notopocorystes testacea* (GAB 37-844); (N) Carapace of *Raninella tridens* (GAB 37-833). (O) Dominant decapod, *Avitelmessus grapsoideus*, of the *Avitelmessus* assemblages, Coon Creek Formation (GAB 139). (A-B), (D-G), (J), (L-N) × 2.0 (Scale 1); (C), (K) × 1.5 (Scale 2); and (H), (O) × 1.0 (Scale 3).

nephrops browni (0.4%); and *Protocallianassa cheyennensis* (~0.5%) and numerous gastropods, bivalves, non-baculitid cephalopods, foraminifera, ostracodes, bony fish, sharks and reptiles. This decapod-dominated community, the *Dakoticancer* Association (see Fig.9A-G), periodically inhabited the Western Interior Cretaceous sea bottom when conditions in this unstable environment were favorable to them. Populations of the crabs, representing opportunistic species, rapidly colonized the bottom over favorable areas. When conditions became unfavorable (for yet unknown reasons) the crab populations rapidly declined to previous low levels. The periodic growth of crab populations (J-shaped) were thus superimposed upon the stable, equilibrium populations of molluscs which characterize the typical fossil assemblages of the Pierre Shale. During non-favorable times the faunal elements of the *Dakoticancer* Association were random elements preserved in the molluscan assemblages. During favorable times the crab populations rapidly increased and became dominant and their record over-printed the molluscan assemblages (Fig.5) as illustrated by Hallam (1972a). The reasons for such rapid population fluctuations are unknown but could be tied to fluctuations in food supply (such as the presence of great numbers of worms), fluctuations in physical conditions (salinity, temperature, oxygen, turbidity, etc.), or other undefined ecological parameters.

The presence of decapod-dominated assemblages, interpreted to be preserved community fractions (Bishop 1981) act as periodic 'pictures of the past' which allow us glimpses of the real evolutionary record of the decapods. Not only are these occurrences biocoenoses, but the large numbers of decapods allow maximum morphologic data retrieval and a good idea of faunal composition not masked by low numbers (Koch 1978) present through the bulk of decapod history. Decapod assemblages hold an inordinate role in deciphering the evolutionary history, the community structure, and the biogeography of the decapods.

Two types of paleobiogeographic records are thus indicated, one based on the fortuitous presence of decapod remains in molluscan assemblages (which will always be limited in numbers of specimens and taxa) and another based on periodic blossoming of opportunistic species of decapods which may form decapod-dominated fossil assemblages. This hypothesis, if true, would mean the geological record of opportunistic species (more capable of forming decapod assemblages) should be better than that of equilibrium species (which would be preserved as random elements in molluscan assemblages and easily 'lost' to science unless large collections are made). Interpretations based on these different types of records could easily distort our understanding of the geologic history of the decapods.

3 BIOGEOGRAPHY OF THE CRETACEOUS DECAPODS

3.1 *Cretaceous crabs of the Gulf Coastal Plain*

The Cretaceous Coastal Plain decapod fauna is large, consisting of 35 genera and 52 species (Figs.6, 7). This diversity is surprising when compared to the Western Interior (17 genera, 20 species) and the Atlantic Coastal Plain (13 genera, 14 species). Upon analysis, this greater diversity becomes explicable by noting that this is the only North American province in which Early Cretaceous fossiliferous marine strata are well developed, well exposed, and well enough collected to produce extensive crab faunas. This Early Cretaceous fauna (18 genera, 22 species), when added to the Late Cretaceous fauna (17 genera, 26 species), enlarges the total fauna.

Table 1. Decapods from the Glen Rose Limestone, Central Texas.

Anomura	*Hillius youngi* Bishop, 1983
Protocallianassa klofi Bishop, 1983	*Dioratiopus scotti* Bishop, 1983
Palaeopagurus banderensis (Rathbun), 1935	*Pseudonecrocarinus stenzeli* Bishop, 1983
Roemerus robustus Bishop 1983	*Prehepatus hodgesi* Bishop, 1983
Brachyura	*Torynomma? densus* Bishop, 1983
Palaeodromites naglei Bishop, 1983	Trace fossils
Diaulax roddai Bishop, 1983	Abundant branching burrows (*Thalassinoides*)

Table 2. Cretaceous decapods from the clastic rocks of North-Central Texas.

EAGLE FORD GROUP (*including Woodbine Fm.)	*Homarus* sp.
Macrura	*Ischnodactylus texanus* Rathbun, 1935
Linuparus grimmeri Stenzel, 1945	Anomura
Linuparus watkinsi Stenzel, 1945	*Eomunidopsis cretaceous* (Stenzel), 1945
Linuparus tarrantensis Davidson, 1968*	*Eomunidopsis? limonitica* (Stenzel), 1945
Linuparus cf. *kleinfelderi* Rathbun, 1931	
Astacodes davisi Stenzel, 1945	Brachyura
Homarus brittonestris Stenzel, 1945	*Notopocorystes punctatus* Rathbun 1935
Homarus davisi Stenzel, 1945	*Necrocarcinus texensis* Rathbun, 1935
Anomura	*Paranecrocarcinus moseleyi* (Stenzel), 1945
Upogebia rhacheochir Stenzel, 1945	*Cenomanocarcinus renfroae* (Stenzel), 1945
*Callianassa aquilae** Rathbun, 1935	*Prehepatus pawpawensis* Rathbun, 1935
	Xanthosia aspera Rathbun, 1935
Brachyura	*Ophthalmoplax comancheensis* Rathbun, 1935
Notopocorystes dichrous Stenzel, 1945	*Caloxanthus americanus* Rathbun, 1935
Cenomanocarcinus vanstraeleni Stenzel, 1945	
Protonecrocarcinus ovalis (Stenzel), 1945	DENTON Fm.
	Brachyura
BUDA LIMESTONE	*Raninella mucronata* Rathbun, 1935
Natantia	*Notopocorystes punctatus* Rathbun, 1935
Enopoclytia walkeri (Whitfield), 1880	*Notopocorystes parvus* Rathbun, 1935
Brachyura	*Necrocarcinus texensis* Rathbun, 1935
Graptocarcinus texanus Romer, 1887	*Paranecrocarcinus graysonensis* Rathbun, 1935
	Cenomanocarcinus armatus (Rathbun), 1935
GRAYSON Fm.	*Xanthosia aspera* Rathbun, 1935
Brachyura	*Xanthosia wintoni* Rathbun, 1935
Rathbunopon polyakron Stenzel, 1945	
	DUCK CREEK Fm.
PAWPAW Fm.	Brachyura
Macrura	*Prehepatus cretaceous* Rathbun, 1935

Three general types of occurrence account for most of the Gulf Coast specimens: (1) Preservation in lagoons behind the reefs in the shallow subtidal carbonate facies of the Early Albian Glen Rose Limestone; (2) Preservation in carbonate and clastic facies (Duck Creek, Pawpaw, Main Street, Grayson) of the late Albian and early Cenomanian Washita Group of northern Texas; (3) Preservation in Late Cretaceous sediment in well defined 'crab beds' (i.e. Britton Formation Eagle Ford Group, of Texas, and Coon Creek Formation of Mississippi) or as isolated remains.

The back-reef carbonates of the Glen Rose Limestone have yielded (Table 1, Fig.6) a diverse and interesting fauna (Bishop 1983b). The specimens are all small, preserved as steinkerns or with a chalky exoskeleton. Additional collecting should enlarge this fauna especially when crabs are collected from contemporaneous reef facies in the Glen Rose or Edwards limestone.

Cyclic sediments associated with fluctuating clastic and carbonate facies (Slocki 1967) or the Late Albian and Early Cenomanian in north Texas have yielded a highly varied decapod fauna (Table 2). Other crab fossils are known from these rocks (Hendrick & Wilson 1967, Perkins & Stewart 1971) suggesting that additional collecting will yield additional taxa. Regrettably, urbanization of the Ft Worth-Dallas area has destroyed many of the localities from which these specimens were collected.

The Late Cretaceous of the Gulf Coastal Plain has yielded crab faunas from a few well-known decapod-rich layers as well as from many isolated localities. The Britton Formation of the Eagle Ford Group near Dallas, Texas, has an abundant decapod fauna studied by Stenzel (1945). The fauna consists mostly of lobsters but includes numerous crabs (Table 2). These fossils are preserved in hematite or ironstone concretions which are found as continuous sheets or as layers of nodules, some of which contain crabs.

The famous fossiliferous clays of the Coon Creek Formation in Mississippi have yielded a numerically impressive fauna (Bishop 1983c) consisting mostly of *Avitelmessus grapsoideus* from its upper facies and a *Dakoticancer* Assemblage from its lower facies (Table 3, Fig.7F-N).

The Prairie Bluff Formation has yielded (Stephenson & Monroe 1940) a sparse but diverse decapod fauna (Table 4, Fig.7A-E). This middle Maastrichtian fauna is the youngest known Cretaceous decapod suite from North America. Being one of the least known faunas, its 12 or so species are still mostly undescribed. If a mass extinction event occurred in the decapods at the end of the Cretaceous, this fauna and superjacent Paleocene faunas will show major faunistic changes.

Faunas from the Cretaceous chalks of the Gulf Coastal Plain exist but are still poorly

Table 3. Decapods from the Coon Creek Formation, Mississippi Embayment.

UPPER COON CREEK Fm.	Brachyura
Brachyura	*Dakoticancer australis* Rathbun, 1935
Avitelmessus grapsoideus Rathbun, 1923	*Tetracarcinus subquadratus* Weller, 1905
	Raninella tridens Roberts, 1962
LOWER COON CREEK Fm.	*Notopocorystes testacea* (Rathbun), 1926
Macrura	*Eryma flecta* Rathbun, 1926
Penaeus wenasogensis Rathbun, 1926	*Cristipluma mississippiensis* Bishop, 1983
Enoploclytia sculpta Rathbun, 1926	*Opthalmoplax stephensoni* Rathbun, 1935
Hoploparia n.sp. (shell crusher)	
Linuparus canadensis (Whiteaves), 1885	Trace fossils
	Simple curved burrows
Anomura	Rehealed predation breaks on *Exogyra*
Protocallianassa mortoni (Pilsbry), 1901	Branching burrows (*Thalassinoides*)
Paguristes? whitteni Bishop, 1983	*Ophiomorpha* sp. (knobby burrows)

Table 4. Decapods from the Prairie Bluff Formation, Mississippi Embayment.

Macrura
Hoploparia n.sp. (shell crusher)
Anomura
Protocallianassa sp.
Paguristes n.sp.
Brachyura
Avitelmessus n.sp.
Prehepatus harrisi Bishop, 1984
Raninoides ovalis Rathbun, 1935
New genus, n.sp.
Raninella n.sp.
Eryma? n.sp.
Portunid
Atelecyelid
Macrophthalmid
Trace fossils
Rehealed breaks on *Exogyra*

Table 5. Decapods from the Austin Chalk, Central Texas.

Macrura
Enoploclytia sp.
Astacodes maxwelli Stenzel, 1945
Mecochirid
Anomura
Protocallianassa pleuralum Beikirch & Feldmann, 1980
Thalassinoides (in burrow)
Brachyura
Notopoides? pflugervillensis Beikirch & Feldmann, 1980
Trace fossils
Burrow (with undeterminable claws)

Table 6. Late Cretaceous decapods from the Northern Atlantic Coastal Plain.

NAVASINK Fm.
Anomura
Protocallianassa mortoni (Pilsbry), 1901
Brachyura
Notopocorystes testacea (Rathbun), 1926
WENONAH Fm.
Macrura
Linuparus sp.
Hoploparia sp.
Anomura
Protocallianassa mortoni (Pilsbry), 1901
Pagurus sp.
Brachyura
Notopocorystes sp.
Xanthosia sp.
Archaeocarabus whitfieldi (Pilsbry), 1901
Dioratiopus n.sp.
Raninella sp.
MERCHANTVILLE Fm.
Macrura
Enopoclytia sp.
Hoploparia gabbi Pilsbry, 1901
Hoploparia gladiator Pilsbry, 1901
Oncopareia sp.
Archaeocarabus? whitfieldi (Pilsbry), 1901
Linuparus richardsoni Roberts, 1962
Anomura
Palaeopagurus pilsbryi Roberts, 1962
Protocallianassa mortoni (Pilsbry), 1901
Protocallianassa praecepta Roberts, 1962
Brachyura
Raninella tridens Roberts, 1962
Notopocorystes testacea (Rathbun), 1926
Homolopsis atlantica Roberts, 1962
Homolopsis dispar Roberts, 1962
Tetracarcinus subquadratus Weller, 1905
Necrocarcinus rathbunae Roberts, 1962
Paranecrocarcinus gamma Roberts, 1962
Prehepatus dilksi Roberts, 1962
Xanthosia elegans Roberts, 1962
Xanthias? lenolensis Rathbun, 1935
MAGOTHY Fm.
Anomura
Protocallianassa cliffwoodensis
Brachyura
Tetracarcinus subquadratus Weller, 1905

known. One fauna (Table 5) is known from the Campanian Austin Chalk near Austin, Travis County, Texas (Beikirch & Feldmann 1980), and another is known from the Demopolis Chalk of Alabama (Gordon Bell, pers. comm.). Fragments of a shell-crushing decapod are known from many outcrops in the Mississippi Embayment Chalks. The fauna described from the Austin Chalk represented by very poorly preserved, fragmental material consists of mud shrimp, lobsters and a crab.

3.2 *Cretaceous crabs of the Atlantic Coastal Plain*

The Atlantic Coastal Plain decapod fauna (Table 6, Fig.8A-S) is severely limited because

---

Figure 8. Late Cretaceous decapods from the Northern Atlantic Coastal Plain. (A) Carapace of *Paranecrocarcinus gamma* partly in concretion (Philadelphia Academy of Natural Sciences, ANSP 20031, holotype); (B) Carapace steinkern assigned to *Paranecrocarcinus gamma* (Roberts 1962: 182) (ANSP 19721, paratype); (C) Partial carapace of *Necrocarcinus rathbunae* (Mendryk Collection H M 21); (D) Carapace of *Xanthosia elegans* (ANSP 20028, paratype); (E) Carapace of *X.elegans* (WFIS 17108, holotype); (F-H) *Tetracarcinus subquadratus*: (F) Carapace from the Woodbury Formation (H M 126); (G) Carapace preserved in hematite concretion and subsequently fractured along dessication fractures, (H M 127); (H) Carapace with exoskeleton, from Merchantville Fm. (H M 3); (I-J) *Raninella tridens*: holotype carapace (I) (ANSP 19237) and carapace (J) with some pereiopods (H M 76); (K) Carapace of *Homolopsis dispar* (ANSP 20030, holotype); (L-M) Carapace and cheliped of *Homolopsis atlantica*; (M) Right branchial chamber inflated by bopyrid isopod? (H M 120 & 48); (N) Right propodus of *Prehepatus* n.sp. (Lauginger Collection EL 20); (O) Right propodus (less fixed finger) and dactylus of *Prehepatus dilksi* (ANSP 19728, holotype); (P) Left propodus of *Palaeopagurus pilsbryi* (Wagner Free Institute of Science WFIS 17095, holotype); (Q) Right partial propodus of *Archaeocarabus? whitfieldi* (ANSP 4693, holotype); (R-S) Right and left propodus of *Xanthias? lenolensis* (ANSP 12804, holotype and paratype). (A-D), (F), (H-J), (M-O), (R), (S) × 2.0 (Scale 1); (E), (K-L), (P) × 1.5 (Scale 2); (G), (Q) × 1.0 (Scale 3).

Figure 9. Late Cretaceous decapods from the Western Interior, USA (see Bishop 1981). (A-G) *Dakotican-cer*Assemblage: (A) Carapace of *Dakoticancer overanus* (US Nat. Mus., USNM 173529); (B) Carapace of *Necrocarcinus pierrensis* (Gale Bishop Coll., GAB 4-509); (C) Carapace of *Sodakus tatankayotankaensis* (US Nat. Mus., USNM 173580, holotype); (D) Anterior of carapace of *Zygastrocarcinus mendryki* (South Dakota School of Mines Museum of Geology 10 001, holotype); (E) Carapace of *Homolopsis punctata* (GAB 2-2); (F) Carapace steinkern of *Raninella oaheensis* partly buried under carapace of *D.overanus* (USNM 173589, holotype); (G) Right propodus (less fixed finger) and proximal dactylus of *Protocallianassa cheyennensis* (GAB 14-33). (H-N) Heart Tail Ranch Assemblage, Butte County, S.D.; (H) Carapace of *Dioratiopus hearttailensis* partly enclosed in apatite concretion (GAB 36-26, paratype); (I) Carapace of *Necrocarcinus davisi* (SDSM 10 033, paratype); (J) Left chela of *Protocallianassa russelli* exposed by break across concretion (GAB 36-24); (K) Carapace steinkern of *Raninella* cf. *oaheensis* (GAB 36-101); (L) Anterior of carapace of *Rugafarius fredrichi* (SDSM 10 029, holotype); (M) Partial cephalothorax, abdomen, and right cheliped of *Hoploparia mickelsoni* (SDSM 10 023, holotype); (N) Carapace steinkern of *Xanthosia elegans occidentalis* (GAB 36-361); (O-S) Western Interior molluscan-dominated assemblages: (O) Carapace of *Plagiophthalamus*? n.sp. (US Geol. Survey D8090), Grand Co., Colo.; (P) Carapace of specimen representing a new species and genus of Dakoticancroid crab, Fox Hills Formation (US Geol. Survey WAC 1) Corson Co., S.D.; (Q) Partial carapace of *Dioratiopus dawsonensis*, Bearpaw Shale, Montana (USNM 173393, holotype); (R) Carapace of *Ekalakia lamberti*, Bearpaw Shale, S.E. Montana (Carter Co. Mus. CCM 6131); (S) Carapace of *Eomunidopsis cobbani*, Larimer Sandstone, Colorado (USNM 370191, holotype). (I-K), (N-S) × 2.0 (Scale 1); (A-H), (L) × 1.5 (Scale 2); (M) × 1.0 (Scale 3).

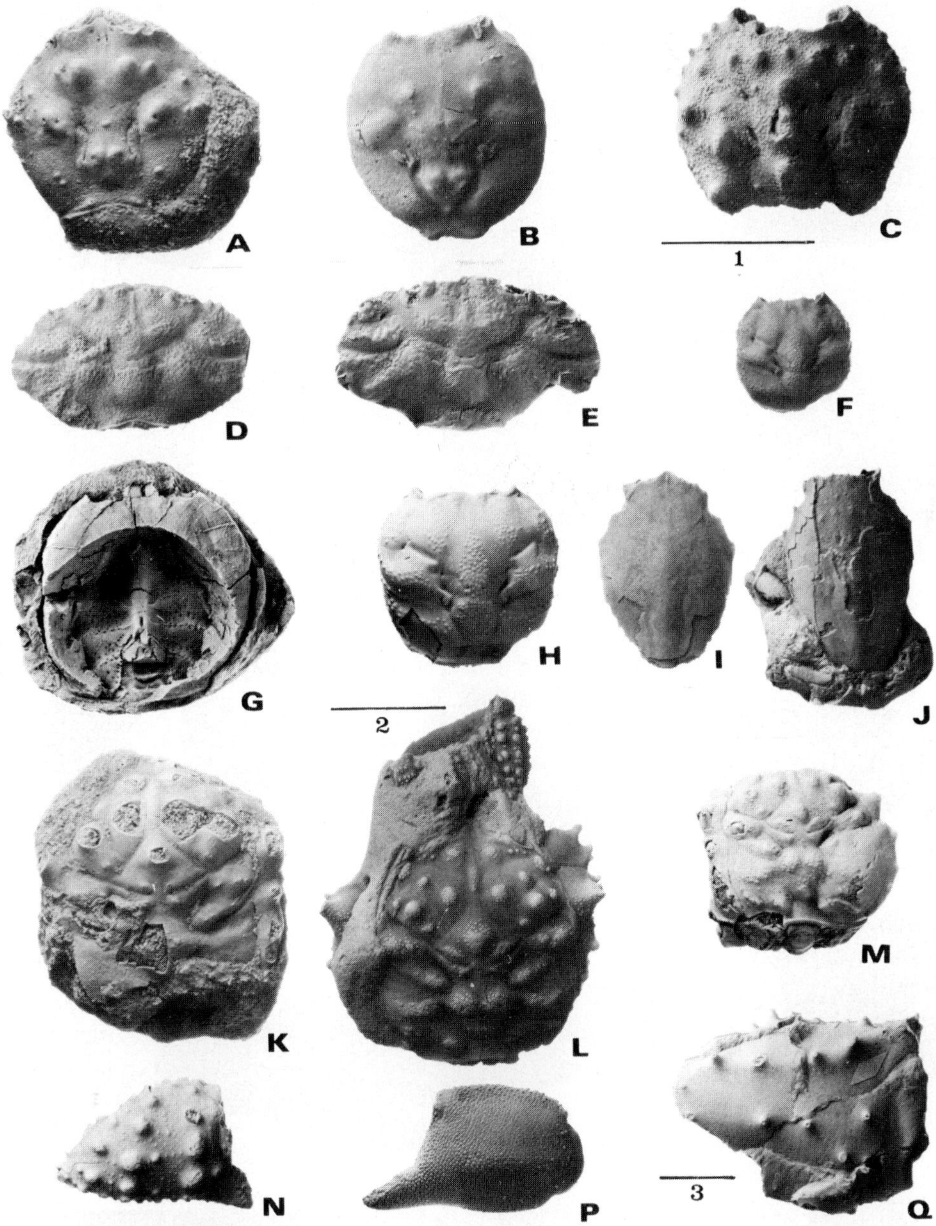
A
B
C
1
D
E
F
G
H
I
J
2
K
L
M
N
P
3
Q
O
R
S

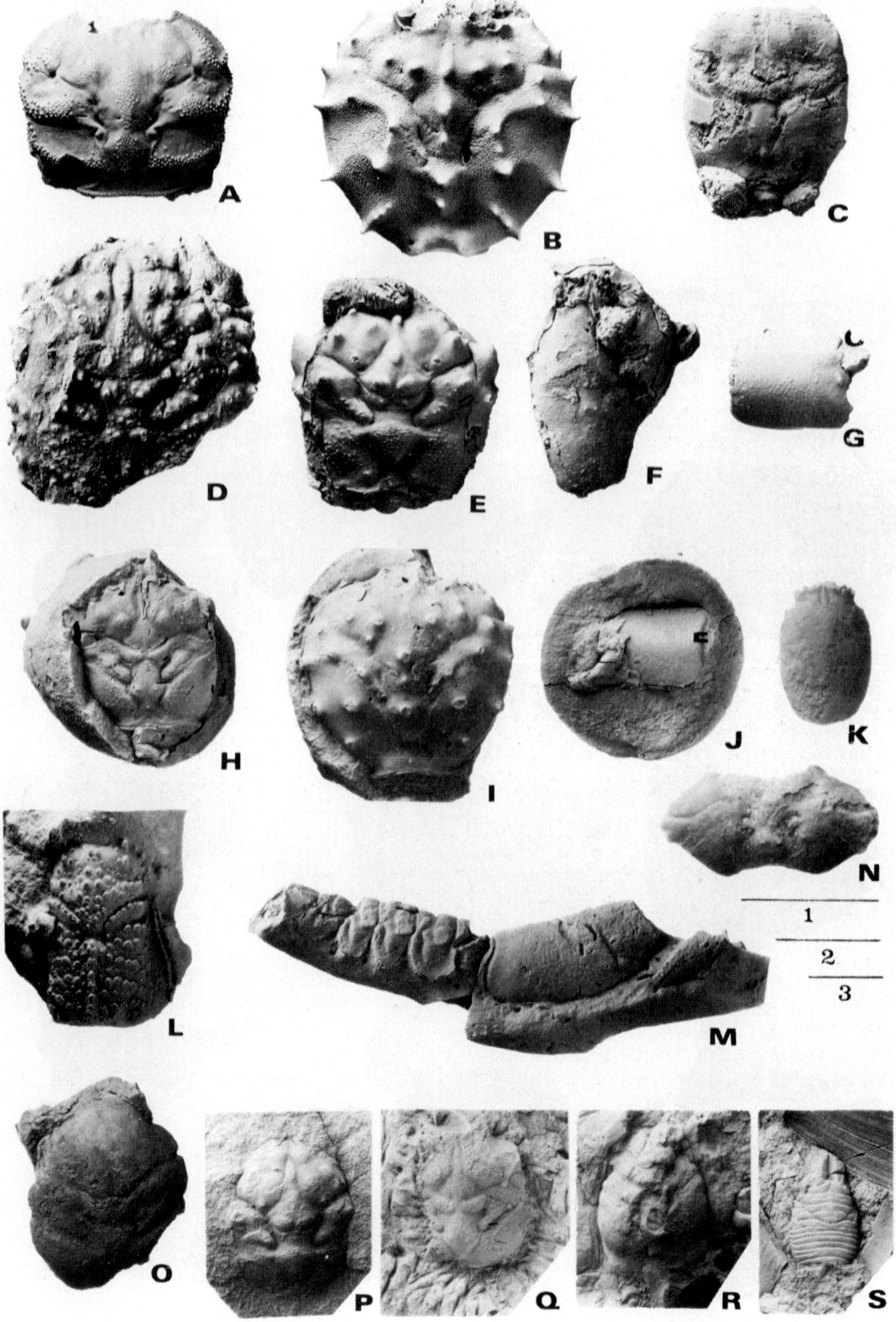
A
B
C
D
E
F
G
H
I
J
K
L
M
N
O
P
Q
R
S
1
2
3

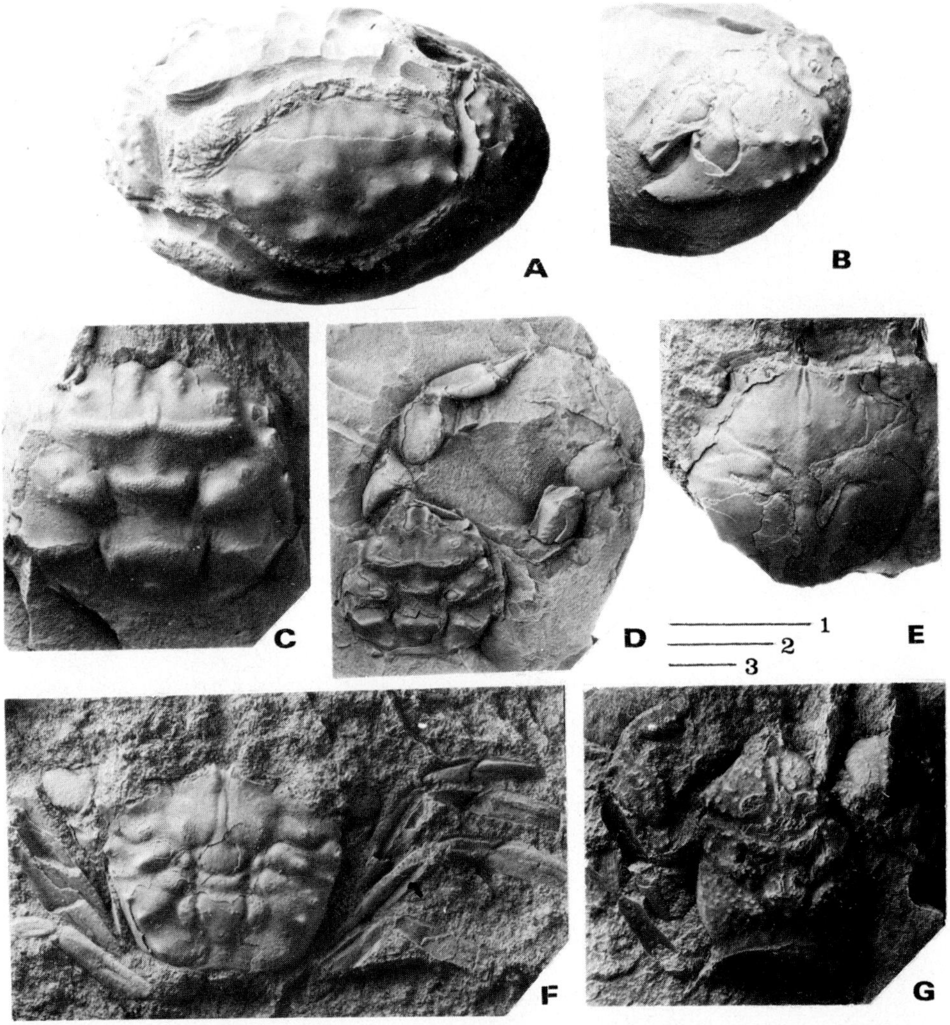

Figure 10. Late Cretaceous decapods from the Pacific Slope of North America. (A-B) Carapace and left chela of *Zanthopsis? sternbergi* from Baja, Mexico (Hennessey Collection EH 1); (C-D) *Archaeopus vancouverensis*from British Columbia, carapace (C) and carapace with chelipeds (D) (Geological Survey of Canada GSC 5974, holotype and Mendryk Collection HM 122); (E) Anterior of carapace of *Notopocorystes harveyi* (GSC 5817, holotype); (F) Nearly complete crab, *Longusorbis cuniculosis* preserved in concretion (GAB 124, topotype); (G) Carapace and pereiopods of *Zygastrocarcinus richardsoni* (GSC 5995, holotype). (C) × 2.0 (Scale 1); (F-G) × 1.5 (Scale 2); (A-B), (D-E) × 1.0 (Scale 3).

Table 7. Decapods from the Cretaceous of the western interior of North America.

DAKOTICANCER ASSEMBLAGE, PIERRE SHALE, SOUTH DAKOTA
Macrura
Palaeonephrops browni (Whitfield), 1907

Anomura
Protocallianassa cheyennensis Rathbun, 1930

Brachyura
Dakoticancer overanus Rathbun, 1917
Necrocarcinus pierrensis (Rathbun), 1917
Homolopsis punctata Rathbun, 1917
Zygastrocarcinus mendryki (Bishop), 1982
Raninella oaheensis Bishop, 1978
Sodakus tatankayotankaensis Bishop, 1978

Trace Fossils
Ophiomorpha type burrows

HEART TAIL RANCH ASSEMBLAGE, PIERRE SHALE, SOUTH DAKOTA
Macrura
Linuparus sp.
Hoploparia mickelsoni Bishop, 1983

Anomura
Protocallianassa russelli Bishop, 1983
Rugafarius fredrichi Bishop, 1983

Brachyura
Necrocarcinus davisi Bishop, 1983
Dioratiopus hearttailensis Bishop, 1983
Xanthosia elegans var. *occidentalis* Bishop, 1983
Raninella oaheensis Bishop, 1978

Trace fossils
Indian Bead type burrows

12 MILE CROSSING LOCALITY, LOWER PIERRE SHALE, SOUTHEASTERN MONTANA
Macrura
Linuparus sp.

Brachyura
Necrocarcinus davisi Bishop, 1983
Zygastrocarcinus cardsmithi Bishop, 1986

CARLILE SHALE LOBSTER ASSEMBLAGE, NORTHERN BLACK HILLS
Macrura
Linuparus canadensis (Whiteaves), 1885
Hoploparia sp.

Brachyura
Necrocarcinus n.sp.
Zygastrocarcinus n.sp.
Notopocorystes sp.

of poor exposures. Most crab fossils have come from man-made outcrops (clay pits, dredge spoil, etc.) and therefore are limited in quantity, quality, and known stratigraphic position. Recent collections made by a cadre of paraprofessional collectors, especially by Mr Harold Mendryk, from fresh spoil banks along the Chesapeake and Delaware Canal on the Delmarva Peninsula (Delaware) have yielded a large number (approximately 2,000) of excellently preserved fossil crabs. Some of these specimens were collected while very fresh and provide some of the first really excellent material for carcinological examination (Fig.8F,H,J,L). These specimens deteriorate rapidly upon exposure, yielding mostly steinkerns for study, which has lead to many misidentifications in the past. An unusual type of preservation (Fig.8G) by pyritization occurs in parts of the Merchantville formation, yielding crabs in concretions exhibiting shrinkage cracks, thus providing fissures along which pyrite forms. These faunas were mostly described by Roberts (1962) in *Cretaceous Fossils of New Jersey* which was, unfortunately, published with virtually unusable photographic plates. The figures of these taxa represent the first time this fauna has been adequately figured.

3.3 *Cretaceous crabs of the Western Interior*

Cretaceous crabs of the Western Interior (Table 7, Fig.9) occur either as part of a decapod assemblage zone (i.e. 'crab beds') or as isolated remains randomly preserved with the other invertebrates.

Table 8. Decapods from the Pacific Slope of North America. From British Columbia unless marked by * = from Baja, Mexico.

Macrura	Brachyura
Phlyetisoma dawsoni (Woodward), 1900	*Archaeopus antennatus* Rathbun, 1908
Glyphea sp.	*Archaeopus vancouverensis* (Woodward) 1896
Linuparus canadensis (Whiteaves), 1885	*'Zanthopsis' sternbergi* Rathbun, 1926*
Linuparus vancouverensis (Whiteaves), 1895	*Longusorbis cuniculosis* Richards, 1975
Enoploclytia minor Woodward	*Zygastrocarcinus richardsoni* (Woodward), 1896
Meyeria? harveyi Woodward, 1900	*Notopocorystes harveyi* (Woodward), 1896
Anomura	
Protocallianassa whiteavesii (Woodward), 1896	

Of the decapod assemblages, the *Dakoticancer* Association (Bishop 1971, 1972a,b,c, 1974, 1977, 1978, 1981a) is the most studied and best known (Table 7). This assemblage (Fig.9A-G) is repeated at several stratigraphic levels within the Pierre Shale of South Dakota. In each occurrence the assemblage is confined to a thin interval of rock (±3 m) and widely distributed (1,500 sq.km).

An older decapod assemblage was discovered in 1976 lower in the Pierre Shale at the Heart Tail Ranch, Butte Co., S.D. This Assemblage (Table 7, Fig.9H-N) is similar to the *Dakoticancer* Assemblage in occurrence and mode of preservation but has a slightly different faunal composition (partly because it predates the migration of *'Dakoticancer'* into the Western Interior).

Specimens in collections of the Carter County Museum and the U.S. National Museum indicates another crab fauna occurs high in the Pierre Shale (= Bearpaw Shale) in eastern Montana south of Baker on Cedar Creek Anticline. This fauna is the source of *Ekalakia lamberti* (Fig.9R) and also contains *Raninella oaheensis.*

Another crab zone in the Bearpaw Shale of north-central Montana has yielded numerous specimens of *Notopocorystes eichhorni* and *Zygastrocarcinus griesi.* The crabs are preserved with molluscs including *Didymoceras nebrascense* in fairly large calcite concretions (Bishop 1983d).

Two other interesting described decapod faunas occur in the Western Interior Cretaceous. Beautifully preserved lobsters in calcite concretions are found on the shores of Fort Peck Reservoir in east-central Montana. The occurrence was described by Feldmann, Bishop & Kammer (1977) and includes *Palaeonephrops browni, Hoploparia bearpawensis,* and *Linuparus pustulosus.* On the north flank of the Black Hills, low in the Pierre Shale, a lobster and crab locality (Table 7) has been located and collected by me. Even lower in the Cretaceous is a Turonian (Carlile Shale) decapod assemblage zone (Table 7) which is found at widely scattered outcrops around the Black Hills uplift. In addition to *Linuparus* sp. two crabs are known, *Necrocarcinus* sp. and *Zygastrocarcinus* n.sp. Many other isolated decapods (Fig.9O-S) are known from the Western Interior Cretaceous molluscan assemblages.

3.4 *Cretaceous crabs of the Pacific Slope*

The Pacific Slope Cretaceous decapod fauna is currently very small (Table 8, Fig.10A-G) and highly questionable in stratigraphic and taxonomic assignment. The known fauna

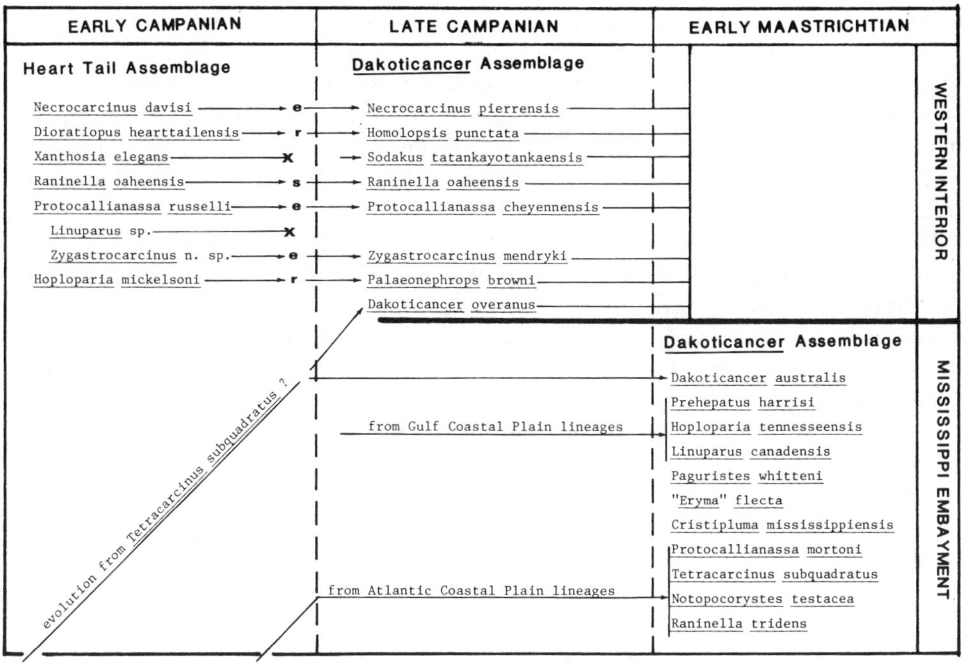

Figure 11. Paleobiogeographic overtones of the evolution of the *Dakoticancer* community in North America. e – evolution; r – ecological replacement; s – stasis; X – extinction.

is included here to indicate that the Pacific slope needs much attention. The presence of fossiliferous strata reported by Jeletzsky (1969) and Ward (1978) and the assurance received verbally from West Coast Cretaceous paleontologists (e.g. David Jones & Jim Fox) substantiate the probable presence of crab fossils.

3.5 *Summary of Cretaceous decapod biogeography*

Cretaceous crabs representing this immense span of time (40 million years) are thus far relatively poorly known from North America. The 47 genera and 95 species known from North America, when compared to the British Cretaceous crab fauna (Early Aptian through Coniacian) of 28 genera and 63 species (Wright and Collins 1972), reveals the relatively equal status of our known of North American and British Cretaceous crabs. Recent faunas consisting of at least 89 genera in the Western Atlantic Carolinian province (Williams 1965) and at least 66 genera and 132 species on the California coast (Schmitt 1921) emphasize the incompleteness of the Cretaceous fossil crab record, especially when we consider the time spans involved.

Differences in crab faunas are decipherable in the regions of the Northern Atlantic Coastal Plain, the Gulf Coastal Plain, the Mississippi Embayment, the Western Interior, and the poorly known Pacific Slope. These provincial differences appear at first only to be artifacts of a poor and geographically and stratigraphically widely scattered record (Figs. 2, 3) through the North American Cretaceous (i.e. the Glen Rose fauna, the North Texas fauna, the Merchantville fauna, the Pierre Shale fauna, etc.). Certainly the disjunct, spotty record does affect our conceptualization of Cretaceous 'crab history' to a significant extent, but upon closer examination the apparent faunal differences are substantiated wherever we have phylogenetically related taxa (e.g. within the Dakoticancridae, Necrocarcinae, Homolidae, or Raninidae) well represented through shorter intervals (Campanian and Maastrichtian) of the Cretaceous. In each of these families or subfamilies, species-level taxa are recognized from separate provinces while some taxa clearly are found in more than one province.

The family Dakoticancridae, known only from North America, is represented on the Atlantic Coastal Plain by *Tetracarcinus subquadratus* and one *Avitelmessus grapsoideus*; in the Mississippi Embayment by *Tetracarcinus subquadratus, Avitelmessus grapsoideus*, and *Dakoticancer australis*; and in the Western Interior by *Dakoticancer overanus* and a few *Tetracarcinus subquadratus.* The near restriction of *Dakoticancer* to the Western Interior, *Tetracarcinus* to the Atlantic Coastal Plain, and *Avitelmessus* to the Mississippi Embayment with some 'missing' particularly to the south (where the Western Interior and Northern Atlantic provinces were connected along the Tethys) strongly suggests provincialism in the Dakoticancridae. Specimens now under examination clearly show *Tetracarcinus* was present in the Western Interior in the Campanian and probably coexisted with *Dakoticancer* which finally became the dominant representative of the family in the Western Interior. Recent work has led to the discovery of a crab assemblage low in the Campanian on the North Flank of the Black Hills which lacks *Dakoticancer* and thus is interpreted to predate migration of *Dakoticancer* into the Western Interior.

The juxtaposition of the Heart Tail Assemblage (Table 7, Fig.9H-N) and the *Dakoticancer* Assemblages (Table 7, Fig.9A-G) in the Pierre Shale of the Western Interior provides an opportunity to compare and contrast (Fig.11) two decapod assemblages, one obviously derived from the other, within a single paleobiogeographic province. Each assemblage is widespread through thin intervals of claystone and similarly preserved as burrowed and pelleted apatite concretions. Both are dominated by decapods and in each case dominated by one to three decapod taxa. The most obvious difference in the Heart Tail Assemblage and the *Dakoticancer* Assemblages is in the identity of the taxa present. The Heart Tail Assemblage is dominated by *Protocallianassa russelli, Necrocarcinus davisi* and *Hoploparia mickelsoni*, but includes the decapods *Dioratiopus hearttailensis, Raninella oaheensis, Xanthosia elegans occidentalis*, and *Rugafarius fredrichi.* The *Dakoticancer* Assemblages are dominated by *Dakoticancer overanus* but include *Necrocarcinus pierrensis, Homolopsis punctata, Sodakus tatankayotankaensis, Raninella oaheensis, Zygastrocarcinus mendryki, Protocallianassa cheyennensis*, and *Palaeonephrops browni.* The changes and similarities in these two assemblages which are observed over the 7 million years between the existence of these community fractions can be explained by: (1) the immigration of *Dakoticancer* into the Western Interior and its subsequent dominance of the repeated *Dakoticancer* Assemblages; (2) the evolution of *Necrocarcinus davisi* into *N.pierrensis* and *Protocallianassa russelli* into *P.cheyennensis*; (3) the replacement of *Dioratiopus hearttailensis* by its possible ecological homeomorph *Homolopsis punctata*; (4) the apparent stasis of the conservative burrowing crab *Raninella oaheensis* which is little changed from one assemblage to the other; (5) the presence and possible extinction of the rare taxa *Rugafarius fredrichi* and *Xanthosia elegans occidentalis* in the Heart Tail Assemblage and the presence of the rare *Sodakus tatankayotankaensis* and *Zygastrocarcinus mendryki* in the *Dakoticancer* Assemblage; (6) the presence of the lobsters *Hoploparia mickelsoni* and *Linuparus* sp. in the Heart Tail Assemblage and a few *Palaeonephrops browni* in the *Dakoticancer* Assemblages;

and (7) the ubiquitous presence of the normal Cretaceous shelf molluscs and trace fossils of deposit feeding worms.

The presence of nearly isochronous *Dakoticancer* Assemblages in two geologic provinces, the Western Interior [*Dakoticancer* Assemblages (Bishop 1981) and the Mississippi Embayment (Table 3, Fig.7F-N) (Bishop 1983c)] provides the opportunity to compare and contrast the effects of evolution of the same community-fraction in distinct paleo-biogeographic provinces. The most obvious similarity of the *Dakoticancer* Assemblages in two provinces is the domination of the decapod-dominated assemblages by *Dakoticancer*. Other similarities are the presence of *Protocallianassa, Raninella*, and a few lobsters in each assemblage as well as the ubiquitous molluscan assemblages which form the 'background' over which the decapod-dominated assemblages are overprinted. The differences in the *Dakoticancer* Assemblages from the two provinces include: (1) different species of *Dakoticancer* are dominant; *D.overanus* in the Western Interior and *D.australis* in the Mississippi Embayment; (2) the Mississippi Embayment Assemblage is less exclusively dominated by *Dakoticancer*; (3) the subdominant (accessory) decapod taxa in the Western Interior province are closely related to species already present in the Heart Tail Assemblage while the subdominant decapod taxa in the Mississippi Embayment province show great similarity to faunas from the Northern Atlantic Coastal Plain and Gulf Coastal Plain; and (4) the modes of preservation are slightly different; the Western Interior Assemblages represent single cycles of population explosion whereas the Mississippi Embayment Assemblage apparently represents at least two episodes or cycles of population explosion (phosphatized and non-phosphatized decapod assemblages).

A similar pattern of morphologically distinguishable species of *Homolopsis* in the Western Interior (*Homolopsis punctata*) and Atlantic Coastal Plain (*Homolopsis dispar, Homolopsis atlantica*) tend to substantiate the hypothesized provincialism.

Other patterns of distribution are equally intriguing. The hermit crabs so abundant in the early Albian of Texas (*Pagurus banderensis, Pagurus trevisensis*) are present during the Late Cretaceous on the Atlantic Coastal Plain (*Pagurus pilsbryi*) and the Gulf Coastal Plain (*Paguristes ouachitensis* and *Petrochirus taylori*) but are not yet known from the Western Interior, the Pacific Slope, or the Arctic Slope. The Western Interior harbors a distinctive 'relict' decapod fauna which includes a series of 'primitive' crabs (*Ekalakia, Plagiothalamus*, and *Dioratiopus*) found long after they reach their peak in abundance and diversity in the English Cretaceous. *Homolopsis* is common on the Atlantic Coastal Plain and in the Western Interior, known from the Pacific Slope, but not yet known from the Gulf Coastal Plain. *Graptocarcinus* is known (Wright & Collins 1972) from Texas, Mexico, England, and Czechoslovakia but not from the Western Interior, Atlantic Coastal Plain, or Pacific Slope. The very small Pacific Slope decapod fauna probably is indicative of the great amount of work that remains to be done in that region.

This preliminary report of on-going research suggests strongly that our knowledge of Cretaceous crabs of North America (except for the Pacific Slope) is approaching a level of usefulness for biogeographic studies. Knowledge of some lineages may also be approaching a level of biostratigraphic usefulness. The record, although admittedly fragmentary, strongly biased, and incompletely known, is sufficiently encouraging to warrant continued research efforts.

The importance of decapod-dominated assemblages is emphasized here because the numbers of specimens and taxa in them are so great that virtually the complete preservable fauna may be represented. These assemblages therefore act as 'windows' to the evolving lineages of Cretaceous decapods giving us opportunity to observe similarities and differences in evolving decapod lineages and communities, the biogeographic distribution of taxa at various times, and the ability to completely describe the gross morphology of some taxa.

As one means of encouraging further investigations of this geologically, geographically, and chronologically widespread fauna, an updated bibliography of North American Cretaceous crabs is included in the References. This is the first such compilation to be published since Rathbun's (1935) paper and is intended to complement Feldmann's (1981) study on North American lobsters and shrimps.

ACKNOWLEDGEMENTS

This study was made possible by direct support from the National Geographic Society (Grant 1629), the Faculty Research Committee of Georgia Southern College, The National Science Foundation (Grant DEB 8011570) and The Department of Geology of Georgia Southern College. Indirect support was received from numerous institutions which loaned specimens. I would like to thank personally for their assistance in procuring specimens W.A.Cobban of the U.S. Geological Survey, Larry Eichhorn of Lewistown, Montana, Harry Mendryk of Harrison, New Jersey, Ralph Harris of Blue Springs, Mississippi, Edward Hennessey of Long Beach, California, and the many landowners who have given me permission to collect. The manuscript was typed by Margaret Perry. Don Hattin and Erle Kauffman edited the initial manuscript.

APPENDIX

Inventory of known Cretaceous decapods. Each taxon is evaluated according to the quantity and quality of specimens and its description.

Superior: Abundant prime material, description essentially complete.
Excellent: Prime material and description.
Good: Material abundant, descriptions satisfactory.
Fair: Material in some fragmentary state, descriptions more or less sufficient.
Poor: Material and descriptions inadequate.
Doubtful: Material and descriptions allow no validation of taxa.

Infraorder Anomura H.Milne Edwards, 1832
Superfamily Paguroidea Latreille, 1803
Family Paguridae Latreille, 1803
Pagurus Fabricius, 1775

1 *P.banderensis* Rathbun, 1935, p.39, Pl.9, Figs.7, 8. Stenzel, 1945, p.435, Pl.45, Figs.7-15. Texas: Bandera County, Burnet County, Blanco County, Travis County, Kendall County. Glen Rose Limestone, early Albian. 34 specimens, claws only. Excellent.

2 *P.travisensis* Stenzel, 1945, p.437, Pl.45, Figs.16-18. Texas: Travis County. Comanche Peak Formation, middle Albian. 1 specimen, claw only. Fair.

3 *P.convexus* Whetstone and Collins, 1982, p.1214, Pl.1, Figs.1-5, 7-8. Alabama: Montgomery County. Eutaw Formation, earliest Campanian. 6 specimens, claws only. Excellent.

4 *P.pilsbryi* Roberts, 1962, p.174, Pl.84, Figs.1-4. New Jersey: Burlington County. Merchantville Formation, mid-early Campanian. 6 specimens. Delaware: New Castle County. Merchantville Formation, mid-early Campanian. 1 specimen. Claws. Fair.

Paguristes Dana, 1851

5 *P.ouachitensis* Rathbun, 1935, p.39, Pl.6, Figs.9-11. Arkansas: Clark County. Nacotoch Sandstone, latest Campanian. 1 specimen. Poor.

6 *P.whitteni* Bishop, 1983, p.420, Figs.3E-F, 4A-I, 5. Mississippi: Union County. Coon Creek Formation, early Maastrichtian. 25 specimens, all chelipeds. Excellent.

7 *P.* n.sp. Bishop, pers. observ. Mississippi: Union County. Prairie Bluff Formation, middle Maastrichtian. 3 specimens, all claws. Good.

Petrochirus Stimpson, 1859

8 *P.taylori* Rathbun, 1935, p.40, Pl.3, Figs.1-3. Texas: Collin County. Taylor Marl, latest-early Campanian. Doubtful.

Roemerus Bishop, 1983

9 *R.robustus* Bishop, 1983, p.42, Pl.3, Figs.10-31, Fig.8G, Table 4. Texas: Travis County and Kendall County. Glen Rose Limestone, early Albian. 2 specimens, both claws. Good.

Superfamily Galatheoidea Samouelle, 1819

Family Galatheidae Samouelle, 1819

Eomunidopsis Via Boada, 1981

10 *E.cretacea* (Stenzel), 1945, p.430, Pl.43, Fig.3. Texas. Tarrant County. Pawpaw Shale, latest Albian or earliest Cenomanian. 1 specimen, carapace. Good.

11 *E.limonitica* (Stenzel), 1945, p.431, Pl.43, Fig.4. Texas: Tarrant County. Pawpaw Shale, latest Albian or earliest Cenomanian. 1 specimen, carapace. Good.

12 *E.cobbani* Bishop, 1985a, p.602, Figs.1-3. Colorado: Larimer County. Larimer Sandstone, late Campanian. 48 specimens, carapaces, chelipeds, and pereopods. Excellent.

Rugafarius Bishop, 1984

13 *R.fredrichi* Bishop, 1985b, p.615, Figs.3,4,7. South Dakota: Butte County, Gammon Ferruginous Member, Pierre Shale, early Campanian. 1 specimen, carapace. Good.

Infraorder Brachyura Latreille, 1803

Section Dromiacea de Haan, 1833

Superfamily Dromioidea de Haan, 1833

Family Prospoidae von Meyer, 1860

Ekalakia Bishop, 1976

14 *E.lamberti* Bishop, 1976, p.399, Pl.1, Figs.1-6, Fig.1. Montana: Carter County and Fallon County. Bearpaw Shale, Campanian. 4 specimens, carapaces. Excellent.

Plagiophthalmus Bell, 1863

15 *P.* n.sp. Bishop, pers. observ. Colorado: Grand County. Pierre Shale, Campanian. 1 specimen, carapace. Good.

Family Dynomenidae Ortmann, 1892

Palaeodromites A.Milne-Edwards, 1865

16 *P.naglei* Bishop, 1983, p.44, Pl.1, Figs.12-17, Figs.8A, 10, Table 5. Texas: Kendall County and Blanco County. Glen Rose Limestone, early Albian. 9 specimens, carapaces. Excellent.

Graptocarcinus Roemer, 1887

17 *G.texanus* Roemer, 1887, p.173, Fig.d. Rathbun, 1935. Pl.10, Figs.13-15. Glaessner, 1969, Figs.301-304. Wright and Collins, 1972, Pl.9, Fig.2a-c, Fig.9h-i. Texas: Travis County and Hays County. Buda Limestone, Earliest Cenomanian. 4 specimens. Good. Mexico: San Luis Potosi. Taninul Limestone, Albian. 1 specimen. Also known from England and Czechoslovakia.

18 *G.muiri* Stenzel, 1944, p.551, Pl.93, Figs.3-5. Wright and Collins, 1972, Pl.9, Fig.2a-c, Fig.9h-i. Mexico: San Luis Potosi. Albian. 1 specimen. Good.

Family Dialulacidae Wright and Collins, 1972

Diaulax Bell, 1963

19 *D.roddai* Bishop, 1983, p.45, Pl.1, Figs.1-2, Figs.8A, 10, Table 5. Texas: Kendall County, Glen Rose Limestone, early Albian. 3 specimens, carapaces. Good.

20 *Diaulax* n.sp. (HM 116) Bishop, pers. observ. Delaware: New Castle County. Merchantville Formation, mid-early Campanian. 1 specimen, carapace, chelipeds, pereopods. Good.

Family uncertain

Rathbunopon Stenzel, 1945

21 *R.polyakron* Stenzel, 1945, p.450, Pl.41, Figs.18-21, Fig.16. Glaessner, 1969, Figs. 301-305. Texas: Travis County. Grayson Formation, lower Cenomanian. 1 specimen, carapace. Good.

Superfamily Homoloidea White, 1947

Family Homolidae White, 1847

Zygastrocarcinus Bishop, 1983

22 *Z.griesi* Bishop, 1983, p.901, Figs.2, 3a-k, 4d. Montana: Valley County. Bearpaw Shale, late Campanian. 4 specimens, carapaces, cheliped, and pereopods. Excellent.

23 *Z.mendryki* Bishop, 1982, p.221, Figs.3-4. South Dakota: Corson County. Mobridge Member, Pierre Shale, early Maastrichtian. 2 specimens, carapaces. Good.

24 *Z.cardsmithi* Bishop, 1986, Montana: Carter County. Gammon Ferruginous Member, Pierre Shale, early Campanian. 1 specimen, carapace. Good.

25 *Z.* n.sp. Bishop, pers. observ. South Dakota: Butte County. Pool Creek Member, Carlile Shale, Turonian. 1 specimen, carapace. Good.

26 *Z.richardsoni* Woodward, 1896, p.224, Fig.3. Rathbun, 1926, p.86, Pl.20, Fig.3. British Columbia: Queen Charlotte Islands. 'Middle Cretaceous'. 1 specimen, carapace. Good.

Homolopsis Bell, 1863

27 *H.punctata* Rathbun, 1917, p.388, Pl.33, Figs.1-3. Bishop, 1981, Figs.13-14c, 13-19b,d. South Dakota: Corson County, Dewey County, Pennington County. Pierre Shale, Late Campanian through early Maastrichtian. 48 specimens, carapaces, chelipeds, and pereopods. Superior.

28 *H.atlantica* Roberts, 1962, p.178, Pl.89, Fig.4. New Jersey: Burlington County. Merchantville Formation, mid-early Campanian. 1 specimen. Fair. (Holotype lost). Delaware: New Castle County. Merchantville Formation, mid-early Campanian. 100 specimens, carapaces, chelipeds, and pereopods. Superior.

29 *H.dispar* Roberts, 1962, p.180, Pl.89, Fig.5. New Jersey: Burlington County. Merchantville Formation, mid-early Campanian. 1 specimen, carapace. Poor.

Superfamily Dakoticancroidea Rathbun, 1917

Family Dakoticancridae Rathbun, 1917

Dakoticancer Rathbun, 1917

30 *D.overanus* Rathbun, 1917, p.386, Pls.32, 33, Figs.6-14. Bishop, 1972, Pl.122, Figs.1-10; 1974, Pl.1, Figs.1-10; 1976, Fig.3; 1981, Figs.13-14A, 9A,C,E,F,I-K,M-T. South Dakota: Dewey County, Corson County, Pennington County, Lyman County, Butte County, Meade County. Pierre Shale, Zone of *Exiteloceras jenneyi* to *Baculites grandis*, mid-Campanian to early Maastrichtian. 5075 specimens. Montana: Dawson County. Bearpaw Shale, early Maastrichtian. 1 specimen. Colorado: Jackson County, Grand County, Boulder County, Larimer County. Pierre Shale, late Campanian. Carapaces, chelipeds, and pereopods. Superior.

31 *D.australis* Rathbun, 1935. p.40, Pl.10, Fig.20. Rathbun *in* Wade, 1926, p.189, Pl.67, Fig.3. Glaessner, 1969, p.R491, Figs.303-2a-b. Bishop, 1983, p.426, Figs.3G-M, 5, 6A-C, 7. Mississippi: Union County, Pontotoc County. Coon Creek Formation, early Maastrichtian. 537 specimens. Tennessee: McNairy County. Coon Creek Formation, early Maastrichtian. 5 specimens. Texas: Bexar County. Navarro Formation. 1 specimen, carapaces, chelipeds, and pereopods. Superior.

Tetracarcinus Weller, 1905

32 *T.subquadratus* Weller, 1905, p.136, Pl.15, Figs.4-6. Weller, 1905, p.328, Figs.4-6. Rathbun, 1935, Pl.10, Figs.16-18. Glaessner, 1969, p.R491, Fig.303, 1. Bishop, 1983c, Fig.3I-J. New Jersey: Monmouth County. Mogothy Formation and Woodbury Formation, late Santonian to mid-late Campanian. Delaware: New Castle County. Merchantville Formation, mid-early Campanian. 32 specimens. Mississippi: Union County. Coon Creek Formation, early Maastrichtian. 27 specimens. Wyoming: Carbon County. Lewis Shale, early-late Campanian. 1 specimen. Good. Carapaces, chelipeds, and pereopods. Superior.

Avitelmessus Rathbun, 1923

33 *A.grapsoideus* Rathbun, 1923, p.404, Pls.101, 102, Fig.4. Rathbun *in* Wade, 1926, Pls. 69-70. Kesling and Reiman, 1957, Pls.1-4. Glaessner, 1969, Fig.303-3a-d. North Carolina: Bladen County. Peedee Formation, late Campanian or early Maastrichtian. 1 specimen. Alabama: Barbour County and Lowndes County. Ripley Formation, late Campanian or early Maastrichtian. 65 specimens. Mississippi: Alcorn County, Union County, Tippah County, Pontotoc County. Coon Creek Formation, early Maastrichtian. 101 specimens. Tennessee: McNairy County. Coon Creek Formation, early Maastrichtian. 15 specimens. Carapaces, chelipeds, and pereopods. Superior.

New genus Bishop, pers. observ.

34 n.sp. Bishop, pers. observ. South Dakota: Corson County. Fox Hills Formation, Maastrichtian. 3 specimens, carapaces. Excellent.

Superfamily Oxystomata H.Milne-Edwards, 1834

Family Dorippidae de Haan, 1841

Torynomma Woods, 1953

35 *T.?densus* Bishop, 1983, p.53, Pl.3, Figs.32-40, Fig.8J, Table 8. Texas: Kendall County. Glen Rose Limestone, early Albian. 2 specimens, claws. Fair.

Sodakus Bishop, 1978

36 *S.tatankayotankaensis* Bishop, 1978, p.608, Pl.1, Figs.1-6, Figs.3, 4. Bishop, 1981, Fig.13-4F. South Dakota: Corson County. Mobridge Member, Pierre Shale, early Maastrichtian. 3 specimens, carapaces. Good.

Hillius Bishop, 1983

37 *H.youngi* Bishop, 1983, p.46, Pl.1, Figs.8-11, Fig.8C. Texas: Kendall County. Glen Rose Limestone, early Albian. 1 specimen, carapace. Fair.

Dioratiopus Wood, 1953

38 *D.dawsonensis* Bishop, 1973, p.19, Pl.1, Figs.1-7. Montana: Dawson County. Bearpaw Shale, Maastrichtian. 1 specimen, carapace. Good.

39 *D.scotti* Bishop, 1983, p.47, Pl.1, Figs.6-7, Fig.8D. Texas: Hays County. Glen Rose Limestone, early Albian. 1 specimen, carapace. Poor.

40 *D.hearttailensis* Bishop, 1985b, p.616, Figs.8-9. South Dakota: Butte County. Gammon Ferruginous Member, Pierre Shale, early Campanian. 6 specimens, carapace, chela, proximal pereopods. Excellent.

41 *Dioratiopus* n.sp. Bishop, pers. observ. New Jersey: Monmouth County. Wenowah Formation, late Campanian. 4 specimens, carapaces, pereopods. Good.

Superfamily Calappoidea de Haan, 1833

Family Calappidae de Haan, 1833

Cenomanocarcinus Van Straelen, 1936

42 *C.oklahomensis* Rathbun, 1935, p.44, Pl.11, Fig.9. Oklahoma: Bryan County. Marietta Formation, late Albian. 1 specimen. Texas: Grayson County. Denton Clay, late Albian. 6 specimens, carapace, pereopods. Good.

43 *C.vanstraeleni* Stenzel, 1945, p.447, Pl.44, Fig.15. Texas: Denton County, Grayson County, Dallas County. Britton Formation, Eagle Ford Group, early Turonian. 12 specimens. New Mexico: Aqua Salado Grant. Mancos Shale, middle Turonian. 1 specimen. Carapaces, chelipeds, pereopods. Excellent.

44 *C.renfroae* Stenzel, 1945, p.443, Pl.41, Figs.13, 15. Texas: Tarrant County. Pawpaw Shale, latest Albian or earliest Cenomanian. 5 specimens, carapaces. Excellent.

45 *C.armata* Rathbun, 1935. p.50, Pl.11, Figs.32, 33. Texas: Tarrant County. Denton Clay, late Albian. 1 specimen, abdomen. Doubtful.

New genus Bishop.

46 n.sp. Bishop, pers. observ. Wyoming: Fremont County. Frontier Formation, Turonian. 1 specimen, carapace. Fair.

Necrocarcinus Bell, 1862

47 *N.texensis* Rathbun, 1935, p.45, Pl.11, Figs.20-22. Texas: Grayson County and Tarrant County. Denton Clay and Pawpaw Formation, late Albian or early Cenomanian. 2 specimens, carapaces. Fair.

48 *N.pierrensis* Rathbun, 1917, p.389, Pl.33, Figs.4, 5. Bishop, 1981, Fig.13-4B. South Dakota: Corson County, Dewey County, Stanley County, Pennington County. Pierre Shale, late Campanian to early Maastrichtian. 186 specimens, carapaces, chelipeds, pereopods. Excellent.

49. *N.rathbunae* Roberts, 1962, p.181, Pl.84, Fig.12, Pl.87, Figs.1, 2. New Jersey: Monmouth County. Woodbury Formation, late-early Campanian. 2 specimens. Delaware: New Castle County. Merchantville Formation, mid-early Campanian. 15 specimens. Carapaces (with sterna). Excellent.

50 *N.siouxensis* Feldmann, Awotua, & Welshenbaugh, 1976, p.985, Pl.1, Figs.1-5. North Dakota: Sioux County. Fox Hills Formation, Maastrichtian. 1 specimen, carapace. Good.

51 *N.davisi* Bishop, 1985b, p.619, Fig.10. South Dakota: Butte County. Gammon Ferruginous Member, Pierre Shale, early Campanian. 59 specimens, carapaces. Excellent.

52 *N.* n.sp. Bishop, pers. observ. South Dakota: Butte County. Poole Creek Shale Member, Carlile Shale, Turonian. 2 specimens, carapaces. Fair.

Paranecrocarcinus Van Straelen

53 *P.graysonensis* Rathbun, 1935, p.45, Pl.11, Figs.23-25. Texas: Grayson County. Denton Clay, late Albian. 2 specimens, carapaces. Fair.

54 *P.moseleyi* Stenzel, 1945, p.441, Pl.41, Figs.12, 15. Texas: Denton County. Pawpaw Shale, latest Albian or earliest Cenomanian. 1 specimen, carapace. Good.

55 *P.gamma* Roberts, 1962, p.182, Pl.85, Figs.7, 8, 11. New Jersey: Burlington County. Merchantville Formation, mid-early Campanian. 1 specimen. Delaware: New Castle County. Merchantville Formation, mid-early Campanian. 2 specimens. Carapaces (with sterna). Good.

Protonecrocarcinus

56 *P.ovalis* Stenzel, 1945, p.442, Pl.41, Figs.7-9, Figs.14, 15. Texas: Dallas County. Britton Formation, Eagle Ford Group, late Cenomanian. 1 specimen, carapace. Fair.

Pseudonecrocarcinus Förster, 1968

57 *P.stenzeli* Bishop, 1983, p.49, Pl.1, Figs.3-5, Fig.8B. Texas: Kendall County. Glen Rose Limestone, early Albian. 6 specimens, carapaces. Good.

Prehepatus Rathbun, 1935.

58 *P.cretaceous* Rathbun, 1935, p.47, Pl.11, Figs.29, 30. Texas: Grayson County. Duck Creek Limestone, late Albian. 1 specimen, claw. Fair.

59 *P.pawpawensis* Rathbun, 1935, p.48, Pl.11, Figs.26-28. Texas: Tarrant County. Pawpaw Shale, late Albian or early Cenomanian. 1 specimen, claw. Fair.

60 *P.dilksi* Roberts, 1962, p.183, Pl.89. Fig.2. New Jersey: Burlington County. Merchantville Formation, mid-early Campanian. 3 specimens, claws. Fair.

61 *P.hodgesi* Bishop, 1983. p.50, Pl.3, Figs.1-19, Figs.8F, 11. Texas: Blanco County and Kendall County. Glen Rose Limestone, early Albian. 30 specimens chelipeds. Excellent.

62 *P.harrisi* Bishop, 1985c, p.1029, Figs.1-2. Mississippi: Union County. Coon Creek Formation and Prairie Bluff Chalk, early to middle Maastrichtian. 4 specimens, claws. Good.

Superfamily Raninoidea de Haan, 1841

Family Raninidae de Haan, 1841

Notopocorystes McCoy, 1849

63 *N.harveyi* Woodward, 1896, p.225, Fig.4. Rathbun, 1926, Pl.20, Fig.4. British Columbia: Vancouver Island. Haslam Formation or Comax Formation, late Santonian or early Campanian. 2 specimens, carapaces. Good.

64 *N.testacea* Rathbun, 1926, p.180, Pl.68, Figs.1-4. Roberts, 1962, Pl.85, Figs.9, 10. Tennessee: McNairy County. Coon Creek Formation, early Maastrichtian. 3 specimens. Mississippi: Union County. Coon Creek Formation, early Maastrichtian. 24 specimens. New Jersey: Monmouth County. Nevasink Formation, latest Campanian or earliest Maastrichtian. 2? specimens. Carapaces, chelipeds, pereopods. Excellent.

65 *N.dichrous* Stenzel, 1945, p.438, Pl.43, Figs.4-7, 13. Glaessner, 1969, Fig.311-7. Texas: Dallas County and Denton County. Britton Formation, Eagle Ford Group, early Turonian. 69+ specimens, carapaces, chelipeds, pereopods. Excellent.

66 *N.eichhorni* Bishop, 1983, p.905, Figs.5, 6A-J. Montana: Valley County. Bearpaw Shale, late Campanian. 34 specimens, carapaces, chelipeds, pereopods. Excellent.

67 *N.?ripleyensis* Rathbun, 1935, p.49, Pl.12, Figs.6-10. Mississippi: Pontotoc County. Coon Creek Formation, early Maastrichtian. 1 specimen, claw. Poor.

Notopoides Henderson, 1888

68 *N.pflugervillensis* Beikirch and Feldmann, 1980, p.322, Figs.5A-B, 11. Texas: Travis County. Pflugerville Member, Austin Formation, early Campanian. 4 specimens, carapace, pereopods. Fair.

Notosceles Bourne, 1922

69 *N.bournei* Rathbun, 1928, p.1, Pl.1, Figs.1-4. Rathbun, 1935, Pl.26, Fig.5. Texas: Bastrop County. Navarro Group, late Campanian or early Maastrichtian. 1 specimen, carapace. Good. Many other occurrences in the Gulf Coast Paleocene rocks.

Raninella A.Milne-Edwards, 1862

70 *R.mucronata* Rathbun, 1935, p.50, Pl.14, Figs.32, 33. Texas: Tarrant County. Denton Clay, late Albian. 1 specimen, carapace. Fair.

71 *R.punctatus* Rathbun, 1935, p.48, Pl.12, Figs.14-16. Texas: Grayson County. Denton Clay, late Albian. 1 specimen, carapace. Good.

72 *R.parvus* Rathbun, 1935, p.48, Pl.12, Figs.11-13. Texas: Tarrant County. Denton Clay, late Albian. 1 specimen, carapace. Fair.

73 *R.tridens* Roberts, 1962, p.187, Pl.88, Figs.5, 6. New Jersey: New Castle County. Merchantville Formation, mid-early Campanian. 1 + specimen, carapace, chelipeds, pereopods. Excellent.

74 *R.oaheensis* Bishop, 1978, p.615, Pl.1, Figs.7-11, Pl.2, Figs.1-20, Fig.5. Bishop, 1981, Fig.13-4E, 9G. South Dakota: Corson County, Dewey County, Stanley County, Lyman County, Butte County, Pierre Shale, early Campanian to early Maastrichtian. 14 specimens. Montana: Carter County. Bearpaw Shale, Campanian or? Maastrichtian. 2 specimens. Carapace, pereopods. Excellent.

75 *R.carlilensis* Feldmann and Maxey, 1980, p.85B, Figs.1-3. Kansas: Ellis County. Blue Hill Shale Member, Carlile Shale, late-middle Turonian. 4 specimens, carapace and cheliped. Excellent.

76 *R.starkvillensis* Rathbun, p.51, Pl.9, Figs.4, 5. Mississippi: Oktibbeha County. Oktibbeha tongue, Prairie Bluff Chalk, Middle Maastrichtian. 1 specimen. *R.armata* Rathbun, 1935 = *Cenomanocarcinus armatus* Rathbun, 1935.

Section Oxyrhyncha Latreille, 1803
Family Majidae Samouelle, 1819
Stenocionops Desmarest, 1823

77 *S.primus* Rathbun, 1935, p.58, Pl.13, Figs.9-12. Arkansas: Howard County. Brownstown Formation, early Campanian. 1 specimen, right manus. Poor.

Section Brachyrhyncha Borradaile, 1907
Superfamily Portunoidea Refinewque, 1815
Family Carcineretes Beurlen, 1930
Carcineretes Withers, 1922

78 *C.woolacotti* Withers, 1922, p.535, Pl.16, Fig.1, Pl.17, Figs.2-6. Jamaica: Chapelton. Rio Minho, Turonian? 3 specimens, carapace, chelae, pereopods. Excellent.

Ophthalmoplax Rathbun, 1935.

79 *O.comancheensis* Rathbun, 1935, p.54, Pl.13, Figs.1-5. Texas: Tarrant County and Grayson County. Pawpaw Formation, latest Albian or earliest Cenomanian. 4 specimens, carapace front, chelae fingers. Poor.

80 *O.stephensoni* Rathbun, 1935, p.52, Pl.13, Figs.13-18, Pl.26, Fig.10. Glaessner, p.R514, Fig.324-1. Mississippi: Union County. 'Ripley Formation', late Campanian or early Maastrichtian. 1 specimen. Alabama: Lowndes County. Ripley Formation, late Campanian or early Maastrichtian. 1 specimen. Texas: Maverick County. Late Cretaceous. 13 specimens. South Carolina: Davis Landing. Peedee Formation, late Campanian or early Maastrichtian. 1 specimen. Carapaces, chelae, pereopods. Excellent.

Longusorbis Richards, 1975

81 *L.cuniculosus* Richards, 1975, p.1857, Figs.4, 5, 7-11. British Columbia: Vancouver Island. Spray Formation, late? Campanian. 83 specimens, carapaces, chelipeds, pereopods. Excellent.

Woodbinax Stenzel, 1952

82 *W.texanus* Stenzel, 1952, p.216, Pl.5, Fig.11, Fig.8. Glaessner, 1969, Fig.324-2. Texas: Grayson County. Woodbine Formation, Cenomanian. 1 specimen, carapace. Fair.

Superfamily Xanthoidea Dana, 1851
Family Xanthidae Dana, 1851
Caloxanthus

83 *C.americanus* Rathbun, 1935, p.56, Pl.11, Figs.12-19. Wright and Collins, 1972, Pl.21, Fig.7, Pl.22, Fig.1a-b. Texas: Tarrant County. Pawpaw Shale, late Albian or early Cenomanian. 14 specimens, carapace, chela. Excellent.

New genus Bishop.

84 n.sp. Bishop, pers. observ. Mississippi: Union County. Prairie Bluff Chalk, middle Maastrichtian. 1 specimen, carapace. Good.

85 N.gen. *sternbergi* (Rathbun), 1926, p.54, Pl.39, Figs.1-4. California: San Diego County. Cretaceous. 1 specimen. Mexico: Baja. California: del Norte. Rosario Formation, late Cretaceous. 3 specimens. Carapaces, chelipeds. Good.

Xanthias Rathbun, 1897

86 *X.lenolensis* Rathbun, 1935b, p.165, Figs.1-4. New Jersey: Burlington County. Merchantville Formation, mid-early Campanian. 2 specimens. Delaware: New Castle County. Merchantville Formation, mid-early Campanian. 2 specimens. Chela. Poor.

Xanthosia Bell, 1863

87 *X.aspera* Rathbun, 1935, p.41, Pl.11, Figs.1-5. Texas: Grayson County. Denton Clay, late Albian. 2 specimens, carapaces. Excellent.

88 *X.wintoni* Rathbun, 1935, p.42, Pl.11, Figs.6-8. Texas: Grayson County, Tarrant County. Denton Clay and Pawpaw Shale, late Albian to ?early Cenomanian. 7 specimens, carapaces. Good.

89 *X.elegans* Roberts, 1962, p.177, Pl.89, Figs.1, 3. Delaware: New Castle County. Merchantville Formation, mid-early Campanian. 2 specimens. New Jersey: Burlington County. Merchantville Formation, mid-early Campanian. 1 specimen. South Dakota: Butte County. Gammon Ferruginous Member, Pierre Shale, mid-early Campanian. 2 specimens. Carapaces. Good.

Menippe de Haan, 1841

90 *M.cretacea* Rathbun, 1935, p.57, Pl.13, Figs.7, 8. Texas: Denton County. Woodbine Formation, Cenomanian. 2 specimens. Mississippi: Oktibbeha County. Oktibbeha tongue (= Prairie Bluff Formation), Maastrichtian. 1 specimen. Claw fragments. Poor.

New genus Bishop.

91 n.sp. Bishop, pers. observ. Mississippi: Union County, Oktibbeha County, Lee County. Demopolis Formation, Coon Creek Formation, Prairie Bluff Formation, early Campanian to middle Maastrichtian. 7 specimens. Alabama: Green County. Selma Chalk, Campanian. 3 specimens. Claw fragments. Fair.

Superfamily Ocypodoidea Rafinesque, 1815

Family Retroplumidae Gill, 1894

Archaeopus Rathbun, 1908

92 *A.antennatus* Rathbun, 1908, p.347, Pl.47, Figs.4-7, Pl.48, Pl.49, Figs.2-4. Rathbun, 1926, Pl.4, Figs.4-7, Pl.5, Pl.6, Figs.2-4. California: San Mateo County. Chico Formation, late Cretaceous. 3 specimens, carapaces, chelipeds. Excellent.

93 *A.vancouverensis* Woodward, 1826, p.226, Figs.5, 6. Whiteaves, 1903, Figs.15, 16. Rathbun, 1926, Pl.20, Figs.1, 2. British Columbia: Hornby Island. Spray Formation or ? Northumberland or Geoffrey Formation, late Campanian or early Maastrichtian. 3 specimens. Vancouver Island. Haslam or Comox Formation, late Santonian or early Campanian. 1 specimen, carapaces, chelipeds, pereopods. Excellent.

Cristipluma Bishop, 1983

94 *C.mississippiensis* Bishop, 1983c, pl.428, Figs.3K, 8A-C. Mississippi: Union County. Coon Creek Formation, early Maastrichtian. 2 specimens, carapaces. Good.

Incertae sedis

95 *Dromia (?) anomala* Rathbun, 1935, p.43, Pl.11, Figs.10, 11. Texas: Grayson County. Denton Clay, late Albian. 1 specimen, carapace and chela. Poor.

96 *Actaea cretacea* Rathbun, 1935, p.55, Pl.9, Fig.3. Texas: Tarrant County. Pawpaw Clay, late Albian or early Cenomanian. 1 specimen, claw. Poor.

97 *Necrocarcinus*? sp., Rathbun, 1935, p.46, Pl.11, Fig.31, Pl.13, Fig.6. Texas: Tarrant County, Pawpaw Clay, late Albian or early Cenomanian. 2 specimens, carapace and a claw. Doubtful.

REFERENCES

Beikirch, W. & R.M.Feldmann 1980. Decapod Crustaceans from the Pflugerville Member, Austin Formation (Late Cretaceous: Campanian) of Texas. *J. Paleont.* 54: 309-324.

Bishop, G.A. 1971. Fossil decapod Crustacea from the Pierre Shale (Upper Cretaceous) of South Dakota. *Diss. Abstr.*

Bishop, G.A. 1972a. Crab Bitten by a Fish from the Upper Cretaceous Pierre Shale of South Dakota. *Bull. Geol. Soc. Am.* 83: 3823-3826.

Bishop, G.A. 1972b. Moults of *Dakoticancer overanus*, an Upper Cretaceous crab from the Pierre Shale of South Dakota. *Palaeontology* 15: 632-636.

Bishop, G.A. 1972c. Preservation of fossil crabs from the *Dakoticancer* Assemblage, Upper Cretaceous Pierre Shale, South Dakota. *Bull. Ga. Acad. Sci.* 30: 80.

Bishop, G.A. 1973. *Homolopsis dawsonensis*: A New Crab (Crustacea, Decapoda) from the Pierre Shale (Upper Cretaceous, Maastrichtian) of Cedar Creek Anticline, eastern Montana. *J. Paleont.* 47: 19-20.

Bishop, G.A. 1974. A Sexually Aberrant Crab (*Dakoticancer overanus* Rathbun, 1917) from the Upper Cretaceous Pierre Shale of South Dakota. *Crustaceana* 26: 212-218.

Bishop, G.A. 1975. Chap.13: Traces of Predation. In R.W.Frey (ed.), *The Study of Trace Fossils.* New York: Springer-Verlag.

Bishop, G.A. 1976. *Ekalakia lamberti* n.gen., n.sp. (Crustacea, Decapoda). From the Upper Cretaceous Pierre Shale of Eastern Montana. *J. Paleont.* 50: 398-401.

Bishop, G.A. 1977. Pierre Feces: A Scatological Study of the *Dakoticancer* Assemblage, Pierre Shale (Upper Cretaceous) of South Dakota. *J. Sedim. Petrol.* 27: 129-136.

Bishop, G.A. 1978. Two new crabs, *Sodakus tatankayotankaensis* n.gen., n.sp. and *Raninella oaheensis* n.sp. (Crustacea, Decapoda), from the Upper Cretaceous Pierre Shale of South Dakota. *J. Paleont.* 52: 608-617.

Bishop, G.A. 1981a. Chap.13: Occurrence and Fossilization of the *Dakoticancer* Assemblage, Upper Cretaceous Pierre Shale, South Dakota. In Gray et al. (eds.), *Communities of the Past.* Stroudsburg: Hutchinson Ross.

Bishop, G.A. 1981b. The lobster *Linuparus* preserved as an attachment scar on the oyster *Exogyra costata*, Ripley Formation (Late Cretaceous) Union County, Mississippi. *Miss. Geol.* 2: 2-5.

Bishop, G.A. 1982. *Homolopsis mendryki*: A new fossil crab (Crustacea, Decapoda) from the Late Cretaceous *Dakoticancer* Assemblage, Pierre Shale (Maastrichtian) of South Dakota. *J. Paleont.* 56: 221-225.

Bishop, G.A. 1983a. A second sexually aberrant specimen of *Dakoticancer overanus*Rathbun, 1917, from the Upper Cretaceous *Dakoticancer* Assemblage Pierre Shale, South Dakota. *Crustaceana* 44: 23-26.

Bishop, G.A. 1983b. Fossil decapod crustaceans from the Early Cretaceous Glen Rose Limestone of Central Texas. *Trans. San Diego Soc. Nat. Hist.* 20: 27-55.

Bishop, G.A. 1983c. Fossil Decapod Crustaceans from the Late Cretaceous Coon Creek Tongue, Ripley Formation, Union County, Mississippi. *J. Crust. Biol.* 3: 417-430.

Bishop, G.A. 1983d. Two new species of crab, *Notopocorystes (Eucorystes) eichhorni* and *Zygastrocarcinus griesi* (Crustacea: Decapoda) from the Bearpaw Shale (Campanian) of north-central Montana. *J. Paleont.* 57: 900-910.

Bishop, G.A. 1985a. A new crab *Eomunidopsis cobbani* from the Pierre Shale of Colorado. *J. Paleont.* 59: 601-604.

Bishop, G.A. 1985b. Fossil Decapod Crustaceans from the Gammon Ferruginous Member, Pierre Shale (Late Campanian), Black Hills, South Dakota. *J. Paleont.* 59: 605-624.

Bishop, G.A. 1985c. A new crab, *Prehepatus harrisi*, from the Coon Creek and Prairie Bluff Formations, Union Co., Mississippi. *J. Paleont.* 59: 1028-1032.

Bishop, G.A. 1986. A new crab, *Zygastrocareinus cardsmithi*, from the lower Pierre Shale, southeastern Montana. *J. Paleont.* 60:

Charig, A. 1979. A new look at the dinosaurs. London: Heinemann.

Feldmann, R.M. 1981. Paleobiogeography of North American lobsters and shrimps (Crustacea, Decapoda). *Geobios* 14: 449-468.

Feldmann, R.M. & M.Maxey 1980. *Raninella carlilensis*, a new Raninid crab from the Carlile Shale (Turonian) of Kansas. *J. Paleont.* 54: 858-861.

Feldmann, R.M. & C.B.McPherson 1980. Fossil decapod crustaceans of Canada. *Geol. Surv. Can. Pap.* 79(16): 1-20.

Feldmann, R.M., E.E.B.Awotua & J.Welshenbaugh 1976. *Necrocarcinus siouxensis* a new species of calappid crab (Crustacea: Decapoda) from the Fox Hills Formation (Cretaceous: Maastrichtian) of North Dakota. *J. Paleont.* 50: 985-990.

Feldmann, R.M., G.A.Bishop & T.W.Kammer 1977. Macrurous decapods from the Bearpaw Shale (Cretaceous Campanian) of Northeastern Montana. *J. Paleont.* 51: 1161-1180.

Gill, J.R. & W.A.Cobban 1966. The Red Bird Section of the Upper Cretaceous Pierre Shale in Wyoming. *Prof. Pap. US Geol. Surv.* 393-A: 1-73.

Gill, J.R. & W.A.Cobban 1973. Stratigraphy and geologic history of the Montana Group and equivalent rocks, Montana, Wyoming, and North and South Dakota. *Prof. Pap. US Geol. Surv.* 776: 1-37.

Glaessner, M.F. 1929. Crustacea Decapoda. In J.F.Pompeckj (ed.), *Fossilium Catalogus* Pt.41. Berlin: W.Junk.

Glaessner, M.F. 1960. The fossil decapod Crustacea of New Zealand and the evolution of the Order Decapoda. *NZ Geol. Surv. Paleont. Bull.* 31: 1-63.

Glaessner, M.F. 1969. Decapoda. In R.C.Moore (ed.), *Treatise on Invertebrate Paleontology*, Part R, Arhtropoda 4, volume 2. Lawrence: Univ. Kansas Press.

Glaessner, M.F. 1980. New Cretaceous and Tertiary crabs (Crustacea: Brachyura) from Australia and New Zealand. *Trans. R. Soc. S.Aust.* 104: 171-192.

Hallam, A. 1972a. Models involving population dynamics. In T.J.M.Schopf (ed.), *Models in Paleobiology*. San Francisco: Freeman & Cooper.

Hattin, D.E. 1962. Stratigraphy of the Carlile Shale (Upper Cretaceous) in Kansas: *Bull. Kan. Geol. Surv.* 156: 1-155.

Hattin, D.E. 1964. Cyclic sedimentation in the Colorado Group of west-central Kansas. In D.F.Marriam (ed.), *Symposium on Cyclic Sedimentation. Bull. Kan. Geol. Surv.* 169: 205-217.

Hendricks, L. & W.F.Wilson 1967. Introduction. In L.Hendricks (ed.), Comanchean (Lower Cretaceous) stratigraphy and paleontology of Texas. *Permian Basin Sect., Soc. Econ. Paleont. and Miner. Pub.* 67-8: 1-5.

Holland, F.D. Jr & A.M.Cvancara 1958. Crabs from the Cannonball Formation (Paleocene) of North Dakota. *J. Paleont.* 32: 495-505.

Jeletzky, J.A. 1969. Marine Cretaceous biotic provinces of Western and Arctic Canada. *Proc. No. Am. Paleont. Conv.* 1: 1638-1659.

Kauffman, E.G. 1962. Mesozoic paleontology and stratigraphy, Huerfano Park, Colorado. *Diss. Abstr.* 22: 1467.

Kauffman, E.G. 1967. Coloradoan macroinvertebrate assemblages, central Western Interior, United States. In E.G.Kauffman & H.E.Kent (eds.), *Paleonenvironments of the Cretaceous Seaway in the Western Interior — A Symposium.* Boulder: Colorado School of Mines.

Kauffman, E.G. 1975. Dispersal and biostratigraphic potential of Cretaceous benthonic Bivalvia in the Western Interior. In W.G.E.Caldwell (ed.), *The Cretaceous System in the Western Interior of North America, Geol. Ass. Can. Spec. Pap.* 13: 163-194.

Kauffman, E.G. 1977. Geological and biological overview: Western Interior Cretaceous Basin. *Mount. Geol.* 14: 75-99

Kesling, R.V. & I.G.Reimann 1957. An Upper Cretaceous crab, *Avitelmesus grapsoideus* Rathbun: *Contr. Mich. Univ. Mus. Paleont.* 14: 1-15.

Koch, C.F. 1978. Bias in the published fossil record. *Paleobiology* 4: 367-372.

Kues, B.S. 1980. A fossil crab from the Mancos Shale (Upper Cretaceous) of New Mexico. *J. Paleont.* 54: 862-864.

Obradovich, J.D. & W.A.Cobban 1975. A Time Scale for the Late Cretaceous of the Western Interior of North America. In W.G.E.Caldwell (ed.), *The Cretaceous System in the Western Interior of North America. Spec. Pap. Geol. Soc. Can.* 13: 31-54.

Owens, J.P., J.P.Minard, N.F.Sohl & J.F.Mello 1970. Stratigraphy of the outcropping Post-Magothy Upper Cretaceous Formations in southern New Jersey and northern Delmarva Peninsula, Delaware and Maryland. *Prof. Pap. U.S. Geol. Surv.* 674: 1-60.

Perkins, B.F. & C.L.Stewart 1971. Stop 12: Amon Carter Park. In B.F.Perkins (ed.), *Trace Fossils, A field guide to selected localities in Pennsylvanian, Permian, Cretaceous, and Tertiary rocks of Texas and related papers. La. St. Univ. Misc. Pub.* 71-1: 80-88.

Pilsbry, H.A. 1901. Crustacea of the Cretaceous formation of New Jersey. *Proc. Acad. Nat. Sci. Phila.* 53: 111-118.

Rathbun, M.J. 1917. New species of South Dakota Cretaceous crabs. *Proc. U.S. Nat Mus.* 52: 385-391.

Rathbun, M.J. 1923. Decapod crustaceans from the Upper Cretaceous of North Carolina. *N. Carol. Geol. Surv.* 5: 407.

Rathbun, M.J. 1926. In J.P.D.Hull (ed.), *Guide notes on the Midway in southwestern Arkansas. Bull. Am. Ass. Petr. Geol.* 9: 167-170.

Rathbun, M.J. 1926. Decapoda. In Bruce Wade (ed.), *The fauna of the Ripley Formation on Cook Creek. Tennessee. Prof. Pap. U.S. Geol. Surv.* 137: 184-191.

Rathbun, M.J. 1926. The fossil stalk-eyed crustacea of the Pacific Slope of North America. *Bull. U.S. Nat. Mus.* 138: 1-155.

Rathbun, M.J. 1928. Two new crabs from the Eocene of Texas. *Proc. U.S. Nat. Mus.* 73: 1-6.

Rathbun, M.J. 1930. Fossil decapod Crustaceans from Mexico. *Proc. U.S. Nat. Mus.* 78: 1-10.

Rathbun, M.J. 1930a. A new *Callianassa* from the Cretaceous of South Dakota. *J. Wash. Acad. Sci.* 20: 1-3.

Rathbun, M.J. 1930b. *Hoploparia westoni* Woodward. *Wash. Acad. Sci.* 20: 180-183.

Rathbun, M.J. 1935. A new Xanthid crab from the Cretaceous of New Jersey. *Proc. Acad. Nat. Sci. Phila.* 87: 165-166.

Rathbun, M.J. 1935b. Fossil Crustacea of the Atlantic and Gulf Coastal Plain. *Geol. Soc. Amer. Spec. Pap.* 2: 1-160.

Rathbun, M.J. 1937. Cretaceous and Tertiary Crabs from Panama and Columbia. *J. Paleont.* 11: 26-28.

Richards, B.C. 1975. *Longusorbis cuniculosus*: A new genus and species of Upper Cretaceous crab; with comments on Spray Formation at Shelter Point, Vancouver Island, British Columbia. *Can. J. Earth Sci.* 12: 1850-1863.

Roberts, H.B. 1962. The Upper Cretaceous decapod crustaceans of New Jersey and Delaware. In H.G.Richards (ed.), *The Cretaceous fossils of New Jersey. Bull. N. Jers. Bur. Geol. & Topo.* 61: 163-191.

Robinson, C.S., W.J.Mapel & M.H.Bergendahl 1964. Stratigraphy and structure of the northern and western flanks of the Black Hills uplift, Wyoming, Montana, and South Dakota. *Prof. Pap. U.S. Geol. Surv.* 404: 1-134.

Roemer, F.A. 1887. *Graptocarcinus texanus*, ein Brachyure aus der oberen Kreide von Texas. *Neues Jb. Miner. Geol. Palaont.* 1: 173.

Rothrock, E.P. 1947. Geology of the Missouri Valley and vicinity near Mobridge. *S.D. Geol. Surv. Rept. Invest.* 58: 1-29.

Russell, E.E., D.M.Keady, E.A.Mancini & C.C.Smith 1983. Upper Cretaceous lithostratigraphy and biostratigraphy in northeast Mississippi, southwest Tennessee and northwest Alabama, shelf chalks and coastal clastics. *Soc. Econ. Paleont. Miner.* spring field trip, 1983.

Schäfer, W. 1951. Fossilisations-Bedingungen brachyurer Krebse. *Abh. senckenb. naturforsch. Ges.* 485: 221-238.

Schäfer, W. 1954. Form und Funktion der Brachyuren-Schere. *Abh. senckenb. naturforsch. Ges.* 489.

Schäfer, W. 1972. *Ecology and Palaeoecology of Marine Environments.* Chicago: Univ. Chicago Press.

Schmitt, W. 1921. The Marine Decapod Crustacea of California. *Univ. Calif. Publs. Zool.* 23: 1-470.

Slocki, S.F. 1967. Physical stratigraphy of the Georgetown Limestone equivalents in Tarrant, Denton, and Cooke Counties, Texas. In Leo Hendricks (ed.), *Comanchean (Lower Cretaceous) stratigraphy and paleontology of Texas: Permian Basin Sect., Soc. Econ. Paleont. Miner. Pub.* 67-8: 183-217.

Stenzel, H.B. 1941. Pers. Comm. dated Nov. 24, 1941, *Dallas Petrol. Geol.*, Geology of Dallas Co., Texas.

Stenzel, H.B. 1944. A new Cretaceous crab *Graptocarcinus muiri*, from Mexico. *J. Paleont.* 18: 550-552.

Stenzel, H.B. 1945. Decapod crustaceans from the Cretaceous of Texas. *Publc. Univ. Tex. Bur. Econ. Geol.* 4401: 401-476.

Stenzel, H.B. 1953. Decapod crustaceans from the Woodbine Formation of Texas. *Prof. Pap. U.S. Geol. Survey* 242: 212-217.

Stephenson, L.W. & W.H.Monroe 1940. The Upper Cretaceous deposits. *Miss. State Geol. Surv. Bull.* 40: 1-296.

Thorson, G. 1957. Bottom Communities (Sublittoral or shallow shelf). In Joel Hedgepeth (ed.), *Treatise on Marine Ecology and Paleoecology. Mem. Geol. Soc. Am.* 67: 461-534.

Wade, B. 1926. The fauna of the Ripley Formation on Coon Creek, Tennessee. *Prof. Pap. U.S. Geol. Surv* 137: 1-272.

Ward, P.D. 1978. Revisions to the stratigraphy and biochronology of the upper Cretaceous Nanaimo Group, British Columbia and Washington State. *Can. J. Earth Sci.* 15: 405-423.

Weimer, R.J. 1959. Upper Cretaceous stratigraphy, Colorado. In *Symposium on Cretaceous Rocks of Colorado and adjacent areas 11th Field Conference, Rocky Mount. Ass. Geol.*

Weimer, R.J. 1960. Upper Cretaceous stratigraphy, Rocky Mountain Area. *Proc. Am. Ass. Petrol. Geol.* 44: 1-20.

Weller, S. 1905a. The fauna of the Cliffwood Clays. *A. Rep. N.Jers. State Geol.* 1904: 133-144.

Weller, S. 1905b. The fauna of the Cliffwood, New Jersey, clays. *N.Jers. Geol. Surv. Ann. Rept. for 1904*: 136-141.

Weller, S. 1905c. The fauna of the Cliffwood (N.J.) Clays. *J. Geol.* 3: 324-337.

Weller, S. 1907. A report on the Cretaceous Paleontology of New Jersey *Geol. Surv. Palaeont.* Ser.4: 846-853.

Whetstone, K.N. & J.S.H.Collins 1982. Fossil crabs (Crustacea: Decapoda) from the Upper Cretaceous Eutaw Formation of Alabama. *J. Paleont.* 56: 1218-1222.

Whiteaves, J.F. 1900. Mesozoic fossils. On some additional or imperfectly understood fossils from the Cretaceous rocks of the Queen Charlotte Islands, with a revised list of species from these rocks. *Geol. Surv. Can.* 1: 263-307.

Whitney, F.L. 1913. Fauna of the Buda Limestone. *Trans. Tex. Acad. Sci.* 12: 27.

Williams, A.B. 1965. Marine decapod crustaceans of the Carolinas. *Bull. U.S. Bur. Comm. Fish.* 65: 1-298.

Williams, G.D. & C.R.Stelck 1975. Speculations on the Cretaceous Palaeogeography of North America. In W.G.E.Caldwell (ed.), *The Cretaceous System in the Western Interior of North America. Geol. Ass. Canada Spec. Pap.* 13: 163-194.

Withers, T.H. 1922. On a new brachyurous Crustacean from the Upper Cretaceous of Jamaica. *Ann. Mag. Nat. Hist.* 10: 534-541.

Withers, T.H. 1927. *Ranina trechmanni*, a new Cretaceous crab from Jamaica. *Geol. Mag.* 64: 176-180.

Woodward, H. 1896. On some Podophthalmatous Crustacea from the Cretaceous Formation of Vancouver and Queen Charlotte Islands. *J. Geol. Soc. Lond.* 52: 221-228.

Wright, C.W. & J.S.H.Collins 1972. British Cretaceous Crabs. *Palaeontogr. Soc. Monog.* 533: 1-114.

Wright, C.W. & J.S.H.Collins 1975. *Glaessnerella* (Crustacea Decapoda Cymonomidae), a replacement name for *Glaessneria* Wright and Collins, 1972 non Takeda in Miyake, 1969. *Palaeontology* 18: 441.

Young, P. 1974. Lower Albian and Aptian Cretaceous ammonities of Texas. In B.F.Perkins (ed.), *Aspects of Trinity Division Geology. Geosci. and Man* 8: 175-281.

Evolution and zoogeography of the Isopoda Bopyridae, parasites of Crustacea Decapoda

JOHN C.MARKHAM
Arch Cape Marine Laboratory, Oregon, USA

ABSTRACT

The isopod family Bopyridae comprises 469 described species, all of them ectoparasites of decapod crustaceans; it accounts for at least 80% of all species of its suborder, the Epicaridea. The family is divided into ten subfamilies, each of which is largely restricted to hosts belonging to a single decapodan infraorder or lower taxon. The most primitive subfamily is Pseudioninae, members of which probably first infested anomurans of the family Galatheidae. All other subfamilies are derivable from the Pseudioninae, some through a series of genera infesting thalassinideans and designated the Thalassinidean Transition. Evidently the two major subfamilies of abdominal parasites arose separately. The center of distribution and probably of origin of the Bopyridae is the tropical Indo-West Pacific, with progressively fewer species being found westward to the eastern Pacific. Most subfamilies follow this overall pattern, but some deviate markedly from it.

1 INTRODUCTION

The family Bopyridae, which contains at least 80% of the described species of the isopod suborder Epicaridea, consists, at present, of 469 described species currently assigned to 129 genera in ten subfamilies. All species are holoparasitic, their definitive hosts being decapod crustaceans. Except for the questionable case of the peculiar species *Entophilus omnitectus*, the bopyrids are undisputed ectoparasites, being attached as larvae to the surfaces of calanoid copepods and as adults within the branchial chambers or clinging to the abdomens of their decapod hosts. Unlike the members of some epicaridean families (whose limits are uncertain and thus whose number is subject to disagreement), the Bopyridae are easily recognizable as isopods. Except for the modification of mouth parts for sucking, they conform well with the definition of the Isopoda of Watling (1983). At the same time, they are clearly modified for parasitic life and could no longer live free: their bodies are distorted, their pereopods incapable of locomotion, their eyes probably functionless even when present, their mouth appendages reduced or lost, and most of their body volume devoted to the production of brood after brood of eggs.

Although there is considerable discussion about the relationship of the Isopoda to other crustaceans, there does seem to be general agreement (Siewing 1963, Hessler 1983, Sieg

Table 1. Subfamilies of the Bopyridae.

Subfamily	Genera	Species	Main hosts	Other hosts	Main region(s)
Pseudioninae	32	152	Anomura	Caridea Thalassinidea Astacidea	All except East Pacific
Argeiinae	5	9	Caridea	Stenopodidea	Indo-West Pacific
Bopyrinae	22	100	Caridea	–	Indo-West Pacific; West Atlantic
Ioninae	31	88	Brachyura	Thalassinidea Palinura	All except East Pacific
Orbioninae	7	35	Penaeoidea	–	Indo-West Pacific
Entophilinae	1	1	*Munida* spp.	–	Pantropical
Bopyrophryxinae	1	1	*Parapagurus*	–	Indonesia
Phyllodurinae	1	1	*Upogebia*	–	North-East Pacific
Athelginae	7	36	Paguroidea	–	North Atlantic
Hemiarthrinae	22	45	Caridea	–	Indo-West Pacific; East Atlantic

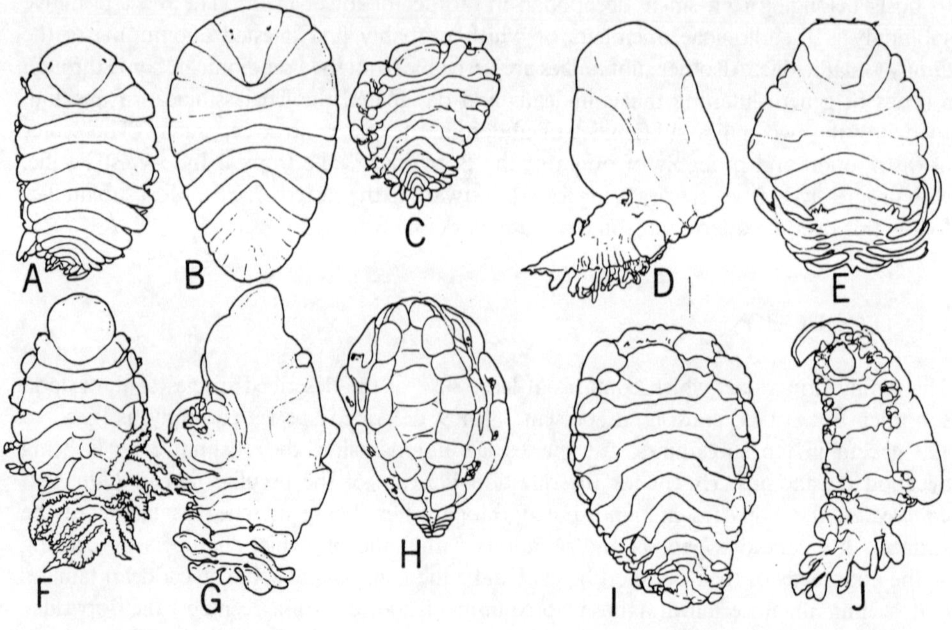

Figure 1. Representatives of 10 bopyrid subfamilies, females of species drawn to different scales (actual lengths 2-10 mm). (A) *Pleurocryptosa megacephalon* (Pseudioninae); (B) *Bopyrione longicapitata* (Bopyrinae); (C) *Argeia pugettensis* (Argeiinae); (D) *Bopyrophryxus branchiabdominalis* (Bopyrophryxinae); (E) *Phyllodurus abdominalis* (Phyllodurinae); (F) *Allokepon goetici* (Ioninae); (G) *Dicropleon processae* (Hemiarthrinae); (H) *Entophilus omnitectus* (Entophilinae); (I) *Orbione halipori* (Orbioninae); (J) *Athelges takanoshimensis* (Athelginae).

1983) that the order is among the most advanced and has evidently given rise to nothing else. Within the Isopoda, the position of the Epicaridea is probably comparable, viz. that of an advanced and terminal group. Its affinities to other isopods are obscure because epicaridean fossils per se are unknown, and only scattered studies (e.g. Strömberg 1971, Anderson & Dale 1981) have been made of the early developmental patterns of epicarideans. Clearly most of the characters which distinguish the Epicaridea evolved with the assumption of a parasitic existence. Ecologically, it is tempting to try to derive the suborder from such part-time parasites as those of the flabelliferan families Cymothoidae (Shiino 1965) or Aegidae or the Gnathidae. Morphologically and embryologically, the question is open. Within the Epicaridea, on the other hand, there is good reason to believe that the Bopyridae are the most primitive and other families are derivable from them.

Whatever the origin of the Epicaridea, it has little direct bearing on the distribution of the extant species. Instead, I want to consider a possible sequence of evolution within the family Bopyridae and relate that to the recorded distribution of its species.

Zoogeography is clearly a function of the distribution of ancestors, but in this case the situation is complicated by the need to consider not only the distribution of the group under discussion but also of its hosts. The early (microniscan) larva of a bopyrid attaches to a copepod (Dale & Anderson 1982) and there develops into the stage (cryptoniscan) which then infests the definitive decapod host. Although host species-specificity is rarely if ever absolute among the Bopyridae, suitable host species of both copepods and decapods must be available, so their past and present distribution affects that of their parasites. Assessment of these factors is not at all helped by the fact that, being uncalcified, the calanoid copepods have left no fossil record (Palmer 1969), and the same applies to the bopyrids themselves (Hessler 1969). Further, of course, the complete account of the distribution of the world's bopyrids (to say nothing of the calanoids and decapods) remains to be written. And finally, there are errors of generic placement which may mislead the user of the bopyrid literature.

2 BOPYRID SUBFAMILIES

The family Bopyridae is divided into ten subfamilies, each of whose members are largely limited to hosts belonging to a single decapodan infraorder. These subfamilies are summarized in Table 1. Most of the species (all of the members of the five subfamilies Pseudioninae, Argeiinae, Bopyrinae, Ioninae and Orbioninae) infest their hosts branchially; they account for 385 species or 82% of the total of the family. The species in the subfamilies Phyllodurinae, Athelginae and Hemiarthrinae are abdominal parasites; the single species of the Entophilinae occurs in its host's visceral cavity, and the single bopyrophryxine is found simultaneously in its host's branchial chamber and on its abdomen. An example of each subfamily is illustrated in Figure 1.

3 FOSSIL RECORD

As already mentioned, the Bopyridae have left no direct fossil record. However, fossilized decapods are known in which branchiostegites are deformed in a manner similar to those of recent decapods harboring branchial bopyrids. Such evidence, though indirect and sparse,

Table 2. Recorded fossil decapods showing bopyrid infestation.

Host superfamily: Family Host species	Age	Location	Author(s)
Nephropoidea: Erymidae			
Palaeasticus sp.	L. Cret.	Moravia	Glaessner 1969
Galatheoidea: Galatheidae			
Galathea antiqua	Jur.	Austria	Remeš 1921 Bachmayer 1955
G.striata	Jur.	Austria	Remeš 1921, Bachmayer 1955
Gastrosacus wetzleri	U. Jur.	Bavaria	Van Straelen 1928
G.zitteli	U. Jur.	Moravia	Van Straelen 1928
G.meyeri	Jur.	Austria	Bachmayer 1955
Gastrosacus sp.	Jur.	Austria	Bachmayer 1955
Mesogalathea striata	Jur.	Moravia	Glaessner 1969
unspecified	Jur.	Unspec.	Balss 1913
Dromioidea: Prosopidae			
Pithonoton marginatum	Jur.	Moravia	Glaessner 1969
	U. Jur.	Poland	Radwański 1972
P.rostratum	U. Jur.	Poland	Radwański 1972
Nodoprosopon heydeni	U. Jur.	Poland	Radwański 1972
Goniodromites polyodon	U. Jur.	Moravia	Remeš 1921, Van Straelen 1928
G.katholickyi	U. Jur.	Moravia	Remeš 1921, Van Straelen 1928
G.bidentatus	U. Jur.	Moravia	Remeš 1921, Van Straelen 1928
unspecified	Jur.	Austria	Bachmayer 1955
Dromioidea: Dromiidae			
Cyclothyreus complanatiformes	U. Jur.	Moravia	Remeš 1921, Van Straelen 1928
C.oxythyreiformes	U. Jur.	Moravia	Remeš 1921, Van Straelen 1928
C.strambergensis	U. Jur.	Moravia	Remeš 1921, Van Straelen 1928
Raninoidea: Raninidae			
Eurycorystes carteri	Cret.	England	Van Straelen 1928
Notopocorystes mantelii	L. Cret.	England	Bell 1863
Xanthoidea: Xanthidae			
Xantho agassizi	L. Cret.	Moravia	Van Straelen 1928

is quite useful. The published records of such parasitized fossil decapods are summarized in Table 2.

Detectable fossilization of infested decapods would, of course, discriminate against the abdominal parasites, which would have left no sign of their presence. Such lightly calcified but presently heavily infested groups as the carideans and paguroids are also unrepresented by parasitized individuals in the fossil record, even though their distant ancestors probably did bear bopyrids. Nonetheless, some interesting conclusions are possible. First, the age of the fossils indicates that bopyrids occurred as far back as the Jurassic, the time of first appearance of nearly all extant decapod groups (Glaessner 1969). Second, the oldest and most numerous species of fossil decapod hosts are in the family Galatheidae. Third, three different infraorders, the Astacidea, Anomura and Brachyura, are represented. And finally, there are seven species of fossil dromioids known as the hosts of bopyrids, in contrast to a single recent species, a *Hypoconcha* sp.

4 EVOLUTION WITHIN THE BOPYRIDAE

It appears that few authors have paid much attention to the possible evoluionary relationships within the Bopyridae. Nierstrasz (1931), in a discussion hedged with assertions that much of what he was saying was purely conjectural, advanced the idea that a reasonable phylogenetic sequence would start with those species most similar to other isopods and then, by progressive loss of body parts, would progress through the Epicaridea as a whole. Unfortunately, he dealt only with generalities and cited no specific examples. Shiino (1952) presented the first detailed account of possible bopyrid phylogeny by using and greatly enlarging on Nierstrasz's idea. He divided the family into six groups of genera corresponding to six of the subfamilies listed above. Later, in an elegant little paper, Shiino (1965) succinctly and clearly summarized his earlier discussion. He stressed the need to consider both males and females of each group in looking at evolutionary progression. However, because sex is epigamically determined in the Bopyridae, the first arriving parasite becoming a female and the next a male (Reinhard 1949, Reverberi & Pitotti 1942, Beer 1951), every male is an incipient female. Moreover, the females display far more phylogenetically usable characters than the males do. Thus I prefer to concentrate on the structure of the females (as in Fig.1) and consider the males only secondarily. Generally the males lend excellent substantiation to the evidence provided by the females.

My scheme for the evolution of the Bopyridae (Fig.2) is essentially that of Shiino (1965, 1966) with the important exception that I consider the Hemiarthrinae to be derived from the Pseudioninae along an entirely different route than are the Athelginae; in other words, the abdominal parasites are not monophyletic. I have applied the basic principle of Hierstrasz (1931) that evolution has led from a form resembling a free-living isopod to those highly modified for parasitism. The characters considered to exhibit such change include progressive reduction in the number of distinct pereomeres and pleomeres, reduction and loss of pereopods and pleopods, elaboration of body margins and pleonal appendages and change of structure of oöstegites. Other characters are useful in particular cases. I have also considered correlations between body form and host selection in the various subfamilies.

Shiino (1952, 1965) regarded the genus *Pseudione* as the most generalized, that is, representative of an evolutionary grade from which most other pseudionines and all of the other subfamilies are derivable. With 49 currently recognized species worldwide (some of which may eventually be removed to new genera if the genus is revised), *Pseudione* is by far the largest genus in the family. It is also the only genus whose species commonly occur as parasites of hosts belonging to more than one infraorder. While most species of *Pseudione* infest anomurans, all pseudionine parasites of carideans but one (Markham 1979b) belong to *Pseudione* the two known parasites of astacideans are in *Pseudione*, and several thalassinidean parasites are also members of *Pseudione*. *Pseudione*, however, is not the most primitive genus, only the most generalized. I agree with Shiino (1965) that *Pleurocryptella* is the most primitive. Its females uniquely retain all seven pairs of oöstegites rather than five pairs as in other bopyrids, and its males have prominent pleopods.

Having decided, on morphological grounds, that *Pleurocryptella* is the most primitive bopyrid and *Pseudione* the most generalized, one can then ask what the original hosts of bopyrids were. Shiino (1965), on the basis of *Pleurocryptella*'s occurrence, believes this was in the anomuran family Galatheidae. I concur, but on the basis of additional evidence as well. In the fossil record (Table 2), galatheids are both the most numerous and the oldest known hosts. Among recent galatheids, at least 55 species in five genera are infested,

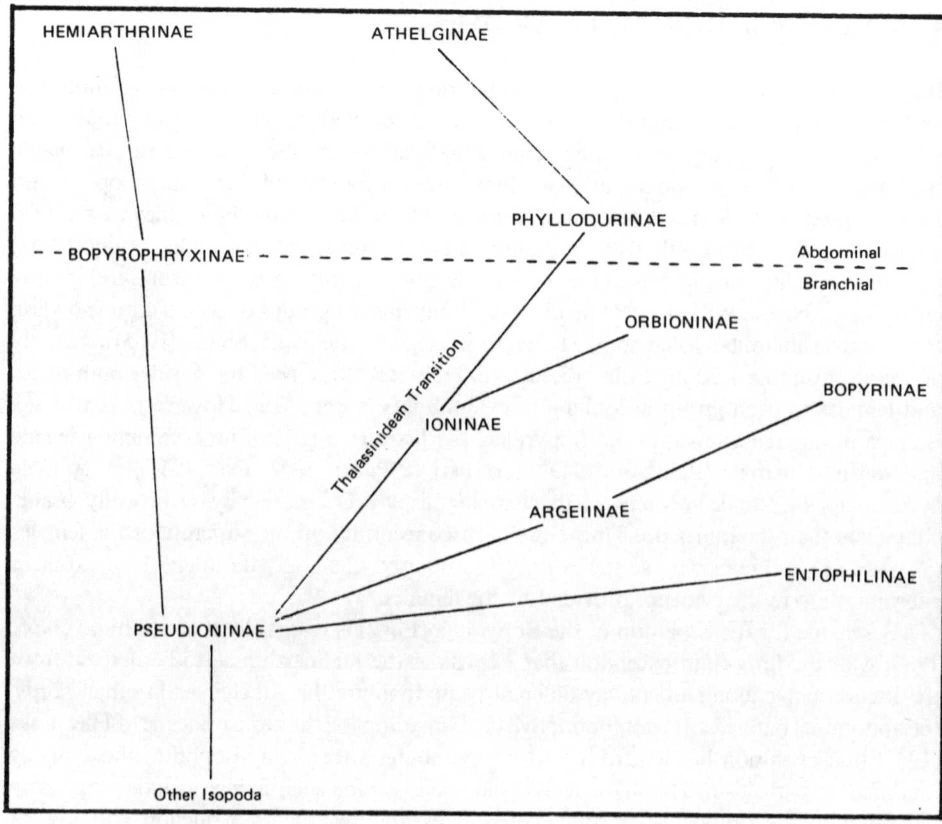

Figure 2. Proposed phylogeny of subfamilies of family Bopyridae. See text for explanation.

a very high proportion of the 225 species in six genera constituting the family, according to Chace (1951). Thirty species of *Munida* ranging from Norway to Madagascar, from British Colombia to the Falkland Islands, and many points between are infested by 39 bopyrid species belonging to 13 pseudionine genera and the lone entophiline species, *Entophilus omnitectus.* Among the genera of *Munida*'s parasites are both *Pseudione* and *Pleurocryptella.* The other common galatheid genus of rather shallow water, *Galathea* (which Ekman 1953, considers, with *Munida* to represent the primitive part of the family), contains eight reported infested species, from Norway to Turkey, Japan to Indonesia and in Mauritius and Florida. Its seven bopyrid species belong to four genera, including *Pseudione. Pleurocryptella* is not among them, but the closely related *Pleurocryptina* is. The third highly infested galatheid genus, *Munidopsis*, which Ekman (1953) considers more advanced and typical of deep water worldwide, contains 13 known host species from middle latitudes in the Atlantic, Pacific and Indian Oceans. Its described parasites belong to seven species in four genera, including *Pseudione* and *Pleurocryptella.*

Within the subfamily Pseudioninae one may trace an evolutionary line from *Pseudione* to other genera and from galatheid hosts to those in other anomuran families. I now want to consider a possible progression to other subfamilies and different hosts (Fig.2). At the evolutionary level of *Pseudione* on *Munida*, one parasite moved from the branchial chamber to the immediately adjacent part of the visceral cavity of its host and, with modification mainly of the pleon, gave rise to *Entophilus omnitectus*, the sole known species of the Entophilinae. Other *Pseudione* spp. moved to other anomurans and a few nonanomurans, namely the astacidean genera *Nephrops* and *Nephropsis* and the thalassinidean genera *Callianassa, Upogebia, Axiopsis* and *Laomedia.*

At a level of development just slightly above that of *Pleurocryptella*, an ancestor of the genus *Asymmetrione* infested hermit crabs (see Markham 1975e, for a discussion of *Asymmetrione*). With some slight modifications, it then would have developed into the subfamily Bopyrophryxinae, exemplified by the single species *Bopyrophryxus branchiabdominalis. Bopyrophryxus branchiabdominalis* is known from a single collection in the Kei Islands, Indonesia, where it infested the hermit crab *Parapagurus monstrosus* Alcock. As the species name indicates, *B.branchiabdominalis* is attached both in its host's branchial chamber and on its abdomen and modified accordingly. A further rather slight modification and transfer entirely to abdominal infestation would lead to the Hemiarthrinae, members of which infest carideans.

In another direction from *Pseudione*, a species infesting a caridean could have given rise to the Argeiinae and thence, through the peculiar monotypic genus *Bathygyge* to the subfamily Bopyrinae. All of the species of both Argeiinae and Bopyrinae are branchial parasites of carideans. Another line from *Pseudione* infesting thalassinideans could have gone through 12 other genera in the Pseudioninae and most primitive part of the Ioninae to give rise to the Ioninae. The most important genera in this sequence are *Pseudione, Progebiophilus, Gyge, Hypocepon, Ione* and *Ionella.* Aside from these species infesting thalassinideans, all species of the Ioninae (with one, perhaps accidental, exception) infest Brachyura. A probable derivative of the Ioninae, perhaps through the genus *Gigantione* Kossmann, is the subfamily Orbioninae, whose species infest peneoid shrimps. Continuing on with the parasites of Thalassinidea, we encounter the monotypic subfamily Phyllodurinae, consisting of *Phyllodurus abdominalis*, which has a structure suggestive of branchial parasites but nonetheless infests its host, *Upogebia pugettensis*, abdominally. I propose calling this link among the Pseudioninae, Ioninae and Phyllodurinae, from *Pseudione* to *Phyllodurus,* the Thalassinidean Transition. This term stresses the fact that all of the species which seem to join the three subfamilies infest hosts in the Thalassinidea. The status of the Thalassinidea as a monophyletic group is unsettled, but, however that issue may eventually be resolved, it should have no effect on this proposed evolutionary scheme for the Bopyridae. Because the pattern of host selection and evolutionary change could be interpreted as either co-adaptation or co-speciation (according to the criteria of Brooks 1979), the concept of the Thalassinidean Transition does not by itself favor either monophyly or convergent evolution in the Thalassinidea.

A final change of form and host would lead from the Phyllodurinae to the Athelginae, whose species are abdominal parasites of paguroids (in the broad sense of McLaughlin 1983). Because of different patterns of asymmetry, pereomere fusion, oöstegite development and other characters, I do not consider the Hemiarthrinae and Athelginae to be close relatives. Thus the completely separate proposed routes to these subfamilies of abdominal parasites do not contradict morphological evidence. Indeed, it appears that the resemblances between hemiarthrines and athelgines are really the result of convergent evolution to meet the requirements of abdominal infestation.

Among other epicaridean families, it is possible that the Entoniscidae arose from an advanced ancestor in the Hemiarthrinae which had lost most of its pereopods and body segmentation. The Dajidae may be derived from the Athelginae and possibly gave rise to the Cabiropsidae. Such conjectures have no present embryological or other support. Nor do they affect arguments concerning the Bopyridae.

This whole scheme, of course, is speculative, and many details of morphological arguments in its favor have been glossed over here for want of space. There is currently little embryological evidence available to bolster or challenge such a phylogeny.

5 ZOOGEOGRAPHIC CONSIDERATIONS

With the above evolutionary scheme in mind, one can now look at the distribution of potential hosts available to the Bopyridae. The newest, though surely rough, estimate of the number of living decapod species is 10,000 in 1200 genera of 105 families (Bowman & Abele 1982). Actually infested by the 459 described and several undescribed species of bopyrids are 568 host species (5.7%) in 173 genera (14.4%) of 45 families (42.9%). (All numbers and distributional data relating to the Bopyridae have been calculated from my own files, which are based on records derived from over 1000 references, so no individual sources are cited for the general summaries presented here.) Unfortunately, there is no recently published count of species or genera in the decapod families available, so I must refer to that of Chace (1951), which was compiled from data published up to 1945. Chace's total of decapod genera is 1001, with 8321 species. Because those numbers are both 20% below those of Bowman and Abele, it is probable that, on average, the other numbers used will be 20% too low. Still, a comparison of bopyrid numbers from 1983 with decapod numbers from 1945 (Table 3) will indicate trends. All of the ratios of species of parasites to species of hosts are close, about 10%, except for the Brachyura, whose ratio is only 1.7%. The biological significance of this is unclear; it seems to reflect some sort of 'undersaturation' among those ionines infesting brachyurans. Related to this low ratio is the fact that, in all of the other infraorders, it is common for one genus of hosts to harbor parasites in several genera, whereas the typical genus of the Ioninae infests brachyuran hosts in several genera or even different families. An extreme of this is the case of *Dactylokepon*, whose eight species infest hosts in eight genera in five families belonging to four superfamilies, including, perhaps accidentally, the only palinuran known to be infested (Markham 1975d). Another manifestation of this low ratio of parasite-to-host species is that, of the 91 species of Brachyura known to be infested, 45 are the sole representatives of their respective genera. In the family Majidae, every one of the 12 infested

Table 3. Numbers of species in various decapod taxa (after Chace 1951) and numbers of species of bopyrid isopods infesting members of those taxa.

Taxon	Total species	Parasite species infesting	Parasite species: Host species
Penaeoidea	318	35	11.0
Caridea	1 590	15	9.9
Thalassinidea	257	34	13.2
Anomura	1 270	123	9.7
Brachyura	4 428	75	1.7

Table 4. Decapod crustaceans from selected localities and rates of bopyrid infestation in those localities and of contained species only elsewhere.

Locality	Author	Total	Infested species		Infested only elsewhere	
			No.	%	No.	%
Arctic and Bering Sea	Rathbun 1904	77	15	19.5	8	10.4
Aleutians-Monterey	Rathbun 1904	197	29	14.7	11	5.6
Peru	del Solar et al. 1970	215	1	0.5	6	2.8
Hudson Bay	Squires 1967	13	1	16.7	9	69.2
Maine-New Jersey	Williams 1974	92	9	9.8	15	16.3
Carolinas	Williams 1965	220	25	11.4	24	10.9
Offshore Carolinas	Wenner & Boesch 1979	145	8	5.5	16	11.0
Bermuda	Markham & McDermott 1981	277	9	13.3	30	10.8
S.E. Florida	Gore 1976	256	21	8.2	51	19.9
Surinam	Holthuis 1959	115	7	6.1	13	11.3
Patagonia, 36-44°S	Boschi 1979	55	0	0.0	5	9.1
S.E. tip of S. America	Boschi 1979	29	2	6.9	0	0.0
Patagonia, deep	Boschi 1979	34	0	0.0	7	20.6
Britain	Plymouth Marine Fauna 1957	91	17	18.7	8	8.8
Iberia	Zariquiey Alvarez 1968	315	7	2.2	45	14.3
E. Mediterranean	Holthuis & Gottlieb 1958	191	4	2.1	34	17.8
S. Africa	Barnard 1950	530	14	2.6	49	9.2

species belongs to a different genus. These facts may be a consequence of extensive speciation in the Brachyura (Rice 1983), perhaps proceeding too fast for the parasites to match.

Information compiled from selected decapod check-lists (Table 4) shows that, in most places, the number of species of decapods actually infested is usually smaller than the number of decapods found there but infested only elsewhere. This evidently reflects the principle of Pielou (1974) that parasites do not range as far as their host species.

5.1 *Geographical distribution of the Bopyridae*

The zoogeography of parasites illustrates several interesting aspects of evolution and dispersal in coordination with and contrary to those of their hosts (see Brooks 1979, 1980, 1981a,b for discussions). Eichler (1949) propounded some useful rules by which parasites in general may be correlated with the evolution and distribution of their hosts, and Brooks (1979) expanded on them. Although most systematists dealing with the Bopyridae have at least implicitly acknowledged that the bopyrid faunae of distant regions are related, Codreanu (1961) was evidently the first researcher to stress the occurrence of congeneric parasites on congeneric hosts in widely separated regions. Brooks (1979: 305) points out that 'parasitologists were among the earliest biologists to support the concept of continental drift . . .', and it is abundantly clear that modern plate tectonic theory explains many aspects of the distribution of the Bopyridae.

A plot of the 469 species of described bopyrids by 5° ranges of latitude in the four major oceanic regions (Fig.3A) shows several general trends. (Because many species occur over more than 5° of latitude and some even in more than one ocean, the total plotted is much more than 469.) In the eastern Pacific, the numbers at all latitudes are low, with much of western South America completely unrepresented. In the western Atlantic,

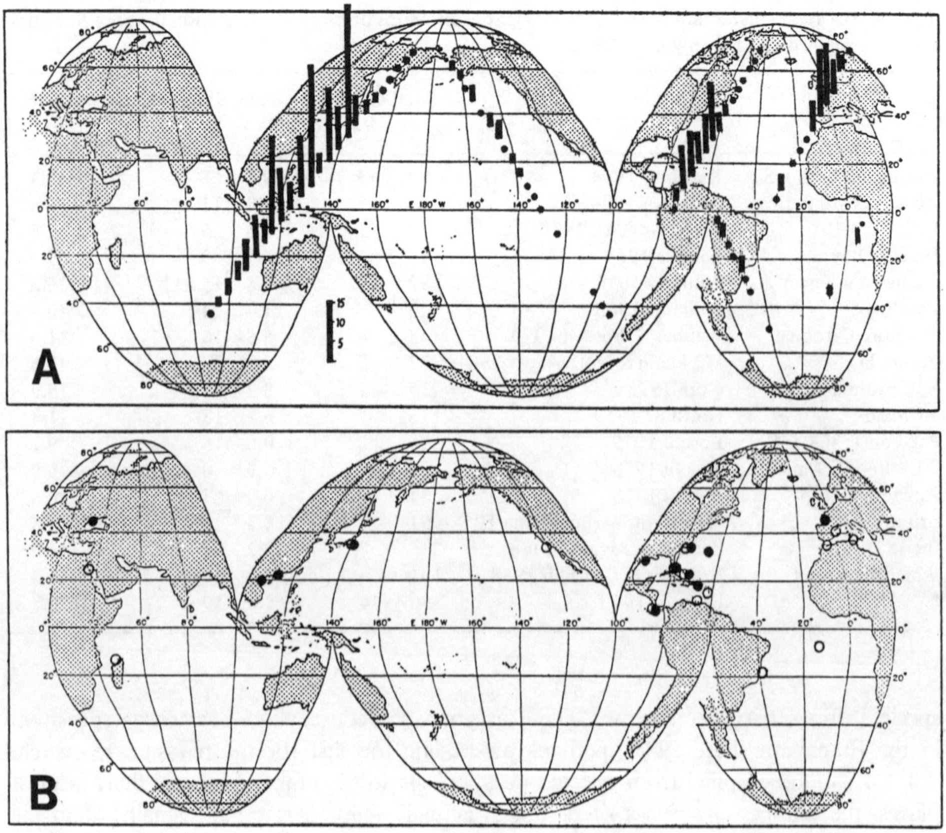

Figure 3. (A) Distribution of 469 species of Bopyridae in four oceanic regions by 5° latitude bands. The height of each column represents the number of species in the 5° interval from the base of that column to 5° farther north. The scale labeled 5, 10, 15 (species) is proportional to the columns. (B) Distribution of seven species of *Asymmetrione* (open circles) and six species of *Bopyrissa* (filled circles).

the peak is from the Carolinas through the Caribbean, i.e. the north temperate and tropical regions, with South America again rather sparsely represented. The eastern Atlantic region has the greatest concentrations of species along the western European and northern Mediterranean coasts; records from western Africa are very sparse. In the Indo-West Pacific, especially from the northern subtropics through the tropics, are found by far the most species of bopyrids. Three maxima represent the faunae of Japan, the Philippines and the coast of India and Indonesia. Because the fauna of China is almost completely unknown (Markham 1982) and that of Australia rich in still-undescribed species (A.J.Bruce, pers. comm.), the Indo-West Pacific, already the region containing half the known species of Bopyridae, may actually harbor nearly twice the number now recorded.

The center of present distribution, though not necessarily the center of origin, for the Bopyridae is thus clearly the subtropical and tropical Indo-West Pacific. The second richest region is the correspondin portion of the western Atlantic from North Carolina through the Gulf of Mexico and the Caribbean. The only concentration in temperate waters is that of Europe. This distribution, with the greatest number of species found in warm waters, runs counter to the generalization of Abele (1982) that isopod species are most numerous in temperate to cold waters. As parasites of decapods, the bopyrids should be expected to reflect the distribution of their hosts, and they do, in that the decapods also reach a peak in the tropics (Abele 1982) and have more than half of their species in the Indo-West Pacific. Nonetheless, as a group whose numbers account for an eighth of all known species of isopods, the epicarideans cannot be overlooked in a discussion of the distribution of the Isopoda.

Bopyrid isopods, despite the great bathymetric range of decapods, are largely restricted to shallow and moderately deep waters. Only one described species has been recorded from more than 4000 m and two others below 2000 m. Many species are common in intertidal regions around the world.

To some unknown degree, of course, the plot presented (Fig.3) is a record not of actual occurrence but of collectors' activities. Thus the fauna of Japan is very well documented because of the work of Shiino (e.g. Shiino 1933, 1934, summarized by Shiino 1972). India was well covered by Chopra (1923, 1930), and Indonesia by Nierstrasz & Brender à Brandis 1929, 1931). More recently, Bourdon (1968) has dealt with the bopyrid fauna of Europe, Madagascar (Bourdon 1976, 1979a) and several other regions. Markham (Markham 1975b, 1979a, in press) has worked in the northwestern Atlantic and elsewhere. Lemos de Castro (1965) has published a series of short reports on the Bopyridae of Brazil. My personal experiences in Hong Kong (Markham 1982) and western Thailand (in prep.) have shown that even a brief collecting effort can turn up numerous species in a place previously known to have but few.

In at least two regions, small recorded numbers may actually reflect a paucity of bopyrids. For example, there are few species known from the western coast of North America, a region quite well surveyed. I have seen specimens which represent five still-undescribed species from southern California, but even their description will leave the count very low. The situation off the coast of South America is much less clear. In the eastern Atlantic, the coast of Africa, it seems, should have far more species, but perhaps it really does not. A recent comprehensive coverage of the Brachyura of West Africa (Manning & Holthuis 1981) by two authors who carefully cite the presence of parasites in their samples, mentions not a single bopyrid among thousands of brachyuran specimens examined.

5.1.1 *Subfamily Pseudioninae*

The Pseudioninae are distributed in a pattern similar to that of the family as a whole but with some marked differences in detail. Fully half of the species occur in the Indo-West Pacific. But rather than showing numerical peaks in the tropics, they reach a maximum of 24 species from 30 to 35°N, mostly near southern Japan. The second greatest number of species is 15 found between 5 and 10°S, mostly in Indonesia. In the eastern Pacific, only about 12 species occur, with three recorded from 45-50°N and at most two species in any other region. Only one species is recorded from south of the equator, off Peru. In the western Atlantic, no pseudionine is known north of 40°N, but within each 5° section from 40°N to 10°N six to nine species occur; thereafter, scattered species range from 5°S to 50°S. In the eastern Atlantic, European species are numerous (12 found at 40 to 45°N), and widely scattered ones occur along western Africa.

Both *Asymmetrione* and *Bopyrissa* (Fig.3B) are worldwide primitive genera of pseudionines which infest hermit crabs and contain, respectively, seven and six species (not all yet

Table 5. Hosts and localities for six species of *Bopyrissa* and seven species of *Asymmetrione.*

Parasite genus	Host species	Locality
Bopyrissa	*Diogenes merguiensis*	Vietnam
	D.pugilator	Black Sea
	D.edwardsi	Japan
	Clibanarius misanthropus	France
	C.bimaculatus	Hong Kong
	C.tricolor	North-West Atlantic
	C.vittatus	North-West Atlantic
	Clibanarius sp.	West Central America
Asymmetrione	*Clibanarius bimaculatus*	Japan
	C.signatus	Red Sea
	C.tricolor	North-West Atlantic
	Clibanarius sp.	West Central America
	Dardanus arrosor	Marocco
	Paguristes occulatus	Mediterranean
	Pagurus spp.	North-West Atlantic

described). Most species of branchially infesting female bopyrids occur in about equal numbers in mirror-image forms in which either the right side (in dextral forms) or left side (in sinistral forms) is the longer, and which occupy their hosts' right or left branchial chambers, respectively. For completely unknown reasons, in *Asymmetrione* and *Bopyrissa*, however, each species has only a dextral or sinistral form (there are two exceptions, one species in each genus in which one form is preponderant but not exclusive). Species of *Bopyrissa* are parasites only of diogenid hermit crabs, while *Asymmetrione* spp. infest both diogenids and pagurids (Table 5). *Clibanarius tricolor* (Gibbes) serves as the host for both *Asymmetrione clibanarii* Markham 1975 and *Bopyrissa wolffi* Markham 1978 in some of the same western Atlantic localities; the former species occurs in its left branchial chamber and the latter only in its right one.

Aporobopyrus (Fig.4A), containing eight species, is the largest genus of parasites of porcellanids. Pantropically distributed, *Aporobopyrus* spp. infest species of *Petrolisthes* (in the Red Sea, northern Atlantic, Brazil, Japan and the Ryukyus), *Alliaporcellana* (Red Sea), *Palyonyx* (Red Sea), *Pachycheles* (California, Philippines) and *Pisosoma* (Ryukyus).

Munidion is entirely amphi-American except for a species from the Philippines and the Java Sea (Markham 1975c). All of the hosts are *Munida* spp. except for *Pleuroncodes planipes* Stimpson from central California to Baja California.

The 49 currently recognized species of *Pseudione* occur in all oceanic regions, and their host selection (discussed above) is so diverse that generalizations about the genus as a whole are of no value. However, two cases of *Pseudione* are worthy of note in a zoogeographic discussion.

Both *Pseudione hyndmanni* and *P.giardi* are parasites of northerly cold-water pagurid hermit crabs. *Pseudione hyndmanni* ranges from Norway and the White Sea and Iceland through Denmark, along all British coasts to northwest France (Bourdon 1968). Its most common host is *Pagurus bernhardus*, others being *P.pubescens, Anapagurus laevis* and *A.chiroacanthus.* It is also recorded as infesting a *Pagurus* sp. in Japan, and *Parapseudione lata*, a parasite of *Pagurus middendorffi* at Hokkaido, may be a synonym. Calman (1898),

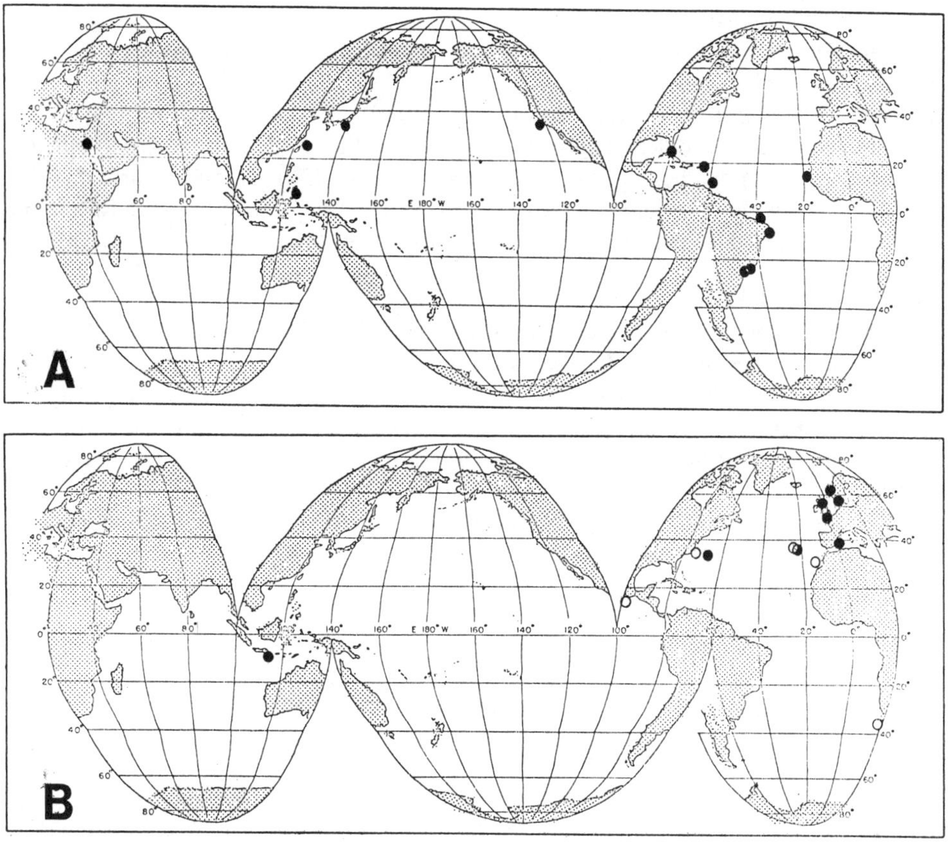

Figure 4. (A) Distribution of ten species of *Aporobopyrus.* (B) World distribution of *Pseudione affinis* (filled circles) and *Bathygyge grandis* (open circles).

in describing *Pseudione giardi* from Puget Sound, Washington, stated that it was difficult to distinguish from *P.hyndmanni*; Bourdon (1968) considered it to be *P.hyndmanni*'s closest relative. Markham (1975a) reviewed the species, considered it to be distinct and summarized its range, from Bristol Bay in the Bering Sea to the San Juan Islands, Washington. Its known hosts are five species of *Pagurus* and, in a single instance, the lithodid *Dermaturus mandti.* It is of interest that one of the pagurid host species of *Pseudione giardi, Pagurus ochotensis*, is similar enough to the European species *P.bernhardus* that it was originally described as a variety of that species (McLaughlin 1974). Thus the distribution of *Pseudione hyndmanni* and *P.giardi* consists of two very closely similar species each infesting several closely related hosts in comparable boreal regions.

Pseudione affinis (Sars) (Fig.4B) is the most commonly encountered species of *Pseudione* infesting carideans. It has been reported from a very great geographical range although infesting rather few species of midwater pandalid shrimps. Its hosts are *Pandalus* spp. and *Dichelopandalus bonnieri* Caullery in western Europe, *Plesionika martia* (A.Milne Edwards) in France and the Java Sea and *Plesionika* spp. in Bermuda, the Java Sea and the Mediterranean.

5.1.2 *Subfamily Entophilinae*

This subfamily is represented by a single species, *Entophilus omnitectus*, originally reported from Hawaii (Richardson 1903). Recently, Bourdon (1976, 1979b) reported finding several specimens in Madagascar and the Azores. In each case, the host, infested viscerally rather than branchially, was a different species of the galatheid genus *Munida.* Bourdon (1979b) also alluded to unpublished records of the occurrence of *E.omnitectus* in Australia and the Philippines, without mentioning their hosts. This is clearly a circumtropical distribution of an extent comparable only to the range of *Pseudione affinis.*

5.1.3 *Subfamily Argeiinae*

This small subfamily of nine species in five genera is concentrated in the Indo-West Pacific, where all but two species occur. One species, *Argeia pugettensis*, is quite a common parasite along the boreal coasts of both the northwestern and northeastern Pacific, where it infests a total of 23 species of crangonids and one hippolytid (Markham 1977a). Another argeiine is also found in the eastern Pacific, but only *Argeia atlantica* known from Newfoundland to the Bahamas in moderately deep water, has been recorded from the Atlantic. This distributional pattern implies an origin in the western Pacific or Indian Ocean and dispersal from there through the Arctic rather than by a circumtropical route.

5.1.4 *Subfamily Bopyrophryxinae*

A single species, *Bopyrophryxus branchiabdominalis* Codreanu is assigned to the Bopyrophryxinae. It is known from a single pair found infesting the hermit crab *Parapagurus monstrosus* Alcock in the Kei Islands, Indonesia (Codreanu 1965).

5.1.5 *Subfamily Bopyrinae*

This subfamily of 100 species of caridean-infesting bopyrids has its greatest representation in the subtropical to tropical Indo-West Pacific, where over half of its species occur. There are 41 species in the tropical Indian Ocean alone. Only five species are known from the Eastern Pacific, none south of 15°N. Five European species and one species each from northwestern and southwestern Africa constitute the Bopyrinae in the eastern Atlantic. In the western Atlantic, there are 14 species in the Carolinian area and 16 in the Caribbean region and southward to Brazil, with none reported south of 25°S. A review (Markham in press) will describe some new species from the northwestern Atlantic. All species occur in temperate or tropical waters except for *Bopyroides hippolytes*, which is restricted to the arctic and circumboreal regions. Dispersal of the subfamily seems to have been circumtropical, through the Tethys sea with one passage through the arctic, probably from the eastern Atlantic.

Bathygyge grandis (Fig.4B) is a very widespread species infesting midwater shrimps of the genus *Glyphocrangon.* Originally recorded from off western Mexico (Hansen 1897), it has since been found off South Africa, in the Azores and Canaries and off Virginia.

Bopyroides hippolytes is circumboreal, occurring on all arctic and boreal coasts of the Atlantic and Pacific. It infests 23 species in the hippolytid genera *Spirontocaris, Eualus, Heptacarpus, Hippolyte* and *Lebbeus* and the pandalid genera *Pandalus* and *Pandalopsis.* It extends south to northern Japan, the San Juan Islands, Cape Cod and Britain. This distribution covers all but the southern limits of the genus *Spirontocaris* (Ekman 1953), 11 species of which are infested by *Bopyroides hippolytes.* Because the only other accepted species of *Bopyroides* occurs in western Europe (Bourdon 1968), it is probable that *B.hippolytes* originated there and then dispersed through the Arctic.

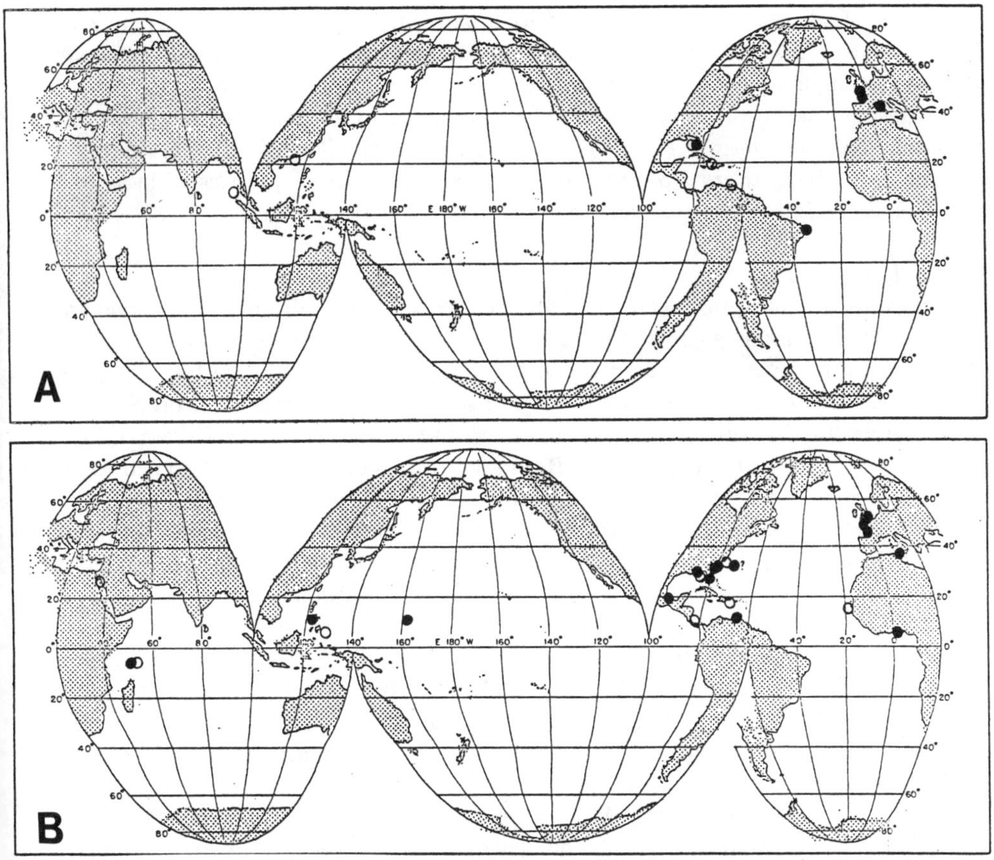

Figure 5. (A) Distribution of four species of *Bopyrione* (open circles) and *Urobopyrus processae* (filled circles). (B) Distribution of five species of *Cancricepon* (filled circles) and eight species of *Dactylokepon* (open circles).

Bopyrione (Fig.5A) contains four species infesting Synalpheus spp. in the Caribbean and the Andaman Islands (Bourdon & Markham 1980) and *Alpheus* spp. in Hong Kong (Markham 1982).

There is evidently only one very widespread species of *Urobopyrus* (Fig.5A), *U.processae* Richardson, which infests processids (all but one in the genus *Processa*) in the Gulf of Mexico, Brazil, the Mediterranean and northern France.

The genus *Probopyrus* has a highly unusual distribution. When species improperly included are removed, the genus becomes quite homogeneous, but species are so plastic and difficult to distinguish that their number is subject to considerable dispute (Markham in press). The genus occurs in the Indo-West Pacific from Japan to India and in the western Atlantic from New Hampshire to Brazil. Its species infest palaemonid shrimps of the genera *Macrobrachium, Palaemon, Palaemonetes* and *Periclimenes*, which tend to be found progressively farther up rivers from the ocean in lower latitudes (Ekman 1953). Though common in brackish waters, *Probopyrus* is the only bopyrid genus known from fully fresh water, being recorded from far up the Mississippi River and in the mountains of the Philippines, among other places.

At least 24 species have been assigned to *Bopyrus*, the type-genus of the Bopyridae, but only three remain recognized (Bourdon 1968). The most widespread is *B.squillarum* found as a common parasite of seven species of *Palaemon* through Europe to the eastern Mediterranean and in the Indo-West Pacific from India to Japan. It is also known to infest the processid *Processa edulis* in southwestern Europe (Bourdon 1968). The other two species of *Bopyrus* are *B.bimaculatus* which infests species of *Leander* in India and New Guinea, and *B.stebbingi* on an unrecorded host in Indonesia.

5.1.6 *Subfamily Ioninae*

The subfamily Ioninae preponderates in the Indo-West Pacific with maxima in southern Japan and northern and southern Indonesia. The tropical western Pacific alone contains 35 species. In the eastern Pacific, three species occur in Hawaii, but only two along the western North American coast; except for three species in Chile, none is recorded south of California. In the western Atlantic, there are five species along the eastern United States and six from southern Florida to Brazil; only one species has been recorded south of 5°S. Species in the eastern Atlantic number six from Europe, especially the Mediterranean, and four scattered along tropical West Africa. This subfamily is almost entirely restricted to warm waters, and most of its hosts are brachyurans. On the other hand, nearly all of the species recorded from temperate waters are those primitive ones parasitic on thalassinideans. Because most of the ionines infest brachyurans, their overall distribution is clearly related to that of their hosts, the Brachyura being far more abundant in the tropics, especially of the Indian and Pacific (Abele 1982). That the Thalassinidean Transition portion of the Ioninae prevails in cooler waters implies that the subfamily originated there, but there is still evidence of considerable Tethyan circumtropical dispersal among genera infesting brachyurans.

The genus *Cancricepon* (Fig.5B) shows a pantropical distribution with the addition of Europe, where *C.elegans* is widespread. The eight other known species and two '*Cancricepon* sp.' occur quite close to the equator, in all regions but the eastern Pacific. Together, they infest members of 12 different genera of xanthid crabs. *Cancricepon choprae*, the only species known from the western Atlantic, infests seven species in six xanthid genera there and the goneplacid *Panoplax depressus* (Markham 1975a).

Dactylokepon (Fig.5B) is also pantropical except for the eastern Pacific. Its eight species, all from subtropical or equatorial regions, infest a total of nine host species, mostly members of different Brachyuran families or even superfamilies (Markham 1979d).

5.1.7 *Subfamily Orbioninae*

The subfamily Orbioninae displays a very unusual distributional pattern. All species are parasites of peneoidean shrimps, their hosts being 34 species in 13 genera of the families Aristeidae, Penaeidae, Solenoceridae and Sicyonidae. The geographical range is exclusively warm Indo-West Pacific from the Red Sea to Japan, between 40°N and 35°S. The only exception is a species from Turkey which evidently entered the eastern Mediterranean with its host species by Lessepsian migration from the Red Sea. On morphological grounds, the orbionines are quite advanced, and their geographical distribution indicates that they must have evolved since the closing of the circumtropical Tethys Sea, thus no later than the Eocene (Cox et al. 1976). Their host group, the Penaeoidea, is considered ancient (Glaessner 1969), and it is very well represented in all temperate and tropical waters. For instance, of the six genera of peneoids known from the shallow Caribbean (Chace

1972), four contain species infested elsewhere. The species *Aristaeomorpha foliacea*is itself circumtropical in moderately deep water (Holthuis 1980), but infested individuals are recorded only from Japan and Australia. Many species of orbionines infest hosts in more than a single genus. Thus the limited geographical range of the subfamily indicates not a shortage of suitable hosts but lack of access to other regions after these hosts had been dispersed.

Some species of the Orbioninae are quite widespread. *Orbione halipori, Epipeneon igens* and *Parapeneon japonica* are recorded from nearly the entire range of the whole subfamily (Markham 1982). Their selection of hosts is also quite broad. Incidence of infestation by some species is occasionally remarkably high, with up to half of a host population parasitized in some areas.

The 35 orbionine species infest only 34 host species for a host-species parasite-species ratio of only 0.97. In contrast, the ratio for the Bopyridae as a whole, 469 species infesting 569 species, is 1.21.

5.1.8 *Subfamily Phyllodurinae*

There is but a single known phyllodurine, *Phyllodurus abdominalis.* Its only recorded host is *Upogebia pugettensis* and its range is restricted to the north-eastern Pacific, in estuaries from southern British Columbia to central California (Markham 1977b). Within that range, however, it is often found in large numbers. *Phyllodurus* is an important member of the Thalassinidean Transition group of genera. An abdominal parasite, it has a body structure more like that of branchial parasites, thereby representing a link between the two modes of infestation (Markham 1977b).

5.1.9 *Subfamily Athelginae*

The subfamily Athelginae contains 36 species in seven genera, all of them dorsoabdominal parasites of paguroid (sensu lato) anomurans. Although present in all oceanic regions, the athelgines are unevenly distributed. In the Indo-West Pacific, species are found from 40°N to 45°S but are nowhere numerous. Only two species occur along the entire length of the eastern Pacific. The greatest number of species is in the Atlantic, with eight species in the Caribbean region and ten around Europe. Although athelgines are known along eastern South America as far south as Brazil, no species in the eastern Atlantic is recorded from south of 35°N.

Of the seven genera of Athelginae, six are polytypic, and each of these genera is represented in more than one ocean. The three species of *Anathelges* are found in the eastern and western Indian Ocean and Brazil. The 11 species of *Athelges* are clustered in Europe (to northern Norway) and range through Japan, in the western Indian Ocean and New Zealand. *Parathelges*, a warm-water genus, has four species between North Carolina and Venezuela and seven others in the Mediterranean, Black Sea, Indonesia and the central tropical Pacific (Markham 1972). *Stegophryxus* is amphi-American, with one species infesting intertidal pagurids from Massachusetts to Curaçao, a second occuring in rather deep water off California and Baja California, and a third recorded from Chile (Markham 1974). *Stegias*, another warm-water genus of three species, occurs in Indonesia, western Costa Rica and Bermuda. The four species of *Pseudostegias* occur in Brazil, Japan and Hong Kong (Markham 1982). Thus, except for *Athelges* and *Stegophryxus*, there is a definite tendency towards equatorial distributions. The subfamily is probably a Tethyan-dispersed group which arose in the Atlantic Ocean. The hosts, both pagurids and diogenids (and one lithodid) are not distributed at all along generic lines of the parasites.

5.1.10 *Subfamily Hemiarthrinae*

In the subfamily Hemiarthrinae the 45 known species in 22 genera are caridean parasites. All species infest their hosts abdominally, typically clinging to the ventral surface of the abdomen among the host's pleopods, though some few species have been found attached otherwise. The probable course of evolution within the Hemiarthrinae is from a female having all body segments distinguishable, the pleopods well-developed and all pereopods present, through a sequence of progressive fusion and loss of segments, pleopods and pereopods. At the most advanced level, distinctions between hemiarthrines and members of other epicaridean families are blurred, so it is likely that some of the latter arose from hemiarthrine ancestors.

As with most other bopyrid subfamilies, the subfamily Hemiarthrinae is centered in the Indo-West Pacific, with 19 species being found in the tropical Indian Ocean alone. All of the species but one are found only in warm waters, those in the Indo-West Pacific from 35°N to 35°S. In other oceans, the distribution is very restricted; the single (undescribed and possibly questionable) warm-water species of the eastern Pacific occurs only in California, the eight warm-water species of the western Atlantic occur no farther south than 10°N, and those of the eastern Atlantic occur only around Europe, between 40° and 50°N. Thus, except in the Indo-West Pacific, the subfamily is not known to occur in or close to the southern hemisphere.

The single species which occurs in cold waters is *Hemiarthrus abdominalis* (Krøyer 1840-1841), whose distribution is completely circumboreal and arctic. Its range and host selection are almost identical to those of *Bopyroides hippolytes*, discussed above. *Hemiarthrus abdominalis* infests a total of 32 different species in the hippolytid genera *Eualus, Spirontocaris* and *Lebbeus*, the most hosts known for any bopyrid species. It is found down to about 45°N in all oceanic regions.

Aside from the distribution of *Hemiarthrus abdominalis*, which occupies a rather advanced evolutionary position in the Hemiarthrinae and so must be secondarily distributed, the pattern of distribution of the Hemiarthrinae is a Tethyan relict which started in the Indian Ocean and extended only as far west as the Caribbean. Because the only other species properly assigned to *Hemiarthrus* occurs in North Carolina (Markham in press), it is a reasonable assumption that *H.abdominalis* originated in the western Atlantic and spread from there through the Arctic Ocean.

Many of the hemiarthrine genera, especially those in the Indo-West Pacific, are monotypic. On the other hand, five of the eight Caribbean species belong to five widespread genera (Markham in press). *Diplophryxus* is otherwise known from Japan, Mozambique, India and the Andamans. The seven species of *Eophrixus* occur in Japan, the Mediterranean and the Caribbean. The four species of *Dicropleon* are known from the Indian Ocean, Hong Kong and the Caribbean, all but one infesting species of *Periclimenes. Hyperphrixus*, with one species each in the Philippines and Caribbean, infests *Periclimenes* and *Periclimenaeus.* And the two species of *Metaphrixus* occur near Zanzibar and in the Caribbean. Except as noted above, the genera of hosts do not correspond well to those of the parasites. However, a particular species, even one with a rather large number of host species, nearly always selects them from a single genus.

5.2 *Host selection and geography in the Bopyridae*

There is some correlation between latitude and numbers of host species for a single species of parasite. All bopyrids of high latitudes infest numerous hosts. The two species with completely arctic-boreal distributions, *Bopyroides hippolytes* and *Hemiarthrus abdominalis*, infest 23 and 32 host species, respectively. *Argeia pugettensis*, an amphi-Pacific boreal species, has at least 23 different hosts. In each case, the host species belong to more than one family. The only other bopyrid species found north of the Arctic Circle, the closely related *Pseudione hyndmanni* and *P.giardi*, each infest five species of *Pagurus.* (*Pseudione hyndmanni* also infests two species of *Anapagurus.*) This tendency may be a result of the small numbers of decapod species in high latitudes (Rathbun 1904, Squires 1967).

In warmer waters, where species of both decapods and bopyrids are much more numerous, most parasites are restricted to three or fewer host species, of which nearly all are congeneric or at least in a single family. In the northwestern Atlantic, parasites of the alpheid genera *Alpheus* and *Synalpheus* present unusual cases. These are the most speciose genera of shallow-water shrimps in the region, *Alpheus* having 24 species and *Synalpheus* 28 (Chace 1972). Of these, five *Alpheus* spp. and 12 *Synalpheus* spp. (those recorded only as 'sp.' not being counted) host bopyrids. Six bopyrid species infest species of *Alpheus*, and two of them infest three different host species apiece. Ten bopyrid species infest species of *Synalpheus*, with three parasites each having three hosts, two parasites having four hosts, and one each having five and six hosts. Only one bopyrid infests both an *Alpheus* and a *Synalpheus*, and no parasite of either host genus is recorded from any nonalpheid host (Markham in press). All but one species of *Alpheus* bear at least two different bopyrid species, and two species of *Synalpheus* each bear four bopyrids, and one each bears five and six bopyrid species. These cases of multiple host and parasite selections may be related to the facts that both *Alpheus* and *Synalpheus* have many variable sympatric species which are so similar as often to be indistinguishable.

Brooks (1979) divides coevolution of hosts and parasites into two separate but interacting phenomena, co-accommodation (the degree to which a parasite species can infest different host species) and co-speciation (the degree to which a parasite species has evolved with a single host species). Species of the Bopyridae reflect both phenomena. The broadest cases of co-accommodation occurred with the evolution of the separate subfamilies as their hosts evolved into different decapodan infraorders. Once established, most of the subfamilies underwent co-speciation (exceptions being the monotypic subfamilies Phyllodurinae and Bopyrophryxinae, which show only co-accommodation and transfer to other hosts with evolution into other subfamilies). The largest and among the oldest subfamilies, the Pseudioninae and Bopyrinae, show the best evidence of cospeciation: they are broadly pantropically distributed, clearly as Tethyan relicts, and many congeneric species infest congeneric hosts around the world. The more recently evolved subfamilies, the Ioninae and Orbioninae among branchial parasites, and the abdominally infesting Hemiarthrinae and Athelginae, show quite low host specificity; a single parasite species typically infests more than one host species, and most genera infest members of different host genera or even families. Of the four latter subfamilies, only the Hemiarthrinae (probably the earliest evolved), exhibit a typical Tethyan distribution; the Ioninae are largely Indo-West Pacific and the Orbioninae exclusively so, and the Athelginae appear to have radiated from the northern Atlantic. Similarly, all of the arcto-boreal species discussed above are evidently secondarily and probably quite recently distributed away from ancestral patterns. Thus there is an evident correlation with incrèased co-speciation and decreased co-adaptation with increasing age of dispersal and evolution.

ACKNOWLEDGEMENTS

Thanks are extended to Dr R.Bourdon (Station Biologique de Roscoff) for remarks on bopyrid phylogeny; to Dr A.J.Bruce (Museums and Art Galleries of the Northern Territory, Darwin, Australia) for information about Australian bopyrids; to Dr R.H.Gore (Academy of Natural Sciences of Philadelphia) for an unpublished checklist of decapods occurring near Ft Pierce, Florida, and for valuable comments on the manuscript; to Dr F.G.Hochberg (Santa Barbara Museum of Natural History) for the loan of bopyrid specimens collected in California; and to Dr L.B.Holthuis (Rijksmuseum van Natuurlijke Historie, Leiden) for current names of caridean species and publications unavailable to me. I am especially indebted to Mrs W.A.Markham for provision of research facilities at the Arch Cape Marine Laboratory, of which this is publication number 9.

REFERENCES

Abele, J.G. 1982. Biogeography. In L.G.Abele (ed.), *The Biology of Crustacea* Vol.1. New York: Academic Press.

Anderson, G. & W.E.Dale 1981. *Probopyrus pandalicola* (Packard) (Isopoda, Epicaridea): morphology and development of larvae in culture. *Crustaceana* 41: 143-161.

Bachmayer, F. 1955. Die fossilen Asseln aus den Oberjuraschichten von Ernstbrunn in Niederösterreich und von Stramberg in Mähren. *S.B. österr. Akad. Wiss. Abt. I*, 164: 255-273.

Baer, J.G. 1951. *Ecology of animal parasites.* Urbana: Univ. Illinois.

Balss, H. 1913. Uber fossile Galatheiden. *Centralbl. f. Minaral. Geol. Palaeont. Jhrg.* 1913: 155-160.

Barnard, K.H. 1950. Descriptive catalogue of South African decapod Crustacea (crabs and shrimps). *Ann. 6th Afr. Mus.* 38: 1-837.

Bell, T. 1863. *A monograph of the fossil malacostracous Crustacea of Great Britain. Part II. Crustacea of the Gault and Greensand.* London: Palaeont. Soc.

Boschi, E.E. 1979. Geographic distribution of Argentinian marine decapod crustaceans. *Bull. Biol. Soc. Wash.* 3: 134-143.

Bourdon, R. 1968. Les Bopyridae des mers Européennes. *Mém. Mus. Natl. Hist. Nat. (Paris) N.S. (A)*50: 77-424.

Bourdon, R. 1976. Epicarides de Madagascar. I. *Bull. Mus. Natl. Hist. Nat. (3): 371, Zool.* 259: 353-392.

Bourdon, R. 1979a. Epicarides de Madagascar. II. *Bull. Mus. Natl. Hist. Nat. (4): 1 Sect.A*(2): 471-506.

Bourdon, R. 1979b. Bopyridae de la campagne Biaçores (Isopoda Epicaridea). *Bull. Mus. Natl. Hist. Nat. (4): 1 Sect.A*(2): 507-512.

Bourdon, R. & J.C.Markham 1980. A new genus and species of bopyrid isopod infesting alpheid shrimps in the genus *Synalpheus* in the western Atlantic Ocean. *Zool. Meded.* 55: 221-230.

Bowman, T.E. & L.G.Abele 1982. Classification of the Recent Crustacea. In L.G.Abele (ed.), *The Biology of Crustacea* Vol.1. New York: Academic Press.

Brooks, D.R. 1979. Testing the context and extent of host-parasite coevolution. *Syst. Zool.* 28: 299-307.

Brooks, D.R. 1980. Allopatric speciation and non-interactive parasite community structure. *Syst. Zool.* 29: 192-203.

Brooks, D.R. 1981a. Raw similarity measures of shared parasites: an empirical tool for determining host phylogenetic relationships? *Syst. Zool.* 30: 203-207.

Brooks, D.R. 1981b. Hennig's parasitological method: a proposed solution. *Syst. Zool.* 30: 229-249.

Calman, W.T. 1898. On a collection of Crustacea from Puget Sound (Puget Sound Zoology, Columbia University Contributions no.9). *Ann. N.Y. Acad. Sci.* 11: 259-292.

Chace, F.A. Jr 1951. The number of species of decapod and stomatopod Crustacea. *J. Wash. Acad. Sci.* 41: 370-372.

Chace, F.A. Jr 1972. The shrimps of the Smithsonian-Bredin Caribbean Expeditions with a summary of the West Indian shallow-water species (Crustacea: Decapoda: Natantia). *Smiths. Contr. Zool.* 98: 1-179.

Chopra, B. 1923. Bopyrid isopods parasitic on Indian Decapoda Macrura. *Rec. Indian Mus.* 25: 411-550.

Chopra, B. 1930. Further notes on bopyrid isopods parasitic on Indian Decapoda Macrura. *Rec. Indian Mus.* 52: 113-147.

Condreanu, R. 1961. Crustacei paraziti indo-pacifice în Marea Neagre. *Hidrobiol., Acad. Rep. Pop. Romîne* 3: 133-146.

Condreanu, R. 1965. Sur un bopyride nouveau, *Bopyrophryxus branchiabdominalis* nov.gen. nov.sp. parasite de *Parapagurus monstrosus* des îles Kei et type de la sous-famille nouvelle de Bopyrophryxinae nov. *C.R. Acad. Sci. Paris* 261: 1895-1897.

Cox, C.B., I.N.Healey & P.D.Moore 1976. *Biogeography, an Ecological and Evolutionary Approach. 2nd Ed.* Oxford: Blackwell.

Dale, W.E. & G.Anderson 1982. Comparison of morphologies of *Probopyrus bithynis, P.floridensis,* and *P.pandalicola* reared in culture (Isopoda, Epicaridea). *J. Crust. Biol.* 2: 392-409.

Del Solar, E.M., F.Blancas & R.Mayta 1970. *Catalogo de Crustáceos del Perú.* Lima: Miranda.

Eichler, W. 1949. Some rules in ectoparasitism. *Ann. Mag. Nat. Hist.* (12)1: 588-598.

Ekman, S. 1953. *Zoogeography of the Sea.* London: Sidgwick & Jackson.

Glaessner, M.F. 1969. Decapoda. In R.C.Moore (ed.), *Treatise on Invertebrate Paleontology, Part R, Arthropoda 4.* Lawrence: Geol. Soc. Am. & Univ. Kansas Press.

Gore, R.H. 1976. Indian River coastal zone study species inventory list November 1976. Unpublished.

Hansen, H.J. 1897. Reports on the dredging operations off the west coast of Central America . . . by the U.S. Fish Commission Steamer 'Albatross' during 1891 . . . XXII The Isopoda. *Bull. Mus. Comp. Zool.* 31: 96-129.

Hessler, R.R. 1969. Peracarida. In R.C.Moore (ed.), *Treatise on Invertebrate Paleontology, Part R, Arthropoda 4.* Lawrence: Geol. Soc. Am. & Univ. Kansas Press.

Hessler, R.R. 1983. A defense of the caridoid facies: wherein the early evolution of the Eumalacostraca is discussed. In F.R.Schram (ed.), *Crustacean Phylogeny.* Rotterdam: Balkema.

Holthuis, L.B. 1959. The Crustacea Decapoda of Suriname (Dutch Guiana). *Zool. Verh.* 44: 1-296.

Holthuis, L.B. 1980. Shrimps and prawns of the world. An annotated catalogue of species of interest to fisheries. *FAC Fish. Synop. (125)* Vol.1: 1-261.

Holthuis, L.B. & E.Gottlieb 1958. An annotated list of the decapod Crustacea of the Mediterranean coast of Israel, with an appendix listing the Decapoda of the eastern Mediterranean. *Bull. Res. Council Israel* 7B: 1-126.

Lemos de Castro, A. 1965. Crustáceos Isópodos Epicarídeos do Brasil. III. Gênero *Bopyrella* Bonnier (Isopoda, Bopyridae). *Acad. Brasil. Cienc. Ann.* 37: 283-288.

Manning, R.B. & L.B.Holthuis 1981. West African brachyuran crabs (Crustacea: Decapoda). *Smiths. Contr. Zool.* 306: 1-379.

Markham, J.C. 1972. Four new species of *Parathelges* Bonnier, 1900 (Isopoda, Bopyridae), the first record of the genus from the western Atlantic. *Crustaceana* Suppl.3: 57-78.

Markham, J.C. 1974. Parasitic bopyrid isopods of the amphi-American genus *Stegophyryxus* Thompson with the description of a new species from California. *Bull. Sth. Calif. Acad. Sci.* 73: 33-41.

Markham, J.C. 1975a. Extension of range and new host records for the parasitic isopod *Pseudione giardi* Calman in the northeastern Pacific. *Wasmann J. Biol.* 32: 195-201.

Markham, J.C. 1975b. Bopyrid isopods infesting porcellanid crabs in the northwestern Atlantic. *Crustaceana* 28: 257-270.

Markham, J.C. 1975c. Biological results of the University of Miami deep-sea expeditions. 113. A review of the bopyrid isopod genus *Munidion* Hansen, 1897, parasitic on galatheid crabs in the Atlantic and Pacific Oceans. *Bull. Mar. Sci.* 25: 422-441.

Markham, J.C. 1975d. New records of two species of parasitic isopods of the bopyrid subfamily Ioninae in the western Atlantic. *Crustaceana* 29: 55-67.

Markham, J.C. 1975e. Two new species of *Asymmetrione* (Isopoda, Bopyridae) from the western Atlantic. *Crustaceana* 29: 255-265.

Markham, J.C. 1977a. Description of a new western Atlantic species of *Argeia* Dana with a proposed new subfamily for this and related genera (Crustacea Isopoda, Bopyridae). *Zool. Meded.* 52: 107-123.

Markham, J.C. 1977b. The status and systematic position of the species of the bopyrid isopod genus *Phyllodurus* Stimpson, 1857. *Proc. Biol. Soc. Wash.* 98: 813-818.

Markham, J.C. 1979a. Epicaridean isopods of Bermuda. *Bull. Mar. Sci.* 29: 522-529.

Markham, J.C. 1979b. A new genus and species of bopyrid isopod infesting the crangonid shrimp *Pontophilus abyssi* Smith in deep water of the northwestern Atlantic Ocean. *Proc. Biol. Soc. Wash.* 92: 769-772.

Markham, J.C. 1982. Bopyrid isopods parasitic on decapod crustaceans in Hong Kong and southern China. In B.S.Morton & C.K.Tseng (eds.), *Proceedings of the First International Marine Biological Workshop: The marine flora and fauna of Hong Kong and Southern China.* Vol.1. Hong Kong: Hong Kong Univ. Press.

Markham, J.C. 1985. A review of the bopyrid isopods infesting caridean shrimps in the northwestern Atlantic Ocean with a special reference to those collected by the Hourglass Cruises in the Gulf of Mexico. *Mem. Hourglass Cruises.* 7(3): 1-156.

Markham, J.C. & J.J.McDermott 1981. A tabulation of the Crustacea Decapoda of Bermuda. *Proc. Biol. Soc. Wash.* 93: 1266-1276.

McLaughlin, P.A. 1974. The hermit crabs (Crustacea Decapoda, Paguridea) of Northwestern North America. *Zool. Verh.* 130: 1-396.

McLaughlin, P.A. 1983. Hermit crabs — are they really polyphyletic? *J. Crust. Biol.* 3: 608-621.

Nierstrasz, H.F. 1931. Die Phylogenie der Epicariden. *Zool. Anz.* 34: 123-128.

Nierstrasz, H.F. & G.A.Brender à Brandis 1929. Papers from Dr.Th.Mortensen's Pacific Expedition 1914-16. XLVIII. Epicaridea I. *Vidensk. Medd. Dansk naturh. Foren.* 87: 1-44.

Nierstrasz, H.F. & G.A.Brender à Brandis 1931. Papers from Dr.Th.Mortensen's Pacific Expedition 1914-16. LVII. Epicaridea II. *Vidensk. Medd. Dansk naturh. Foren.* 91: 147-226.

Palmer, A.R. 1969. Copepoda. In R.C.Moore (ed.), *Treatise on Invertebrate Paleontology. Part R. Arthropoda* 4. Lawrence: Geol. Soc. Am. & Univ. Kansas Press.

Pielou, E.C. 1974. Biogeographic range comparisons and evidence of geographic variation in host-parasite relations. *Ecology* 55: 1359-1367.

Plymouth Marine Fauna 1957. Third edition. Plymouth: Marine Biological Association.

Radwański, A. 1972. Isopod-infected prosoponids from the Upper Jurassic of Poland. *Acta Geol. Polonica* 22: 499-506.

Rathbun, M.J. 1904. Decapod crustaceans of the northwest coast of North America. In C.H.Merriam (ed.), *Harriman Alaska Ser.*10: 3-20.

Reinhard, E.G. 1949. Experiments on the determination and differentiation of sex in the bopyrid *Stegophryxus hyptius* Thompson. *Biol. Bull.* 96: 17-31.

Remeš, M. 1921. Excroissance des crustacés du Tithonique de Stramberk. *Internatl. Acad. Sci. Bohême. Bull.* 23: 36-37.

Reverberi, G. & M.Pitotti 1942. Il ciclo biologico e la determinazione fenotipica del sesso di *Ione thoracica* Montagu, bopiride parassita di *Callianassa laticauda* Otto. *Pubbl. Staz. Zool. Napoli* 19: 111-184.

Rice, A.L. 1983. Zoeal evidence for brachyuran phylogeny. In F.R.Schram (ed.), *Crustacean Phylogeny.* Rotterdam: Balkema.

Richardson, H. 1903. Isopods collected at the Hawaiian Islands by the U.S. Fish Commission Steamer Albatross. *Bull. U.S. Fish. Comm.* 23: 819-826.

Shiino, S.M. 1933. Bopyrids from Tanabe Bay. *Mem. Coll. Sci., Kyoto Imper. Univ.* (B)8: 249-300.

Shiino, S.M. 1934. Bopyrids from Tanabe Bay II. *Mem. Coll. Sci., Kyoto Imper. Univ.* (B)9: 258-297.

Shiino, S.M. 1952. Phylogeny of the family Bopyridae. *Ann. Rept. Pref. Univ. Mie. Sect.2, Nat. Sci.* 1: 33-56.

Shiino, S.M. 1965. Phylogeny of the genera within the family Bopyridae. *Bull. Mus. Natl. Hist. Nat. Paris* (2)37: 462-465.

Shiino, S.M. 1966. Phylogeny of the genera within the family Bopyridae. In A.Corradetti (ed.), *Proc. First Internat. Cong. Parasit.* 2. Oxford: Pergamon.

Shiino, S.M. 1972. [The Epicaridea (list of species) from Japan.] *Kansai Shizenkagaku* 24: 7-10 [in Japanese].

Sieg, J. 1983. Evolution of Tanaidacea. In F.R.Schram (ed.), *Crustacean Phylogeny.* Rotterdam: Balkema.

Siewing, R. 1963. Studies in malacostracan morphology: Results and problems. In H.B.Whittington & W.D.I.Rolfe (eds.), *Phylogeny and evolution of Crustacea.* Cambridge: Mus. Comp. Zool.

Squires, H.J. 1967. Decapod Crustacea from *Calanus* collections in Hudson Bay in 1953, 1954, and 1958-61. *J. Fish. Res. Bd. Canada* 24: 1873-1904.

Strömberg, J.-C. 1971. Contribution to the embryology of bopyrid isopods with special reference to *Bopyroides, Hemiarthrus,* and *Pseudione* (Isopoda, Epicaridea). *Sarsia* 47: 1-46.

Van Straelen, V. 1928. Contribution à l'étude des isopodes méso- et cénozoiques. *Mém. Acad. Roy. Belgique. Cl. Sci.* (2)9: 1-68.

Watling, L. 1983. Peracaridan disunity and its bearing on eumalacostracan phylogeny with a redefinition of eumalacostracan superorders. In F.R.Schram (ed.), *Crustacean Phylogeny.* Rotterdam: Balkema.

Wenner, E.L. & D.F.Boesch 1979. Distribution of epibenthic decapod Crustacea along the shelf-slope coenocline, middle Atlantic Bight, U.S.A. *Bull. Biol. Soc. Wash.* 3: 106-133.

Williams, A.B. 1965. Marine decapod crustaceans of the Carolinas. *Fish. Bull.* 65: 1-298.

Williams, A.B. 1974. Marine flora and fauna of the northeastern United States. Crustacea: Decapoda. *NOAA Tech. Rept. NMFS Circ-389*: 1-50.

Zariquiey Alvarez, R. 1968. Crustáceos Decápodos Ibéricos. *Inv. Presq.* 32: 1-510.

Distribution of the Tanaidacea: Synopsis of the known data and suggestions on possible distribution patterns

JÜRGEN SIEG
Abteilung Vechta, Universität Osnabrück, Federal Republic of Germany

ABSTRACT

As in other biological disciplines zoogeography can be divided into a descriptive and an analytic section. A synopsis of the known, but scattered, data allows remarks on faunistics as well as on systematic, ecological, and phylogenetic zoogeography.

Within faunistics, an overview of the established zoogeographic regions is given. The tanaidacean fauna of the main regions (cold water areas of the Antarctic and Arctic, northern and southern temperate waters, tropical warm water) are discussed. The Antarctic cold water area is characterized by having a much higher percentage of endemic elements (approximately 63%) than the Arctic (about 10%). There are only two or three species in these regions that show a bipolar distribution. Only a limited comparison of the tanaidacean faunas of the northern temperate and tropical warm water areas can be made, and relies mainly on data from the Atlantic Ocean, an area much better explored than any other region of the world. It can be shown that there exists a much closer relationship between the northern temperate waters and the tropical warm water than with the southern temperate area in either the Pacific and the Atlantic. On the other hand the main areas of the tropical warm water regions are each characterized by an independent tanaidacean species-composition. Only the Indo-West Pacific may be called (with some reservation) an interceding area. In total the Atlantic tanaidacean fauna shows a closer relationship to the Pacific than to the Indian Ocean.

The shallow-water family Tanaidae probably has a southern 'evolution center'. It is then shown that the phylogenetic method also allows an analysis of the geographic origin of the members of a special fauna. This is exemplified by the tanaid fauna of the Gulf of Mexico. Finally, the relationship between body size and depth is discussed. An analysis shows that the order Tanaidacea is characterized by a constant size reduction during evolution. Therefore, there is a much stronger correlation between size and the phylogenetic age of a taxon than with depth. Probably all known large deep-sea or cold water species of a particular taxon are plesiomorphic compared to the small members occurring in the same region.

1 INTRODUCTION

Although there has been much taxonomic work done on Tanaidacea in the past, this order is still unjustifiedly ignored for the most part in discussions of special biological fields, e.g. ecology or zoogeography. This is mainly a result of the scattered literature, a fact which makes it nearly impossible to determine all relevant data for a particular aspect. There also is a great lack of synopses as well as compilations for fields other than taxonomy. To rectify this, major works have been done recently at the familial or ordinal level. The family Neotanaidae has been revised by Gardiner (1975) and the Pseudotanaidae, as well as the Tanaidae, by Sieg (1977, 1980). In addition to these monographs there was a compilation of all known references to the order Tanaidacea (Sieg 1983). Concomitant with these efforts an EDP-data base has been established at the Universität Osnabrück, Abt. Vechta. One of the main goals of this data base is to give assistance in analyzing distributional patterns (Sieg 1984). In some cases first attempts allow both recognition and understanding of distributional patterns within a certain family (Sieg 1980, 1983a), or a species-composition within a given area (Sieg & Heard 1983). But there is still much 'undiscovered' information hidden in the literature. Therefore, one of the main tasks of this paper was to compile all data available and discuss their informational value. This, in some cases, provides a sound basis for discussion of other newly revealed details.

As in other biological disciplines biogeography can be divided into a descriptive and an analytical section. Descriptive zoogeography commonly has a statistical character and consists of the fields chorology, faunistics, and biocenotic and systematic zoogeography. Chorology, which is the description of all habitats settled by a species world wide, is unimportant in connection with this paper. More information is available within faunistics, which is primarily a 'statistic' of all species so far known and their distributional records. In doing such statistics only for a special taxon, e.g. a family or genus, we deal mainly with systematic zoogeography. Last, but not least, biocenotic zoogeography is the investigation of the distribution and dynamics of ecological systems. In this field there is nearly no information available within the Tanaidacea; thus we are forced to discuss only faunistics and systematic zoogeography in this section.

Analytic zoogeography, on the other hand, is subdivided into an ecological and a historical (or better phylogenetic, Brundin 1975) phase. Both views are linked and complement each other. There is no doubt that the recent distributional pattern of a species within a given region is limited by its ecological needs. But this indication may be quite restrictive, because it does not allow an explanation of the fact that within different regions characterized by the same abiotic factors a species may not always be present. In this case only phylogenetic zoogeography can give an answer. Thus, both points of view are not mutually exclusive and questions raised in connection with a recognized distribution pattern often can only be satisfactorily explained when taking into consideration ecological as well as phylogenetic zoogeography. Unfortunately, in both fields, available information for tanaidaceans is meager, but should increase in the future as our understanding of the phylogeny of the order increases.

2 DESCRIPTIVE ZOOGEOGRAPHY OF TANAIDACEANS

Because of the scattered information on tanaids there has been almost no compilation of species diversity for different areas. To fill this gap the Tanaidacea data base was established (Sieg 1984). In addition to the exact geographical co-ordinates for each location, a

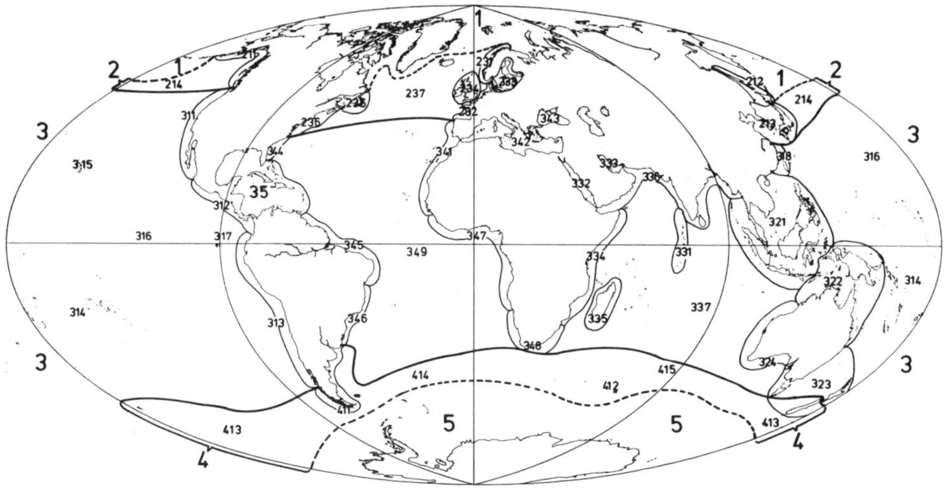

Figure 1. Zoogeographic regions proposed for the Tanaidacea-data base: 1 Arctic cold water area. 2 Arctic transition area; 2.1 North Pacific transition area; 2.1.1 Shelf of the west coast of Canada and Alaska; 2.1.2 Shelf of the east coast of Asia, Japan and southern Kurile Islands; 2.1.3 Deep water bottom of the North Pacific transition area; 2.2 North Atlantic transition area; 2.2.1 Scandinavian shelf (Norway, Denmark); 2.2.2 European shelf (Germany to Portugal); 2.2.3 Baltic Sea; 2.2.4 British Isles; 2.2.5 Gulf of St Lawrence and coasts of Newfoundland; 2.2.6 East coast of the USA; 2.2.7 Deep water bottom of the North Atlantic transition area. 3 Tropical warm water area; 3.1 Tropical warm water area of the Pacific Ocean: 3.1.1 Shelf of the west coast of the USA and northern Mexico; 3.1.2 Shelf of the west coast of Central America; 3.1.3 Shelf of the west coast of South America; 3.1.4 Island of Polynesia; 3.1.5 Hawaiian Islands; 3.1.6 Deep bottom water of the Pacific; 3.1.7 Galapagos Islands and islands near the coast of South and Central America; 3.1.8 Coast of Asia; 3.2 Indo-West Pacific (Islands of Melanesia, New Zealand and Australia); 3.2.1 Shelf of south-east Asia, Indonesia and the Philippine Islands; 3.2.2 Shelf of New Guinea and North Australia; 3.2.3 Shelf of south-east and eastern Australia, Tasmania and New Zealand; 3.2.4 Shelf of south-west and western Australia; 3.3 Tropical warm water area of the Indian Ocean; 3.3.1 Laccadive and Maldive Islands; 3.3.2 Red Sea; 3.3.3 Gulf of Iran and Oman; 3.3.4 Shelf of the east coast of Africa; 3.3.5 Shelf of Madagascar; 3.3.6 Shelf of the Arabian Peninsula, India, Bangladesh, and Burma; 3.3.7 Deep bottom water of the Indian Ocean; 3.4 Tropical warm water of the Atlantic Ocean; 3.4.1 Shelf of the Iberian Peninsula and North Africa; 3.4.2 Mediterranean Sea; 3.4.3 Black Sea; 3.4.4 Shelf of the east coast of the USA; 3.4.5 Shelf of the east coast of South America (northern part); 3.4.6 Shelf of the east coast of South America (southern part); 3.4.7 Shelf of the west coast of Africa; 3.4.8 Shelf of the southern part of Africa; 3.4.9 Deep bottom water of the Atlantic; 3.5 Shelf of the east coast of Central America and the West Indies. 4 Antarctic transition area; 4.1 Shelf of southern South America; 4.2 Kerguelen and other nearby subantarctic islands; 4.3 Subantarctic area of the Pacific; 4.4 Subantarctic area of the Atlantic; 4.5 Subantarctic area of the Indian Ocean. 5 Antarctic cold water area.

characterization by a zoogeographic province was necessary. For this purpose a 'zoogeographic code' was introduced (Fig. 1) which followed the concept of marine zoogeographic provinces as proposed by Ekman (1935) and De Lattin (1967). The great variety of combinations of these two specifications allows a very detailed analysis. Yet even when we are now able to compare the different zoogeographic regions, in many cases this proves a senseless undertaking, owing to the very meager knowledge available from most of these areas (see below). Therefore, only some relatively more interesting cases will be considered here. As specified above, these will be related to two main fields, i.e. faunistics, and systematic zoogeography.

Table 1. Comparison of species composition of Tanaidacea in Arctic and Antarctic cold waters.

Arctic cold water		Antarctic cold water	
Family	Species	Family	Species
Sphyrapidae	*Spyrapus anomalus*	Apseudidae	*Apseudes spectabilis*
	Pseudosphyrapus serratus		*A.paragracilis*
Agathotanaidae	*Agathotanais ingolfi*		*A.unicus*
Anarthruridae	*Paranarthrura insignis*	Whiteleggiidae	*Leviapseudes galatheae*
	P.subtilis		*L.gracillimus*
Leptognathiidae	*Collettea cylindrata*		*L.shiinoi*
	C.minima	Neotanaidae	*Carolangia mirabunda*
	Leptognathia amdrupi		*Neotanais affinis*
	L.arctophylax		*N.americanus*
	L.armata		*N.antarcticus*
	L.brachiata		*N.hadalis*
	L.breviremis		*N.hessleri*
	L.glacialis		*N.kurchatovi*
	L.gracilis		*N.magnificus*
	L.inermis		*N.tricarinatus*
	L.latiremis	Agathotanaidae	*Agathotanais hanseni*
	L.linearis	Anarthruridae	*Paranarthrura abbreviata*
	L.longiremis		*P.insignis*
	L.polita	Leptognathiidae	*Collettea antarctica*
	L.subaequalis		*Exspina typica*
	L.tenella		*Filitanais rebainsi*
	L.tuberculata		*Leptognathia antarctica*
	L.ventralis		*L.armata*
	L.vicina		*L.breviremis*
	Typhlotanais cornutus		*L.elongata*
	T.finmarchicus		*L.gallardoi*
	T.inermis		*L.gracilis*
	T.irregularis		*L.paraforcifera*
	T.mixtus		*L.robusta*
	T.mucronatus		*L.vanhoeffeni*
	T.spinicauda		*Libanius monacanthus*
	T.variabilis		*Mirandotanais vorax*
Paratanaidae	*Heterotanais groenlandicus*		*Tanaopsis antarctica*
Pseudotanaidae	*Cryptocopoides arcticus*		*Typhlotanais armatus*
	Pseudotanais abyssi		*T.greenwichensis*
	P.affinis		*T.macrocephala*
	P.forcipatus		*T.parangularis*
	P.lilljeborgi		*T.rectus*
	P.macrocheles	Nototanaidae	*Nototanais antarcticus*
	P.oculatus		*N.dimorphus*
			Protanaissus longidactylus
		Paratanaidae	*Pseudonototanais werthi*
		Pseudotanaidae	*Cryptocopoides arcticus*
			Pseudotanais abyssi
			P.gaussi
			P.longisetosus
			P.nordenskioldi
		Tanaidae	*Allotanais hirsutus*
6	40	10	48 Total

Table 2. Species composition of Tanaidacea in the northern Pacific temperate waters.

Family	Species	
Apseudidae	*Apseudes littoralis*	
	A. nipponicus	
	Parapseudes latifrons	
Kalliapseudidae	*Kalliapseudes tomiokaensis*	
Metapseudidae	*Apseudomorpha albida*	
	Synapseudes setoensis	
Whiteleggiidae	*Carpoapseudes serratispinosus*	
Neotanaidae	*Neotanais barfoedi*	
	N. hadalis	
Anarthruridae	*Anarthruropsis langi*	
Leptognathiidae	*Arthrura shiinoi*	
	Leptognathia birsteini	
	L. brevimanus	
	L. gracilis	
	L. grevae	
	L. longiremis	
	L. parelegans	
	L. robusta	
	L. rotundicauda	
	Typhlotanais magnificus	
	T. mucronatus	
	T. rectus	
	T. tuberculatus	
Nototanaidae	*Metatanais cylindricus*	
Paratanaidae	*Heterotanais groenlandicus*	
	Heterotanoides ornatus	
	Leptochelia dubia	
	Pseudonototanais werthi	
Pseudotanaidae	*Pseudotanais oculatus*	
Tanaidae	*Arctotanais alascensis*	
	Sinelobus stanfordi	
	Tanais vanis (?)	
	(Tanais loricatus)	
	Zeuxo coralensis	
	Z. kurilensis	
	Z. normani	
11	35 (36)	Total

2.1 *Faunistics*

2.1.1 *The cold temperate water regions*

In connection with planned studies of Antarctic tanaidaceans it was of some interest to have a comparison of species diversity in both polar provinces. Such a compilation reveals a nearly identical number of species (Table 1), but the families represented are considerably different. For the Arctic cold water area only six families are recorded while we find ten in Antarctic waters. The same holds true for the number of genera (10 vs. 20, Table 1). A similar result is obtained even in species richness (Sieg, unpublished data for a great number of samples). This diversity depends not on ecological, but more probably on strictly evolutionary factors. As investigations on the Tanaidae have shown (see 3.2.1)

Table 3. Species composition of Tanaidacea in the northern Atlantic temperate waters.

Family	Species	Family	Species
Apseudidae	*Apseudes acutifrons*	Leptognathiidae	*L.longiremis*
	A.graziliae		*L.manca*
	A.grossimanus		*L.multiserrata*
	A.latreilli		*L.ochracea*
	A.rotundifrons		*L.paramanca*
	A.setiferus		*L.polita*
	A.spinosus		*L.profunda*
	A.talpa		*L.subaequalis*
	A.tenuis		*L.unguicillata*
	A.vicinius		*L.ventralis*
	Francapseudes debilis		*L.voeringii*
	F.uniarticulatus		*Macrinella clavipes*
	Tuberapseudes echinatus		*Tanaopsis gracilioides*
Whiteleggiidae	*Carpoapseudes kudinovae*		*Typhlotanais aequiremis*
	C.simplicirostris		*T.assimilis*
	Leviapseudes gracilis		*T.brevicornis*
	L.gracillimus		*T.cornutus*
	L.segonzaci		*T.eximius*
	L.zenkevitchioides		*T.finmarchicus*
Sphyrapidae	*Pseudosphyrapus anomalus*		*T.inermis*
	P.centobi		*T.irregularis*
	P.serratus		*T.macrocephalus*
	Sphyrapus malleolus		*T.microcheles*
	(S.stebbingi)		*T.mananensis*
	S.tudes		*T.mixtus*
Neotanaidae	*Neotanais affinis*		*T.pinicillatus*
	N.americanus		*T.proctagon*
	N.hastiger		*T.richardi*
	N.insolitus		*T.tenuicornis*
	N.laevispinosus		*T.tenuimanus*
	N.micromorpher		*T.variabilis*
	N.peculiaris	Nototanaidae	*Tanaissus lilljeborgi*
	N.sandersi		*T.psammophilus*
Agathotanaidae	*Agathotanais ingolfi*	Paratanaidae	*Hargeria rapax*
Anarthruridae	*Anarthrura simplex*		*Heterotanais oerstedi*
	Siphonolabrum mirabile		*Leptochelia dubia*
Leptognathiidae	*Collettea cylindrata*		*Pseudoleptochelia filum*
	Exspina typica		*Paratanais limicola*
	Leptognathia angusta		*P.rigidus*
	L.arctophylax	Pseudotanaidae	*Cryptocope abbreviata*
	L.armata		*Pseudotanais affinis*
	L.brevimanus		*P.forcipatus*
	L.breviremis		*P.jonesi*
	L.dentifera		*P.lilljeborgi*
	L.filiformis		*P.macrocheles*
	L.gracilis		*P.similis*
	L.indivisa	Tanaidae	*Hexapleomera robusta*
	L.inermis		*Parasinelobus chevreuxi*
	L.latiremis		*Tanais dulongi*
		11	97 (98) Total

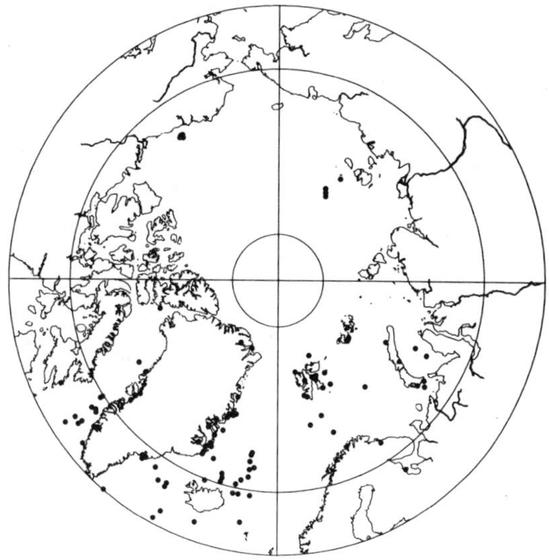

Figure 2. Map for all known tanaidacean records of the Arctic cold water area.

this seems (again) to indicate that the shallow water regions of the southern hemisphere have been something like an 'evolutionary center' for the Tanaidacea. We also have some hints that the majority of families may have evolved during Cretaceous times (Sieg 1980a, 1983a, Schram & Sieg in press), so that geological data on continental drift also support this opinion. I believe this concept is supported as well by the observed different species composition for both regions.

Of the 40 tanaid species so far known from the Arctic (Table 1) at least 28 (= 70%) also occur in the transitional areas in both the Atlantic and the Pacific. The majority (24 species, Table 3) are found in the northern temperate waters of the Atlantic, while just four species (Table 2) occur in the North-Pacific. Of the remaining 12 species one (*Collettea minima*) shows an 'indifferent' type of distribution and seven others (= 17.5%) (of which some are doubtful taxa) are known exclusively from their type localities. As there exists for *Leptognathia brachiata* two, and for *L.vicina* three, nearby loci typici, a broader range is suggested. The same seems to hold true for another species (*Leptognathia amdrupi*) which so far has been recorded from only two localities (Just 1970, Hansen 1913). All these examples show that the Arctic cold water area is characterized by a very low percentage of endemic species (maybe less than 10%). In this connection it might be interesting to recall that only three species (*Leptognathia armata, L.gracilis, Sphyrapus anomalus*) have reached very high northern latitudes. Although there have been various expeditions even into pack ice areas the overwhelming majority of records are found outside the polar basin (Fig.2). Presumably, the time for the arrival of the Tanaidacea from the south was far too short to conquer the marine geological barriers. This also seems to hold true in several cases for other peracaridan groups (Just 1980).

Finally, three of the species occurring in Arctic cold waters have been assumed to show a 'bipolar' type of distribution, but recent researches indicate that at least *Paranarthrura insignis* and (with some doubt) *Cryptocopoides arcticus*, may both have a wider range (Kudinova-Pasternak & Pasternak 1978). Therefore, at the moment only *Pseudotanais abyssi* is left to occur exclusively in Arctic and Antarctic cold waters.

Table 4. Species composition of Tanaidacea in the southern temperate waters.

Family	Species	
Apseudidae	*Apseudes antarcticus*	
	A.crozetensis	
	A.spectabilis	
Metapseudidae	*Cyclopoapseudes diceneon*	
	Metapseudes aucklandiae	
	Synapseudes ideos	
Neotanaidae	*Neotanais antarcticus*	
Leptognathiidae	*Leptognathia arnaudi*	
	L.australis	
	L.glandiceps	
	L.gracilis	
	L.lineata	
	L.luykeni	
	L.parabrevimanus	
	Tanaopsis antarctica	
	T.kerguelenensis	
	Typhlotanais greenwichensis	
	T.kerguelenensis	
	T.longus	
Nototanaidae	*Nototanais dimorphus*	
	Tanaissus lilljeborgi	
Paratanaidae	*Leptochelia barnardi*	
	L.dubia	
	Paratanais oculatus	
	Pseudoleptochelia antarctica	
	Pseudonototanais werthi	
Pseudotanaidae	*Paraiungentitanais longidigitatus*	
	Pseudotanais guillei	
	P.longisetosa	
	P.nordenskioldi	
Tanaidae	*Allotanais hirsutus*	
	Hexapleomera robusta	
	Langitanais bifidirostris	
	L.magnus	
	L.willemoesi	
	Pancoloides litoralis	
	Sinelobus stanfordi	
	(Tanais dulongi?)	
	Zeuxo phytalensis	
	Zeuxoides helleri	
	Z.ohlini	
	Z.pseudolitoralis	
8	41 (42)	Total

In comparison to the Arctic we have a much higher percentage of endemism for the Antarctic tanaid fauna, a fact which supports the idea that the fauna of the Antarctic is older and not recent as assumed by Kudinova-Pasternak & Pasternak (1981). Of the 48 species recorded (Table 1) at least 21 (= 44%) are endemic to the Antarctic. Of these, six species are known exclusively from their type localities. Within the remaining 27 species

14 (= 29%) show a less restricted type of distribution. It is one of the numerous peculiarities of tanaidacean dispersal that with the exception of the bipolar *Leptognathia gracilis*, none of these 14 species so far has been found in temperate waters of the southern oceans. In the case of *L.gracilis* we are really not sure at the moment that the specimens from the southern hemisphere are conspecific with those described from the northern hemisphere. Three others of the 48 species may also show bipolar distribution (see above). To these, *Typhlotanais macrocephalus* might be added, because it has been recorded from various places in the northern temperate waters of both Atlantic (Hansen 1913) and Pacific (Kudinova-Pasternak 1978), and from a single station in the Antarctic region (Vanhöffen 1914). Finally, there exist nine additional species occurring in cold (i.e. polar) as well as in southern temperate waters (adding *L.gracilis* would make the total about 21%). As none of these last mentioned species occur outside the regions in question [in his last paper Shiino (1978) no longer considered his *Apseudes nipponicus* as a synonym to *A.spectabilis*] they all may be considered as endemic to the southern hemisphere. Thus the Antarctic/Subantarctic endemism can be calculated as nearly 63%. This result is similar to the percentage calculated for isopods (66.3%, Kussakin 1967).

Taking into consideration the facts listed below, a characterization of endemism in the southern and northern temperate zones seems to be useless. Therefore, only an overview of the recorded species is given (Tables 2-4).

The transitional areas of the North Atlantic and the North Pacific have nearly nothing in common. Only four species occur in both regions, *Leptognathia brevimana, L.gracilis, L.longiremis* and *Leptochelia dubia.* These are distributed so widely that they must be considered separately. On the other hand such a great gap in species richness between the North-Pacific (35, Table 2) and North Atlantic (98, Table 3) can only be accepted as a result of our ignorance in the Pacific (in particular the Japanese) fauna.

The tanaidacean fauna of the southern temperate region also is very poorly known (Table 4), especially in the southern part of South America and the New Zealand area. Principal studies for the latter islands have shown that there are at least 60 species (Sieg unpublished data). This number is about 50% higher than the total number of species so far recorded for the whole zoogeographic province. However, much more astonishing is the fact that nearly two-thirds of these species are new to science.

2.1.2 *The tropical warm water regions*

The tropical warm water area may be subdivided into the Atlantic, Indian, Indo-West Pacific and Pacific subregion (Fig.1). In general this area is characterized by being much more diverse than the northern or southern temperate zones. Consequently this supports the idea that, in contrast to isopods and amphipods (Abele 1982), tanaidaceans may have their greatest number of species within the tropics. In addition, all four subregions are more or less independent in regard to species composition (Tables 5-8). This probably can be interpreted as a result of their special postmarsupial biology (Sieg 1983a). Consequently the number of species occurring throughout the tropical region is quite low. For the moment we know only three species having a broad distribution: *Leptochelia dubia, Sinelobus stanfordi*, and within some limits *Zeuxo coralensis.* The first two species also penetrate into southern and northern temperate waters.

As in the case of the temperate zones the impression again is that the Atlantic subregion has the most diverse tanaidacean fauna (172 species, Table 8). But as stated above this probably is a result of the much better exploration of this part of the world.

Table 5. Species composition of Tanaidacea in tropical warm Pacific waters.

Family	Species	Family	Species	
Apseudidae	*Apseudes diversus*	Anarthruridae	*Paranarthrura similis*	
	A.galapagensis		*P.undulata*	
	A.garthi		*Siphonolabrum mirabile*	
	A.meridionalis	Leptognathiidae	*Arthrura andriashevi*	
	A.rikiteanus		*Collettea minima*	
	A.seurati		*Exspina typica*	
	A.tropicalis		*Libanius longicephalus*	
	Parapseudes latifrons		*L.pulcher*	
	Pseudoapseudes cedroensis		*Leptognathia armata*	
	P.pernix		*L.birsteini*	
Kalliapseudidae	*Kalliapseudes crassus*		*L.forcifera*	
	K.viridis		*L.gracilis*	
Metapseudidae	*Apseudomorpha glebosus*		*L.longiremis*	
	A.magdalenensis		*L.microcephala*	
	A.oahuensis		*L.parabranchiata*	
	Cyclopoapseudes indecorus		*L.robusta*	
	Synapseudes dispina		*Typhlotanais angularis*	
	S.hancocki		*T.brachyurus*	
	S.intumescens		*T.compactus*	
	S.minutus		*T.elegans*	
	S.rudis		*T.grandis*	
Pagurapseudidae	*Hodometrica australis*		*T.kerguelensis (?)*	
	H.prolixa		*T.magnificus*	
	Pagurapseudes laevis		*T.rectus*	
Whiteleggiidae	*Carpoapseudes bacescui*		*T.setosus*	
	C.longissimus	Nototanaidae	*Nesotanais lacustris*	
	C.menziesi		*Teleotanais gerlachi*	
	C.oculicornutus	Paratanaidae	*Bathytanais bathybrotes*	
	C.serratispinosus		*(Heterotanais oerstedi)*	
	Leviapseudes aberrans		*Leptochelia dubia*	
	L.galatheae		*(L.erythraea)*	
	L.wolffi		*L.minuta*	
	L.zenkevitchi		*Pseudoleptochelia filum (?)*	
Neotanaidae	*Herpotanais birsteini*	Pseudotanaidae	*Cryptocopoides arcticus*	
	H.kirkegaardi		*Pseudotanais affinis*	
	Neotanais americanus		*P.malayensis*	
	N.armiger		*P.nordenskioldi*	
	N.barfoedi		*P.vitjazi*	
	N.calcarulus	Tanaidae	*Anatanais lineatus*	
	N.hadalis		*A.pseudonormani*	
	N.insignis		*Hexapleomera robusta*	
	N.pfaffi		*Pancolus californiensis*	
	N.robustus		*Sinelobus stanfordi*	
	N.sandersi		*Synaptotanais notabilis*	
	N.triangulocephalus		*Tanais gayi*	
	N.tuberculatus		*T.macrocheles*	
	N.vemae		*T.vanis*	
	N.wolffi		*Zeuxo coralensis*	
	N.zenkevitchi		*Z.marmoratus*	
Agathotanaidae	*Agathotanais hanseni*		*Z.normani*	
	Paragathotanais typicus		*Z.paranormani*	
Anarthruridae	*Anarthruropsis galatheae*		*Z.seurati*	
		13	102 (104)	Total

Even if all four tropical subregions are relatively independent at the species level we might expect the Indo-West Pacific fauna to be characterized as intermediate in numbers of species. Therefore, it is not surprising that the Indian Ocean (71 known species, Table 7) and the Pacific (102 known species, Table 5) only share three species, and these three occur nearly all over the world. There are also listed two other common species, *Pseudoapseudes pernix* and *Leptochelia minuta*, but the records of these from the Red Sea (Makkaveeva 1971) are doubtful and the specimens certainly belong to new and undescribed species. Therefore, without question, the Indian and the Pacific regions are each characterized by a totally different tanaidacean fauna. On the other hand, the intermediate position of the Indo-West Pacific fauna is well illustrated by the fact that this area shares nine of the 51 species so far recorded (= 17.5%) (Table 7) with the Pacific region. The relationship between the Indian Ocean fauna and that of the Indo-West Pacific is less striking. There exist at most five species (= 9.8%) that occur in both subregions. Of these, two species, *Leptochelia dubia, Sinelobus stanfordi*, are widely distributed, and a third, *Apseudes gallardoi*, is represented by a different subspecies in each area.

The Atlantic region, which includes the Caribbean, Gulf of Mexico and Mediterranean Sea, seems to have a closer zoogeographic relationship to the Pacific than to the Indian Ocean. At the moment there are 22 known species which occur in the tropical warm waters of both the Pacific and the Atlantic. Of these, five species are distributed widely and three more may eventually prove to have a broad distribution. Five of the remaining 14 species shared in the Atlantic have only been recorded from the Caribbean region. This supports the idea that in several cases colonization of that area took place from the Pacific (see 2.2.2). Of greater impact is the fact that the Atlantic tanaidacean fauna seems to have 16 species in common with that of the Indian Ocean. But this may give a misleading impression because at least six species appear to be doubtful taxa. That is, the specimens collected in Indian Ocean waters probably differ from those reported for the Atlantic and, therefore, may represent undescribed species. Three of the shared species also have a nearly world wide distribution. One species, *Cristapseudes omercooperi* occurs in the eastern part of the Mediterranean Sea as well as in the northern part of the Indian Ocean and thus may be interpreted as a Tethyan-relict.

Another interesting point is found when the zoogeographic affinities of the tropical warm water region are compared with the southern as well as the northern temperate areas.

For the Pacific Ocean fauna we find a significantly different percentage of species shared with the northern and southern transitional zone. Of the 102 reported species 15 (= 15%) also occur in northern temperate waters (Geo-code 2100) while at most four (= 4%) species are known from the southern transitional area. Of these, three species *Leptognathia gracilis, Leptochelia dubia, Sinelobus stanfordi* have a broad distribution, whereas of the fourth one (*Hexapleomera robusta*) we have only doubtful records from the southern temperate zone (Giambiagi 1923).

Similar results are revealed for the Atlantic. At the moment 32 of the 172 recorded species (= 19%) from tropical warm waters are also known from the northern temperate region. However, for the southern temperate zone only ten species are counted. Two of these ten species are again doubtful records (*Leptochelia barnardi, Paratanais oculatus*) and four additional species are characterized as belonging to those members occurring world wide. Three more species belonging to the genus *Zeuxoides*; *Z.helleri, Z.ohlini, Z.spongicola* may not really be a part of the tropical warm water fauna. All three have

Table 6. Species composition of Tanaidacea in tropical Indo-West Pacific waters.

Family	Species	
Apseudidae	*Apseudes caeruleus*	
	A.diversus	
	A.estuarius	
	A.gallardoi gallardoi	
	A.lagenirostris	
	A.latus	
	A.mussauensis	
	A.nagae	
	A.nhatrangensis	
	A.nigrifrons	
	A.setosus	
	Pakistanapseudes tenuicorporeus	
	Pseudosphyrapus dispar	
Gigantapseudidae	*Gigantapseudes adactylus*	
Kalliapseudidae	*Kalliapseudes obtusifrons*	
	K.primitivus	
Metapseudidae	*Apseudomorpha timaruvia*	
	Metapseudes aucklandiae	
	Synapseudes australianus	
Pagurapseudidae	*Hodometrica australis*	
	Macrolabrum boeri	
	Pagurapseudes abrucei	
	P.spinipes	
	Pagurapseudopsis gracilipes	
	P.conspicuus	
Whiteleggiidae	*Leviapseudes sibogae*	
	L.weberi	
	Pseudowhiteleggia typica	
	Whiteleggia multicarinata	
	W.stephensoni	
Neotanaidae	*Neotanais giganteus*	
	N.insignis	
	N.pfaffioides	
	N.triangulocephalus	
Anarthruridae	*Paranarthrurella caudata*	
Leptognathiidae	*Leptognathia longiremis*	
Nototanaidae	*Nototanais dimorphus*	
Paratanaidae	*Bathytanais bathybrotes (?)*	
	Leptochelia dubia	
	L.ignota	
	L.longimana	
	L.minuta	
	L.mirabilis	
	Paratanais elongatus	
	P.linearis	
	P.tenuis	
Pseudotanaidae	*Pseudotanais nordenskioldi*	
Tanaidae	*Anatanais novaezelandiae*	
	Sinelobus stanfordi	
	Synaptotanais abyssorum	
	Tanais dulongi	
	T.tenuicornis	
13	51 (52)	Total

Table 7. Species composition of Tanaidacea in tropical Indian waters.

Family	Species	Family	Species
Apseudellidae	*Apseudella typica*	Metapseudidae	*Tanzanapseudes elegans*
Apseudidae	*Apseudes agulhensis*		*T.langi*
	A.austroafricanus (?)		*T.longiseta*
	A.babelmandebensis	Pagurapseudidae	*Indoapseudes brycesoni*
	A.chilkensis		*Macrolabrum trichopteroides*
	A.espinosus (?)		*Pagurapseudes dactylifrons*
	A.djiboutiensis		*P.tricolae*
	A.grossimanus		*P.varians*
	A.portonovensis		*Pagurapseudopsis gymnophobia*
	A.srilankaensis		*P.iranica*
	A.tuberculatus	Tanapseudidae	*Tanapseudes ormuzana*
	A.unicus	Whiteleggiidae	*Leviapseudes gracillimus*
	Halmyrapseudes killaiyensis		*L.shiinoi*
	H.thaumastocheles		*L.sibogae*
	Msangia larvoides	Neotanaidae	*Neotanais affinis*
	Pakistanapseudes leptochelatus		*N.bacescui*
	P.shiinoi	Leptognathiidae	*Leptognathia dissimilis*
	Pseudoapseudes pernix		*L.paraforcifera*
Cirratodactylidae	*Bacescapseudes adenaicus*		*Typhlotanais peculiaris*
Kalliapseudidae	*Cristapseudes omercooperi*	Nototanaidae	*Tanaissus lilljeborgi (?)*
	Kalliapseudes borceai	Paratanaidae	*Leptochelia barnardi*
	K.makrothrix		*L.dubia*
	K.makrothrixoides		*L.erythraea*
	K.mauritanicus (?)		*L.minuta (?)*
Metapseudidae	*Apseudomorpha hirsuta*		*L.mirabilis (?)*
	A.veleronis		*Paratanais monodi*
	Cryptapseudes acutifrons		*Pseudoleptochelia anomala (?)*
	C.sankarankuttyi		*P.crassicornis*
	Cyclopoapseudes estafricanus		*P.mergellinae (?)*
	Curtipleon carinatoides	Pseudotanaidae	*Pseudotanais gerlachi*
	C.carinatus		*P.oculatus (?)*
	Synapseudes acroporae	Tanaidae	*Sinelobus stanfordi*
	S.comorinensis		*Zeuxo coralensis*
	S.menziesi		*Z.kurilensis*
	S.makkaveevae		*Z.westwoodiana*
	S.violaceus		
		14	62 (71) Total

been exclusively recorded from the coasts of South Africa, a zoogeographic area which belongs much more to the temperate zone than to the tropics (De Lattin 1967).

Finaly, the Indian Ocean only shares two common species, *Leptochelia dubia* and *Sinelobus stanfordi*, with the southern temperate zone.

2.2 *Systematic zoogeographic aspects*

The taxonomy of Tanaidacea is still so unstable that only limited statements can be made in this field. In many cases, such as *Leptognathia* or *Apseudes* we have at the moment

Table 8. Species composition of Tanaidacea in tropical warm Atlantic waters.

Family	Species	Family	Species
Anuropodidae	*Anuropoda francispoori*	Kalliapseudidae	*K.* sp.
Apseudidae	*Apseudes acutifrons*		*Psammokalliapseudes granulosus*
	A. africanus		*P. mirabilis*
	A. agulhensis	Metapseudidae	*Apseudomorpha avicularia*
	A. armatus		*Metapseudes* sp.
	A. austroafricanus		*Synapseudes cystoseirae*
	A. bermudeus		*S. heterocheles*
	A. browni		*S. mediterraneus*
	A. caeca		*S. shiinoi*
	A. cooperi	Pagurapseudidae	*Pagurapseudes bouryi*
	A. deltoides		*P. guitari*
	A. digitalis		*P. largoensis*
	A. elisae		*Parapagurapseudopsis carinatus*
	A. espinosus	Sphyrapidae	*Sphyrapus malleolus*
	A. graciloides	Whiteleggiidae	*Carpoapseudes auritocheles*
	A. grossimanus		*Leviapseudes leptodactylus*
	A. holthuisi		*L. sibogae*
	A. inermis		*L. wolffi*
	A. intermedius		*L. zenkevitchi*
	A. latreilli		*Paraleiopus typicus*
	A. minutus		*Whiteleggia multicarinata*
	A. misarai	Neotanaidae	Carolangia mirabunda
	A. paulensis		*Neotanais affinis*
	A. propinquus		*N. armiger*
	A. robustus		*N. dinotomer*
	A. sicilianus		*N. giganteus*
	A. spinosus(?)		*N. hadalis*
	A. talpa		*N. hastiger*
	A. tenuimanus		*N. mesostenoceps*
	A. triangulata		*N. micromopher*
	A. uncidigitatus		*N. persephone*
	Atlantapseudes nigrichela		*N. sandersi*
	Calozodion wadei		*N. tricarinatus*
	Discapseudes holthuisi		*N. tuberculatus*
	D. surinamensis		*N. vemae*
	Fageapseudes retusifrons		*Venustricus glandurus*
	Halmyrapseudes bahamensis	Agathotanaidae	*Agathotanais hanseni*
	H. cubanensis		*A. ingolfi*
	H. spaansi		*Paragathotanais typicus*
	Parapseudes latifrons	Anarthruridae	*Anarthrura simplex*
	Trichapseudes tridens		*Paranarthrura insignis*
	Tuberapseudes echinatus		*P. kurchatovi*
	Typhlapseudes nereus		*P. similis*
Cirratodactylidae	*Cirratodactylus floridensis*		*P. subtilis*
Kalliapseudidae	*Cristapseudes omercooperi*	Leptognathiidae	*Collettea cylindrata*
	Hemikalliapseudes hanstroemi		*C. vermiformes*
	Kalliapseudes bahamensis		*Exspina typica*
	K. dentatus		*Leptognathia abyssorum*
	K. gianucai		*L. angustocephala*
	K. magnus		*L. armata*
	K. mauritanicus		*L. brevimana*
	K. schubarti		*L. breviremis*

Table 8. (continued).

Family	Species	Family	Species
Leptognathiidae	*L.dissimilis*	Paratanaidae	*Pseudoleptochelia anomala*
	L.filiformis		*P.inermis*
	L.forcifera		*P.longidactyla*
	L.gracilis		*P.magnus*
	L.indivisa		*P.mercantilis*
	L.longiremis		*P.mergellinae*
	L.manca		*P.mortenseni*
	L.unguicillata		*P.provincialis*
	Paratyphlotanais typicus	Pseudotanaidae	*Iungentitanais primitivus*
	Pseudoparatanais batei		*Pseudotanais borcei*
	Tanaopsis graciloides		*P.kurchatovi*
	Typhlotanais kussakini		*P.macrocheles*
	T.longimanus		*P.mediterraneus*
	T.messinensis		*P.mortenseni*
	T.richardi		*P.nordenskioldi*
	T.rotundirostris		*P.unicus*
	T.spiniventris	Pseudozeuxidae	*Pseudozeuxo belizensis*
Nototanaidae	*Androtanais beebei*	Tanaidae	*Hexapleomera robusta*
	Teleotanais gerlachi		*Monoditanais helenensis*
Paratanaidae	*'Gammarus heterocheles'*		*Parasinelobus chevreuxi*
	'Hoplocope oculatus'		*Sinelobus stanfordi*
	Hargeria rapax		*Singularitanais annectens*
	(Heterotanais groenlandicus)		*'Tanais brasiliensis'*
	Heterotanais oerstedi		*Tanais dulongi*
	Heterotanoides capensis		*T.grimaldii*
	Leptochelia barnardi		*Zeuxo coralensis*
	L.dubia		*Z.exsargasso*
	L.forresti		*Z.fresii*
	L.'longichelipes'		*Z.kurilensis*
	L.tenuicola		*Z.simonsiensis*
	Mesotanais dubius		*Z.vanhoeffeni*
	Paratanais euelpis		*Zeuxoides helleri*
	P.oculatus(?)		*Z.ohlini*
	P.spinanotandus		*Z.spongicola*
		17	164 (172) Total

only general ideas of the characters which may actually define these genera. Future research may show that these taxa have to be split into more than two genera. Therefore, depending on the potentially high probability of dealing with poly- or paraphyletic taxa, nearly nothing can be said on distribution patterns for most of the genera. Conversely, after descriptions of many new families during the last three decades the taxonomy on this level now seems to be much more stable. So it is not surprising that the only certain facts within the systematic zoogeography of the tanaidaceans refer to families.

Logically, there are two ways to look at distribution patterns: the horizontal and vertical viewpoint. Because the Anuropodidae, Apseudellidae, Gigantapseudidae, Pseudozeuxidae, and Tanapseudidae are monotypic, these are out of the question. As Cirratodactylidae are very nearly related to Kalliapseudidae (Sieg 1983a) both families are treated together here.

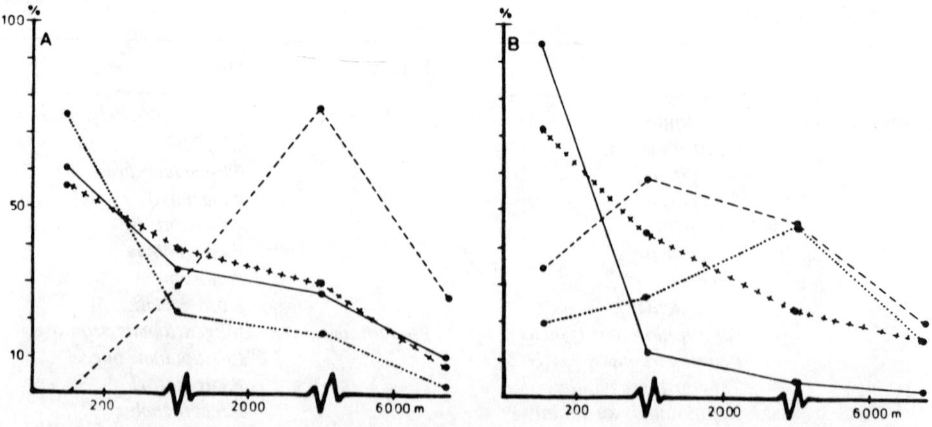

Figure 3. Depth distribution of various taxa within the Tanaidacea. (A) Depth distribution of the order and suborders; ——— Tanaidacea; ...Apseudoidea; — Neotanaidomorpha; +++ Tanaidomorpha. (B) Depth distribution of various families within the Tanaidomorpha; ——— Tanaidae/Paratanaidae; ——— Leptognathiidae; +++ Pseudotanaidae; ...Anarthruridae.

Table 9. Depth distribution of tanaidacean suborders. A. Compilation demonstrating overlapping depth-range of species; B. Compilation for the commonly used bathymetric zones. In this case compilation may total more than 100 % because those species occurring in more than one bathymetric zone are added to each zone separately.

A

Depth	Apseudoidea	Neotanaidomorpha	Tanaidomorpha	Percentage
0 - 200	118	–	98	45.5
200 - 2 000	15	3	35	11.1
2 000 - 6 000	17	18	39	15.6
>6 000	–	5	13	3.8
0 - 2 000	16	–	37	11.2
0 6 000	–	–	8	1.7
200 - 6 000	8	5	12	5.3
0 - >6 000	1	–	6	1.5
200 - >6 000	–	2	6	1.7
2 000 - >6 000	2	2	9	2.7
Σ	177	35	263	

B

Depth	Tanaidacea No.	%	Apseudoidea No.	%	Neotanaidomorpha No.	%	Tanaidomorpha No.	%
0 - 200	284	59.8	135	76.3	0	0.0	149	56.6
200 - 2 000	157	33.0	40	22.6	10	28.6	104	39.5
2 000 - 6 000	134	28.2	28	15.8	27	77.1	80	30.4
>6 000	46	9.7	3	1.7	9	25.7	21	8.0

2.2.1 *Distribution with depth*

The only tanaidacean taxon commonly mentioned so far in connection with vertical distribution is the family Neotanaidae (Wolff 1956, Gardiner 1975, Abele 1982). However, Sieg (1983a) recently gave a broader overview on several additional tanaidomorph families.

Generally the sea bottom is arbitrarily divided into shallow water (0-200 m), bathyal (200-2000 m), abyssal (2000-6000 m) and hadal (below 6000 m) zones. For the majority of crustaceans there is a decrease of species richness with increasing depth (Abele 1982: 251). Some exceptions are the harpacticoid copepods (Coull 1972), desmosomatid isopods (Hessler 1970) and, again, the Neotanaidae (Gardiner 1975) within the Tanaidacea. As always, this is a rough generalization and upon closer examination this opinion, especially within tanaidaceans, must be used conditionally.

Considering the total number of tanaidaceans there also is a negative correlation between number of species and depth (Fig.3, Table 9). Most of the species so far described occur in shallow waters (about 60%), but in depths greater than 6000 m we still find 46 species (about 10%).

Even the three recent suborders show different patterns (Fig.3, Table 9). The Apseudoidea seem to be mainly a shallow water group, the Neotanaidomorpha are classically deep-sea taxa, whereas the Tanaidomorpha show a pattern similar to the whole order. However again, subordinal taxa (e.g. families) are each characterized by a typical bathymetrical distribution.

Within the Apseudoidea the families Metapseudidae, Kalliapseudidae, and Pagurapseudidae are shallow-water forms. The Sphyrapidae probably occur in bathyal and abyssal depths (Fig.4). Within the Whiteleggiidae the subfamily Leviapseudinae is a typical deep water group (the systematic position of the only known shallow water species, *Paraleiopus macrochelis*, is uncertain); in contrast to this the subfamily Whiteleggiinae are restricted to shallow waters. This might provide substantiation for raising both subfamilies to familial level (Gutu 1972, 1982). However, another family, the Apseudidae, seems to extend from shallow waters down to nearly 6000 m. But it must be cautioned that many of the species recorded from the greater depths may not really belong to this family. If so, this taxon may occur only down to bathyal depths.

The lone family Neotanaidae, of the suborder Neotanaidormorpha, so far has been considered as 'the typical' deep-sea taxon. The recognized distributional pattern has been discussed first by Wolff (1956) and then in more depth by Gardiner (1975). Because of this only the estimated figures are listed here (Table 9).

The data for Tanaidomorpha (Fig.4) have been reviewed recently (Sieg 1983a). Tanaidae, Paratanaidae and Nototanaidae are (with some exceptions) shallow water taxa. Pseudotanaidae and Anarthruridae reach from shallow waters to hadal depths, and Agathotanaidae occur from bathyal to hadal depths. The Leptognathiidae show increasing species richness progressing into bathyal depths which gradually decreases again in abyssal and hadal depths (Table 10). This pattern is similar to that observed for Neotanaidae, so there is no doubt that the Leptognathiidae have to be characterized as a typically deep-sea family.

The synopsis also shows that the 200 m line is no ecological limit (Ekman 1935). As easily seen from Table 9, the number of species occurring from 200-2000 m is nearly the same as the number of species distributed in depth from 0-2000 m. This holds true for both the Apseudoidea (15/16 species) and Tanaidomorpha (35/37 species).

A relatively large number of species (contrary to most other crustaceans) also appear to have broad depth ranges. So far, six species (*Collettea cylindrata, Leptognathia breviremis,*

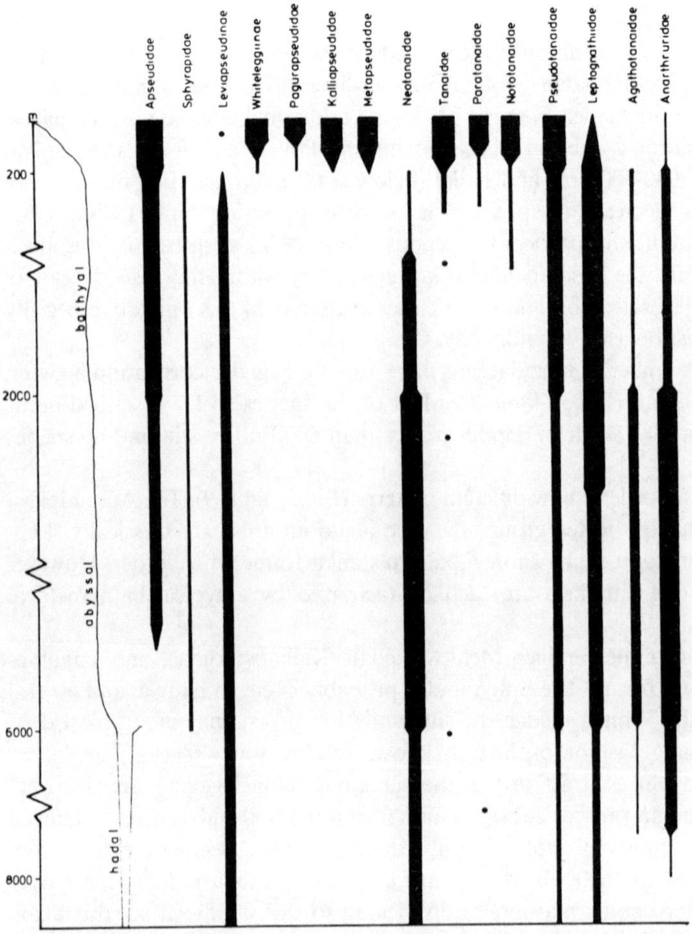

Figure 4. Depth distribution of major families of recent Tanaidacea.

Table 10. Depth distribution of tanaidomorphan families.

Depth	Tanaidae No.	Tanaidae %	Leptognathiidae No.	Leptognathiidae %	Pseudotanaidae No.	Pseudotanaidae %	Anarthruridae No.	Anarthruridae %	Paratanaidae No.	Paratanaidae %
0 - 200	38	95.1	44	33.8	18	72.0	3	20.0	40	95.2
200 - 2 000	5	12.1	77	59.2	11	44.0	4	26.7	4	9.5
2 000 - 6 000	2	4.8	61	46.9	6	24.0	7	46.7	0	0.0
>6 000	1	2.4	26	20.0	4	16.0	3	20.0	1	2.4

L.dentifera, L.gracilis, L.longiremis and *Pseudotanais affinis*; = 1.5%) are known to occur in both shallow waters as well as in hadal depths. An additional six species (*Langitanais willemoesi, Collettea antarctica, Leptognathia forcifera, L.ventralis, Typhlotanais irregularis* and with some doubt *Bathytanais bathybrotes*; = 1.7%) are known from depths of 0-6000 m. At least 25 other species occur in depths between 200 and 6000 m. In total, therefore, nearly 13% of all known species inhabit more than one depth range (Table 9).

2.2.2 *Remarks on the origin of the tanaidacean deep-sea bottom assemblage*

The origin of the deep-sea crustacean fauna has always been of special interest. Some believe that the deep sea may be a refugium for archaic forms. Beurlen (1931) supported the idea that the colonization of bathal and abyssal depths by decapods is the result of interaction of regression and transgression of the seas. On the other hand, Bruun (1956) thought that the decline of temperature at the end of the Tertiary period led to the gradual extinction of the deep-sea fauna, followed by a recolonization from shallow water areas. The data from the Tanaidacea seem to support Beurlen's hypothesis but also support facts recorded by Menzies & Imbrie (1958) which were critically discussed by Zenkevitch & Birstein (1960). In contrast to Wolff (1956), who also thought the deep-sea to be a refugium for archaic forms, Sieg (1983a) showed that the tanaid fauna is (as in decapods) a composite of plesiomorphic (archaic) and apomorphic (young) taxa.

Even if our knowledge of the phylogenetic relationships within the superfamily Apseudoidea is very preliminary (Sieg 1983a) it is certain that this taxon has colonized the deep-sea several times. For example, the Gigantapseudidae is an archaic group while the Leviapseudidae, Sphyrapidae, and Apseudidae are relatively apomorphic forms. In this connection, the Leviapseudinae are less apomorphic than the Apseudidae or Sphyrapidae.

The deep sea, however, can be suggested as a refugium for neotanaidomorphs (Wolff 1956). This phylogenetically old suborder has been replaced in shallow waters by the more advanced Tanaidomorpha (Sieg 1983a). Even within the Tanaidomorpha it is easily shown that the deep sea has been colonized by forms derived from shallow water species. Analysis of phylogenetic relationships in this group leads to the opinion that the Leptognathiidae is an apomorphic family in comparison to Tanaidae or Paratanaidae.

In comparing the relative age of the shallow water against the deep-sea tanaid fauna it must be admitted that both areas contain a nearly equal percentage of archaic and young taxa. There is thus no indication that abyssal and hadal depths are a special refuge for tanaidacean taxa.

2.2.3 *Geographical distribution of tanaidacean families*

At first glance horizontal distribution may be less informative than distribution with depth, because all families seem to occur more or less world wide (but see 3.2.1). In some cases a taxon reaches physically colder water areas (not higher latitudes) in the northern Pacific than in the northern Atlantic. This might be a consequence of the relatively late geological formation of the latter ocean.

Within the Apseudoidea the families Kalliapseudidae and Pagurapseudidae only inhabit the shallow areas in the tropical warm water region. This is probably true even though both taxa have not yet been recorded from several tropical zoogeographic areas in which our knowledge remains very scanty for the whole order. With further studies members of these families will certainly be detected. The Pagurapseudidae for example have so far not been recorded from the Pacific coast of South America and from the entire Atlantic coast of Africa. In the Kalliapseudidae, on the other hand, there exist scattered records for nearly all areas except the islands of the North and South Pacific as well as the Indian Ocean. It is worth mentioning here that the Kalliapseudidae are also known from one species (*Kalliapseudes tomiokaensis*) in Southern Japanese waters, an area which belongs to the North Pacific transitional zone. This holds true also for the Metapseudidae (*Synapseudes setoensis*) and demonstrates the exceptional conditions of that part of the world. The Metapseudidae otherwise seem to occur mainly in shallow waters of the tropical region. But the Metapseudidae differ from the two above mentioned families in that it is also recorded from several places in the southern transitional area (e.g. southern part of New Zealand, southern South America). In this connection it is noteworthy that Shiino (1978) did not record any member of the Metapseudidae from the Kerguelen Islands and other southern Indian Ocean islands. Moreover, within the Whiteleggiidae the subfamily Whiteleggiinae seem to be restricted to the shallow waters of the east coast of Australia and to South Africa. Again the subfamily Leviapseudinae is found in deeper waters of the world oceans except the most northern part of the Atlantic. Little is known of the newly established family Sphyrapidae. This taxon is relatively common in deeper waters of the North Atlantic and has been additionally recorded from the Mediterranean, Tasman Sea and sporadically from the coast of South Africa. The Apseudidae probably show the widest distribution within Apseudoidea. Members are found all over the world (except Arctic cold waters) from shallow waters down to abyssal depths.

The Neotanaidae also show a world-wide distribution (except the shallow water areas of the continents). Because of the relatively great age of this group and the late geological development of the polar basin no members were apparently able to settle within that area.

Of the Tanaidomorpha the families Tanaidae, Leptognathiidae, Paratanaidae, and probably Pseudotanaidae have a nearly world-wide distribution. The Tanaidae impressively demonstrate that tanaids reach much colder waters in the Pacific than in the Atlantic. There is no record for that family above the North Atlantic transitional area, whereas at least three species (*Zeuxo kurilensis, Arctotanais alascensis* and *Sinelobus stanfordi*) occur in the Kuriles or Aleutian Islands, both of which are situated in Arctic cold waters. Similar facts are known from the Paratanaidae. Two species (*Heterotanais modestus* [*Pseudonototanais werthi*] and *Paratanais impressus*) are known from the southern Kuriles while only *Heterotanais groenlandicus* reaches the west coast of Greenland. The Leptognathiidae are found all over the world including both polar regions. Occurring from sublittoral waters down to hadal depths there are a few areas from which this family has not yet been recorded. For the Pseudotanaidae there are also scattered records for most parts of the world. It has been recorded from Arctic as well as from Antarctic cold waters (Sieg 1977) so the family surely has a world wide distribution. Missing records for several zoogeographical areas may be collecting artifacts resulting from losing the relatively small sized specimens (0.8-1.5 mm) through too large net mesh-width or their being overlooked during the sorting process.

The Agathotanaidae, Anarthruridae and Nototanaidae are characterized by a very tenuous distribution pattern at the moment, but it might be that they also occur world wide. The Nototanaidae may be a special case, having a concentration of species in the Antarctic cold waters as well as in the southern transitional area.

3 ANALYTIC ZOOGEOGRAPHY OF TANAIDACEA

3.1 *Ecological zoogeography*

Nearly nothing is known in this field and much remains to be done. Because of poor taxonomic literature, tanaidaceans are normally not treated in ecological research even if they play an important role. However, by looking through the literature we can recognize that members of many families which have nearly the same depth range and occur in the same area still occupy different ecological niches (Sieg 1983a). For example, within the shallow water tanaidomorphs the Tanaidae are mainly associated with algal mats and hydroid colonies. The Paratanaidae are found primarily in sandy bottoms and when recorded from algal mats these contain a great deal of sand or detritus. The Nototanaidae, the third shallow water group, has a preference for muddy bottoms. This also holds true for the Apseudoidea. Kalliapseudids are filter feeders on sandy or muddy bottoms, metapseudids inhabit mainly coral reefs or hydroid colonies, and pagurapseudids use gastropod shells.

Finally a short consideration of eye development and distribution with depth should be done. There is no doubt that the percentage of crustacean species with eyes decreases with depth, but the pattern observed in tanaidaceans seems to be slightly different from that of isopods (Menzies et al. 1968). First, there exists a strong relationship between taxonomic categories and presence or absence of eyes. Within the Apseudoidea the Apseudidae, Whiteleggiinae, Pagurapseudidae, Kalliapseudidae, and Metapseudidae have representatives in intertidal waters. And of tanaidomorphs the Tanaidae, Paratanaidae, Nototanaidae and Pseudotanaidae occur more or less frequently in the same depth. Of these the Tanaidae, Paratanaidae, Whitelegiinae, Pagurapseudidae, and Metapseudidae generally have well-developed eyes. The Kalliapseudidae is the sole family in intertidal waters having no visual elements, whereas in the Apseudidae, Nototanaidae, and Pseudotanaidae we find both species having, and species lacking, eyes.

Correlation between taxonomical categories and presence of eyes is also clearly demonstrated by two other shallow water tanaidomorph families. Some members of the Tanaidae which have been caught in great depths (*Synaptotanais abyssorum*, 1301 m, Nierstrasz 1913; *Langitanais willemoesi*, 2925 m and *L.magnus* 3025 m, both Shiino 1978; *Protanais birsteini*, 6090-6135 m, Kudinova-Pasternak 1970) in all cases had well-developed eyes (although in *P.birsteini* there is still some doubt whether there exist visual elements or not). This also holds true for the Paratanaidae. The sole species so far recorded from a depth greater than 250 m (*Heterotanoides ornatus*, 7370 m) has well-developed facets (Kudinova-Pasternak 1977:117, Fig.1). Thus this is the only known hadal crustacean having eyes.

Can it be argued from the Apseudidae, Nototanaidae, and Pseudotanaidae that there is a correlation between eyes and depth range? The observed patterns are confusing and can only be satisfactorily understood by suggesting a relationship to depth which is obscured by other factors, such as reduction of eyes in shallow water species or secondary immigration of eyeless species into intertidal waters. First of all, within the three above mentioned families no species with well-developed eyes has been recorded from depths greater than 1500 m, which corresponds to the results of Menzies & George (1967:718, Table 7) for isopods. On the other hand in all taxa there occur eyeless species in intertidal waters. For the Pseudotanaidae eight species with eyes have been described so far. These all occur in shallow waters (0-125 m). Of the 16 additionally known species with reduced eye lobes, only five have been mentioned exclusively from depths greater than 200 m. The remaining 11 species all occur in very shallow waters, but sometimes show a remarkably broad depth range (*Cryptocopoides arcticus* 18-5631 m, *Pseudotanais affinis* 40-6890 m, Sieg 1977, 1983). This holds true also for the Nototanaidae. So far only two genera (*Protanaissus, Tanaissus*) of this imperfectly known family do not have eyes and again, one species (*Tanaissus lilljeborgi*) is characterized by a relatively broad depth range (intertidally, Holdich & Jones 1983; to 1105-1660 m?, Kudinova-Pasternak 1975). In the

Apseudidae (which is a very heterogeneous group and which probably must be split into three families, Sieg 1983a) we get similar results. *Halmyrapseudes* and *Discapseudes* demonstrate this in a very good manner. Both genera live in the same intertidal habitat (sometimes these occur together at one and the same spot, Bacescu & Gutu 1975), but *Halmyrapseudes* has well-developed eyes while these are lacking in *Discapseudes.* Within the complex genus *Apseudes* the recorded number of intertidal species without visual elements is low (about four) compared to those having facets (about 16).

Thus the final conclusion for tanaidaceans is that the percentage of species with eyes decreases markedly with depth, but for the intertidal region it is distinctly less (only about 80-85%) compared to the 100% of eyed forms that have been compiled for isopods (Menzies et al. 1968). Yet even in greater depths (abyssal or sometimes hadal depths), sporadic occurrence of eye-bearing species can be observed. This also seems to be true for isopods in abyssal depths (Menzies et al. 1968, Fig.1).

3.2 *Phylogenetic zoogeography*

Owing to a much better knowledge of the evolutionary process within the Tanaidacea a first attempt for a clearer understanding of observed geographical distribution patterns can now be done. This holds true both for the distribution of a particular taxon (e.g. Tanaidae) or the analysis of the origin of species found in a certain area (e.g. Gulf of Mexico). The phylogenetic approach also allows a first interpretation of correlation of size distribution (vertical as well as horizonal).

3.2.1 *Distribution of Tanaidae*

Within the Tanaidacea the family Tanaidae recently has been shown to be a model for geographic vicariance (Sieg 1980, 1983a). Generally, all stages of tanaids, even mancas, are tube dwellers or tube builders. Thus, these tubes are left only occasionally, and specimens seldom occur in the water column, even at night (Hobson & Chess 1976, 1978, Valentin 1978). Therefore, it would require a great deal of time to extend the area of distribution of any particular species over greater distances. For this reason geographical vicariance provides an important method for explaining distributions. Distribution modes which mainly depend on the influence of secondary factors, e.g. water current, pelagic larval stages, etc. may therefore be less important.

There is no doubt that the Tanaidae is a 'shallow water' family distributed world wide between the polar circles, although there exist a few records for depths greater than 200-300 m (see above, Sieg 1983a). In infrafamilial taxa, e.g. subfamilies, vicariance may also explain distributions. Phylogenetically old taxa are only recorded from the southern cold water or transitional areas, whereas phylogenetically young taxa occur mainly in tropical warm water regions (Fig.5). This holds true even for different genera within the tribe Anatanaini (Fig.6).

However, it has also been questioned whether the observed distribution of the three described species of *Anatanais* also follows this rule (Sieg 1983a). The data are weak and need to be tested by the study of additional material.

3.2.2 *Origin of the tanaidaceans of the Gulf of Mexico*

In connection with recent studies of the tanaidaceans of the Gulf of Mexico, it was necessary to develop some ideas on the origin of the species so far recorded. This proposal, described

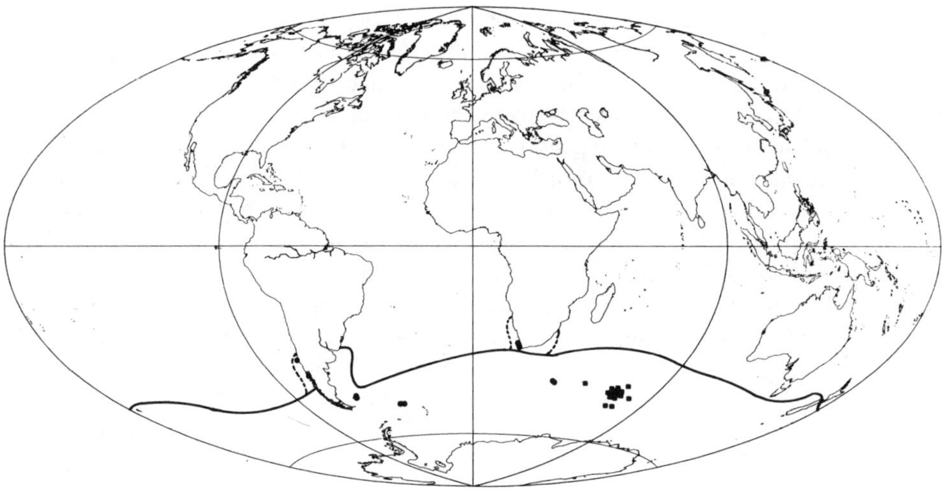

Figure 5. Distribution of plesiomorphic subfamilies within the Tanaidae: ● Archaeotanainae; ■ Langitanainae. The line indicates the border between tropical warm water and southern temperate water; note, the west coast of South Africa may be an exception because of the influence of the Benguela current.

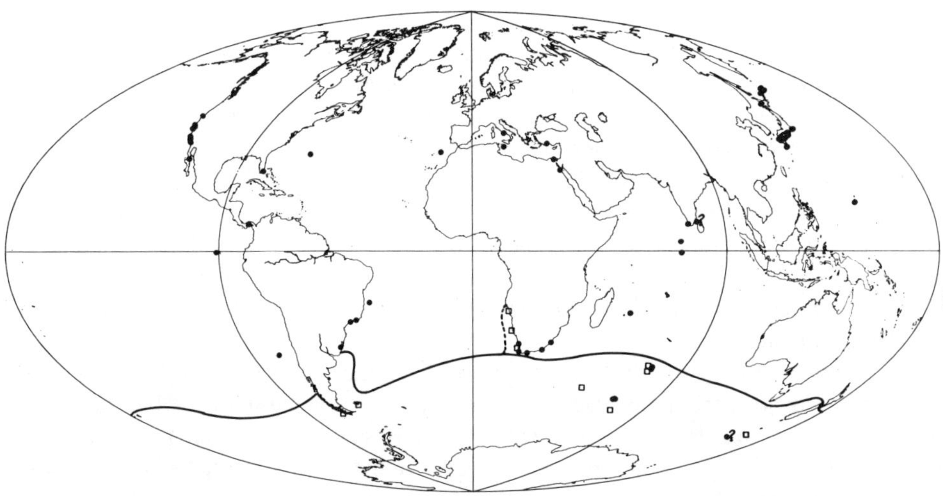

Figure 6. Distribution of the plesiomorphic genus *Zeuxoides* and the apomorphic genus *Zeuxo* of the tribe Anatanaini: ● *Zeuxo*; □ *Zeuxoides*; ? indicates records which may belong to *Zeuxo* or *Zeuxoides*.

below (see also Sieg & Heard 1983) must be accepted as a working hypothesis and needs to be tested by further studies.

Each fauna can be divided into endemic and non-endemic elements. In the case of endemic species information about geographic origin can only be suggested if the sister-group is known. For species having a broader distribution, the pattern itself allows conclusions regarding from which area the immigration may have taken place.

It is necessary to keep in mind that the Gulf of Mexico has had a very fluctuating geological history. Recently, a short review of the different ideas has been given by Hedges (1983). More complex and complete analyses are found in Pilger (1981). According to the available facts the first contact between the Gulf of Mexico and the Pacific was in the lower Jurassic. This primary exchange was undisturbed at last until the Upper Cretaceous. The eastward movement of the 'proto-Antillean archipelago' along the Cayman trough possibly initiated limited exchange with the Tethys Sea. In later times immigration could only take place through the isthmus of Panama and along the Yucatan Peninsula. This last connection was definitively closed in late Pliocene times.

As immigration of species into the Gulf of Mexico took place at various geological times from different geographic areas a complex species composition could be expected. In addition to species showing a nearly world-wide distribution (e.g. *Leptochelia dubia, Sinelobus stanfordi* etc.) at least four additional elements were detected: 1) Species which have a dispersion restricted mainly to the eastern Pacific (along the coast of the Americas) and in the Gulf of Mexico. As a result of the geological events, two subgroups should be expected: a) phylogenetically old genera and b) phylogenetically young species. 2) Species which reached the Gulf of Mexico from the South Atlantic. 3) Species which can be considered as members of a former 'Tethys-fauna'.

Although the facts available at the moment are relatively meager we have hints that all these groups are represented in the recent tanaid fauna. However, a lot of additional faunstic and taxonomic work has to be done for verifying these suggested distributional patterns.

Neotanaidae are considered to be a phylogenetically old taxon, one which probably evolved during Jurassic times. Their characteristic occurrence only in the deep sea may be interpreted as a secondary event because the family probably originally was distributed in shallow water. Looking through the excellent compilation given by Gardiner (1975) we find that so far five species have been recorded from the Gulf of Mexico/Caribbean Sea. Of these, one species (*N.hadalis*) shows a scattered, but world-wide distribution, and a second one (*N.persephone*) is only known from the Puerto-Rican Trench. Compared to these, *Neotanais armiger* shows a dispersion which seems to agree with point 1 (Fig.8A). For the two remaining species (*N.tuberculatus, N.vemae*) we must await the collection of additional material.

We are more successful when searching for examples of phylogenetically young taxa which have invaded from the Pacific. Again, we can distinguish two subgroups. First of all, there should exist species for which the time since immigration took place was too short for speciation. Thus the same species should be found on the Pacific coasts (at least of Central America) and in the Gulf of Mexico/Caribbean Sea. So far, I can only list an undescribed species of the genus *Paratanais.* However, there are several additional species in the Gulf of Mexico which may be identical with those described from different regions of the Pacific coast of Central America (Heard pers. comm.). A second group should contain taxa more or less endemic to the Caribbean Sea/Gulf of Mexico, and have their sister-group on the Pacific coasts of the Americas. These species may be called 'geminate species'. A very characteristic example of this group is found within the genus *Kalliapseudes.* The subgenus *Mesokalliapseudes* was previously known to consist of two species (*Kalliapseudes viridis, K.crassus*) from the Pacific coast of Central America. Recently a new species (*K.bahamensis*) has been described from the Bahamas (Sieg 1983) and is now recognized as having a wide distribution within the eastern Gulf (Heard pers. comm.).

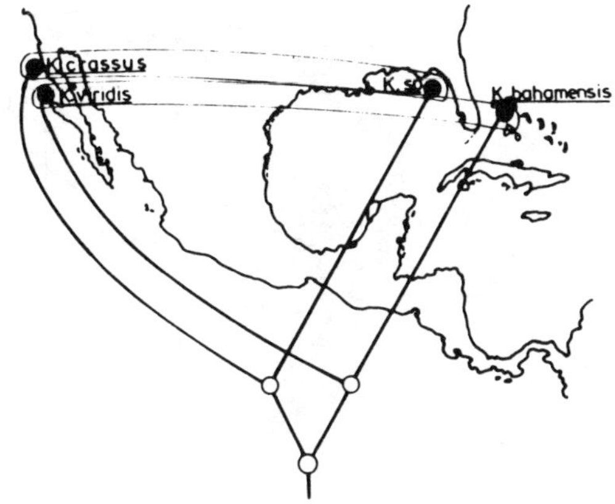

Figure 7. Distribution of the subgenus *Meso kalliapseudes* (genus *Kalliapseudes*).

This species is the nearest relative to *K.viridis* (Fig.7). The subgenus becomes much more interesting when it is known that the second Pacific species (*K.crassus*) is most nearly related to a species occurring in the Gulf of Mexico (this species is also undescribed). This pattern (Fig.7) only becomes understandable if one accepts a geologically late immigration. The same seems to hold true for *Pagurapseudes leavis/P.largoensis* (McSweeny 1982) and several other species (unpublished data, personal observation).

Because the 'Antillean archipelago' has always been a barrier for deeper water species from the Atlantic or Tethys Sea, there will only be shallow water species by which the two remaining points from our list may be verified. During preparation of a revision of the genus *Paratanais* it became clear that *P.euelpis* may be an example of a species which could have invaded from the South Atlantic (Fig.8B). Finally, *Tanais dulongii* very recently recorded from the eastern Gulf (unpublished data) might be called a 'Tethyan element', but there exist some records outside the main area which make this questionable (Fig.8C).

3.2.3 *Distribution and size*

Two main patterns are discussed: the relationship between latitude and size, and between depth and size. For the latter, sometimes the expression 'deep-sea gigantism' has been used (Wolff 1956). Before discussing the facts within the Tanaidacea it is necessary to give some general remarks.

First of all, I believe that there is no possibility for applying data from the species level to higher taxonomic categories. This simply means that the postulated phenomena for the correlation between body size and latitude (Abele 1982:254-256) which in many species undoubtedly exists, does not automatically explain an analogous observation at the family level. All the mainly physiological reasons invoked for the explanation of larger size of organisms at high latitude have no effect on the evolutionary pattern within a family. Against this it might be argued that there exist some examples (Amphipoda, Barnard 1962; Ostracoda, Poulsen 1965) which show a striking relationship between latitude and size, but caution is necessary in all these cases. As we will see, tanaidaceans also would fit into this pattern if we neglected phylogenetic relationships as has been done in those groups listed above. Genealogy has mostly been neglected when attempting to correlate an increase in body size with increasing depth (Belyaev 1972, Wolff 1956a,b). The situation in this case is especially contradictory because Menzies & George (1967) denied 'deep-sea gigantism' for isopods in the first place.

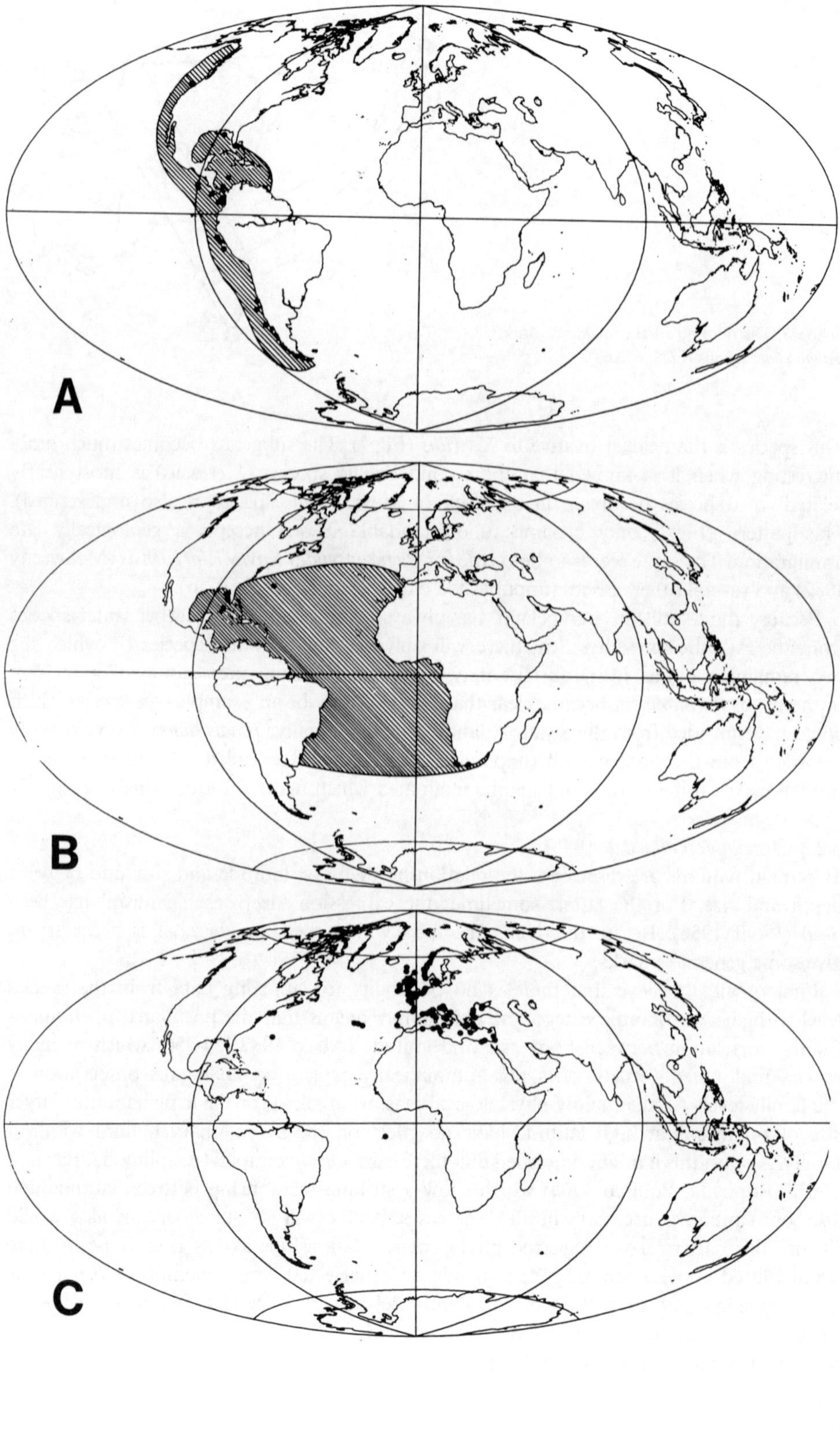
A
B
C

Gradual increase of body size has usually been discussed in connection with orthogenesis, a term introduced by Haacke (1893). The idea of increasing size as an evolutionary process in many taxa is mainly referred to in the works of Cope (1896), Gaudry (1896), and Deperet (1909). Rensch (1947), Remane (1952), and Osche (1966) have discussed this problem critically. Rensch especially drew attention to the insight that an increase of size leads to a better efficiency which obviously gives advantage in natural selection to an organism. Some smaller groups were mentioned which seem to be an exception (e.g. some insects) but it was additionally stated that in most cases the tiny insects may have been derived from larger forms (Rensch 1947:199). Remane's (1952:301-303) statements were much more critical. He gave, especially for arthropods, a long list of exceptions. Within insects, he was able to show for nearly every order cases which contravened the hypothesis. In the taxon Chelicerata the mites are among the most important representatives and (again) these are characterized by a decrease of body size. For Crustacea Remane mentioned groups within Copepoda, Ostracoda, and Cladocera which are contrary to the hypothesis. Later Siewing (1963) added the amphipod family Ingolfiellidae.

Normally within these listed taxa only the smaller subgroups stand opposite to Cope's rule. However, this seems to be totally different in the Tanaidacea where the whole order shows size reduction during evolution. Although this statement may be less positive, based on studies of the suborder Tanaidomorpha, future researches certainly will confirm the general affirmation.

For example, at this time we need only to ascertain that all known fossil forms are larger than any recent species. However, even if we are not able to clarify the final relationship of the presently established apseudoidean families, the taxon considered as the most plesiomorphic (Gigantapseudidae) nonetheless contains the largest species.

As the Neotanaidae is a phylogenetically old taxon (Sieg 1980, 1983a), it is not surprising that it is characterized by large or very large species. The average body size is about 10.0 mm, and the largest species *Herpotanais kirkegaardi* measures 25.0 mm (Gardiner 1975). Even the smallest taxon with an adult length of 4.4 mm (*Neotanais hessleri* with 3.8 mm total body length is only represented by a manca-stage) is larger than the average individual within any tanaidomorphan family. Gardiner (1975:249) discussed very critically 'deep-sea gigantism' and 'overgrowth' in Antarctic waters. As he has shown there possibly may be a correlation between depth and size, but a definitive statement must await the collection of additional material. Unfortunately, Gardiner (1975) has not given any information on the phylogenetic relationships within the Neotanaidae. Therefore, the assumption given above that body size is only correlated to the phylogenetic age of a taxon cannot be definitely proved for this family at the moment. So far as we know *Herpotanais* seems to be the most plesiomorphic genus. On the other hand, the genera *Venustricrus* and *Carololangia* are not only apomorphic compared to *Neotanais*, they are more or less small. With a body size of 8.9 mm or 11.8 mm, these species fit within the average size of individuals in *Neotanais.* If this rule holds true the phylogenetic analysis

Figure 8. (A) Distribution of *Neotanais armiger* (Wolff 1956). Hatching indicates possible distribution pattern, represented by phylogenetically old-plesiomorphic-taxa; (B) Distribution of *Paratanais euelpis* Barnard, 1920. Hatching indicates possible distribution pattern of a species which has invaded from the south; (C) Distribution of a possible Tethyan-element *Tanais dulongii* (Audouin 1826).

of the relationship of *Neotanais* species should reveal that the large species (e.g. *N.robustus, N.hadalis, N.pfaffi*) are plesiomorphic.

Within the Tanaidomorpha we also observe a continuous decrease in body size from phylogenetically old toward phylogenetic young taxa. This holds true not only at the family level but within a family as well. Measurements show that the average body length is about 4-5 mm for Tanaidae, 3-4 mm for Paratanaidae, and about 2-3 mm for Leptognathiidae. Finally, the Pseudotanaidae only show a body size of 0.8-3 mm (average 1.5 mm).

Research on the phylogenetic relationships below family level have only been done for the Tanaidae (Sieg 1980) and Pseudotanaidae (Sieg 1977). The members of the Pseudotanaidae are already so small that measurements of length are relatively inaccurate. No definitive statement, therefore, can be made. In this respect, then, the Tanaidae differ. The family can be subdivided into two main subfamilial lines (Sieg 1980) which both show continuous size reduction.

One branch is represented by the Archaeotanainae, Sinelobinae and Tanainae. The largest species are *Allotanais hirsutus* (10-18 mm) and *Singularitanais annectens* (8-10 mm). Both belong to the most plesiomorphic subfamily Archaeotanainae. *Singularitanais* has to be considered as apomorphic compared to *Allotanais.* Within the Sinelobinae a similar situation is found for *Sinelobus stanfordi* (4-6 mm) and *Parasinelobus chevreuxi* (6-8 mm). In the Tanainae, *Synaptotanais* (6-8 mm) and *Protanais* (8.7 mm) are plesiomorphic compared to *Tanais.* So far as we know at the moment, even within the genus *Tanais*, we can observe this tendency. *Tanais dulongii* is the largest species, followed by *T.grimaldii* and *T.vanis.* This corresponds exactly with the phylogenetic relationship suggested by Sieg (1980:83).

The second line is represented by Langitanainae and Pancolinae. Again the plesiomorphic Langitanainae, represented by *Langitanais* (16-20 mm) and *Arctotanais* (8-10 mm), are distinctly larger than the Pancolinae (3-5 mm). Whether this also holds true for all pancolinian genera cannot be clarified at this time, but within the subgenus *Zeuxo sensu stricto*, the plesiomorphic species *Z.simonsiensis* and *Z.vanhoeffeni* are much larger than the apomorphic *Z.fresii.* These examples impressively show that correlation of body size and distribution can be clarified when phylogenetic relationships are also taken into consideration.

Because plesiomorphic taxa of Tanaidae occur in high latitudes of the southern hemisphere, one could easily and falsely suppose an Antarctic 'overgrowth' for this family. Therefore, before further speculations on the 'deep-sea gigantism' of Neotanaidae are made the phylogenetic relationship of the species has to be defined. When this takes place we might be able to obtain results analogous to those known from deep-sea isopods (Menzies & George 1967).

ACKNOWLEDGEMENTS

The author expresses his deepest appreciation to Dipl.-math. W.Bauch who did most of the programming for the computerized documentation of the Tanaidacea including the subroutine TANMAP. The ideas contained herein are the result of many years of ongoing study, and have profited by discussion with innumerable colleagues. These studies were mainly supported by grants from the Deutsche Forschungsgemeinschaft, whose help is gratefully acknowledged. Special thanks are also due to the editors, R.H.Gore and K.L.Heck, who critically read the manuscript and offered many helpful suggestions.

REFERENCES

Abele, L.G. 1982. Biogeography. In D.E.Bliss (ed.), *The Biology of Crustacea.* Vol.1. New York: Academic Press.

Bacescu, M. & M.Gutu 1975. A new genus (*Discapseudes* n.g.) and three new species of Apseudidae (Crustacea, Tanaidacea) from the north-eastern coast of South America. *Zool. Mededel.* 49: 95-113.

Barnard, J.L. 1962. South Atlantic abyssal amphipods collected by R.V.Vema. In J.L.Barnard et al. (eds.), *Abyssal Crustacea.* New York: Columbia University Press.

Belyaev, G.M. 1972. *Hadal bottom fauna of the world ocean.* Inst. Oceanogr. Acad. Sci. USSR (Isr. Program Sci. Transl. Jerusalem).

Beurlen, K. 1931. Die Besiedlung der Tiefsee. *Natur und Museum* 61: 269-278.

Brundin, L. 1975. Phylogenetische Biogeographie als Glied der Evolutionsbiologie. *Verh. dtsch. Zool. Ges.* 1972: 372-380.

Bruun, A.F. 1956. The abyssal fauna: Its ecology, distribution and origin. *Nature* 177: 1105-1108.

Cope, E.E. 1896. *The primary factors of organic evolution.* Chicago: Univ. Chicago Press.

Coull, B.C. 1972. Species diversity and faunal affinities of meiobenthic Copepoda in the deep sea. *Mar. Biol.* 14: 48-51.

De Lattin, G. 1967. *Grundriss der Zoogeographie.* Jena: Fischer.

Déperét, C. 1909. *Die Umbildung der Tierwelt.* Stuttgart: Schweizerbart.

Ekman, S. 1935. *Tiergeographie des Meeres.* Leipzig: Akad. Verlagsgesellschaft.

Gambiagi, D. 1923. Dos isopodos nuevos pare la fauna Argentina. *Physis* 6: 363-364.

Gardiner, L.F. 1975. The systematics, postmarsupial development, and ecology of the deep-sea family Neotanaidae (Crustacea: Tanaidacea). *Smithson. Contr. Zool.* 170: 1-264.

Gaudry, A. 1896. *Essai de paléontologie philosophique.* Paris: Masson Editeurs.

Gutu, M. 1972. Phylogenetic and systematic considerations upon the Monokonophora (Crustacea, Tanaidacea) with suggestion of a new family and several new subfamilies. *Rev. Roum. Biol.* 17: 297-305.

Gutu, M. 1982. A new contribution to the systematics and phylogeny of the suborder Monokonophora (Crustacea, Tanaidacea). *Trav. Mus. Hist. nat. 'Grigore Antipa'* 23: 81-108.

Haacke, W. 1893. *Gestalt und Vererbung. Eine Entwicklungsmechanik der Organismen.* Leipzig: T.O.Weigel Nachfolger.

Hansen, H.J. 1913. Crustacea Malacostraca I. *Danish Ingolf Exped.* 3(3): 1-145.

Hedges, S.B. 1983. Caribbean biogeography: Implications of recent plate tectonic studies. *Syst. Zool.* 31: 518-522.

Hessler, R.R. 1970. The Desmosomatidae (Isopoda, Asellota) of the Gay Head-Bermuda transect. *Bull. Scripps Inst. Oceanogr.* 15: 1-185.

Hobson, E.S. & J.R.Chess 1976. Feeding oriented movements of the atherinid fish *Pranesus pinguis* at Majuro Atoll, Marshall Islands. *Fish. Bull.* 71: 777-786.

Hobson, E.S. & J.R.Chess 1978. Trophic relationship among fishes and plankton in the lagoon at Enewetak Atoll, Marshall Islands. *Fish. Bull.* 76: 133-153.

Holdich, D.M. & J.A.Jones 1983. The distribution and ecology of British shallow-water tanaid crustaceans (Peracarida, Tanaidacea). *J. Natur. Hist.* 17: 157-183.

Just, J. 1970. Decapoda, Mysidacea, Isopoda, and Tanaidacea from Jørgen Brølund Fjord, North-Greenland. *Medd. Grønl.* 194(9): 1-32.

Just, J. 1980. The malacostracan fauna of the abyssal North Polar Basin (Abstract). In J.G.Griffin (ed.), *International Conference. Biology and Evolution of Crustacea, 19-23 May, 1980.* Sydney: Australian Museum.

Kudinova-Pasternak, R.K. 1970. Tanaidacea kurilokamciatkogo jeloba. *Trudy Inst. Okeanol.* 86: 341-380.

Kudinova-Pasternak, R.K. 1973. Tanaidacea (Crustacea, Malacostraca) collected on the R/V 'Vitjaz' in regions of the Aleutian trench and Alaska. *Trudy Inst. Okeanol.* 91: 141-168.

Kudinova-Pasternak, R.K. 1975. Tanaidacea (Crustacea, Malacostraca) from the Atlantic sector of Antarctic and Subantarctic. *Trudy Inst. Okeanol.* 103: 194-228.

Kudinova-Pasternak, R.K. 1977. Glubokovodnaja donnaja fauna tichoyo okeana. *Trudy Inst. Okeanol. (Sirsoova)* 99: 115-125.

Kudinova-Pasternak, R.K. 1978. Tanaidacea (Crustacea, Malacostraca) from the deep-sea trenches of the western part of the Pacific. *Trudy Inst. Okeanol.* 108: 115-135.

Kudinova-Pasternak, R.K. & F.A.Pasternak 1978. Deep-sea Tanaidacea (Crustacea, Malacostraca) collected in the Caribbean sea and Puerto Rico Trench during the 16th cruise of R/V 'Akademic Kurchatov' and the resemblance between the fauna of deep-sea Tanaidacea of the Caribbean region and the Pacific. *Trudy Inst. Okeanol.* 113: 178-197.

Kudinova-Pasternak, R.K. & F.A.Pasternak 1981. Tanaidacea collected by the Soviet Antarctic expedition. *Trudy Inst. Okeanol.* 115: 108-125.

Kussakin, O.G. 1967. Isopoda and Tanaidacea from the coastal zones of the Antarctic and Subantarctic. *Explor. Fauna Seas* 4: 220-380.

Makkaveeva, E.B. 1971. Kacestvenii sostav i kolicestvennoe raspredelemie Tanaidovik rakov v Krasnom more. In Anonymous (ed.), *Bentos sel'fra krasnogo morja.* Kiev.

McSweeney, E.S. 1982. A new *Pagurapseudes* (Crustacea, Tanaidacea) from Southern Florida. *Bull. Mar. Sci.* 32: 455-466.

Menzies, R.J. & R.Y.George 1967. A re-evaluation of the concept of hadal or ultra-abyssal fauna. *Deep-sea Res.* 14: 703-723.

Menzies, R.J. & J.Imbrie 1958. On the antiquity of the deep-sea bottom fauna. *Oikos* 9: 192-201.

Menzies, R.J., R.Y.George & G.Rowe 1968. Vision index for isopod Crustacea related to latitude and depth. *Nature* 219: 93-95.

Nierstrasz, H.F. 1913. Die Isopoden der Siboga-Expedition. I. Isopoda Chelifera. *Siboga Exp.* 32(A): 1-56.

Osche, G. 1966. Grundzüge der allgemeinen Phylogenetik. In L.Bertalanffy & F.Gessner (eds.), *Handbuch der Biologie* 3: 817-906.

Pilger, R.H. (ed.), 1981. *The origin of the Gulf of Mexico and the early opening of the central North Atlantic Ocean.* Houston: Houston Geological Society, Continuing Education Series.

Poulsen, E.M. 1965. Ostracoda-Myodocopa. Part II: Cypridiniformes-Rutidermatidae, Sarsiellidae and Asteropidae. *Dana Rept.* 12(65): 1-484.

Remane, A. 1952. *Die Grundlagen des natürlichen System, der vergleichenden Anatomie und der Phylogenetik.* Leipzig: Akademische Verlagsgesellschaft.

Rensch, B. 1947. *Neuere Probleme der Abstammungslehre. Die transspezifische Evolution.* Stuttgart: Enke Verlag.

Shiino, S.M. 1978. Tanaidacea collected by French scientists on board the survey ship 'Marion Dufresne' in the regions around the Kerguelen Islands and other subantarctic Islands in 1972, '74, '75, '76. *Sci. Rept. Shima Marineland* 5: 1-122.

Sieg, J. 1977. Taxonomische Monographie der Pseudotanaidae Sieg (Crustacea, Tanaidacea). *Mitt. zool. Mus. Berlin* 53: 4-109.

Sieg, J. 1980a. Taxonomische Monographie der Tanaidae Dana, 1849 (Crustacea, Tanaidacea). *Abh. Senckenberg. naturf. Gesell.* 537: 1-267.

Sieg, J. 1980b. Sind die Dikonophra eine polyphyletische Gruppe? *Zool. Anz.* 200: 401-416.

Sieg, J. 1983a. Pars 6: Tanaidacea. In L.B.Holthuis & H.E.Gruner (eds.), *Catalogus Crustaceorum.*

Sieg, J. 1983b. Evolution of Tanaidacea. In F.R.Schram (ed.), *Crustacean phylogeny.* Rotterdam: Balkema.

Sieg, J. 1984. Fact documentation and literature database for the crustacean order Tanaidacea. *The Systematics Association, Special Volume* 26: 125-136.

Sieg, J. & R.W.Heard 1983. Distribution patterns of Tanaidacea in the Caribbean Sea and the Gulf of Mexico (Abstract). *Assoc. Southern Biol. Bull.* 30: 225.

Siewing, R. 1963. Zur Morphologie der aberranten Amphipodengruppe Ingolfiellidae und zur Bedeutung extremer Kleinformen für die Phylogenie. *Zool. Anz.* 171: 76-91.

Valentin, C. 1978. Zur Biozönotik und Substratpräferenz von Benthoscrustaceen: Ein Beitrag zur Biologie und Autökologie der Cumaceen aus dem Küstenbereich der Insel Ischia (Golf von Neapel/Italien). Thesis Christian-Albrechts-Universität Kiel.

Wolff, T. 1956a. Crustacea Tanaidacea from depths exceeding 6000 meters. *Galathea Rept.* 2: 187-241.

Wolff, T. 1956b. Isopoda from depths exceeding 6000 meters. *Galathea Rept.* 2: 85-156.

Wolff, T. 1962. The systematics and biology of bathyal abyssal Isopoda Asellota. *Galathea Rept.* 6: 7-320.

Zenkevitch, L.A. & J.A.Birstein 1960. On the problem of the antiquity of the deep-sea fauna. *Deep-sea Res.* 7: 10-23.

Caribbean biogeography and a biological calendar for geological events

JAN H.STOCK
Institute of Taxonomic Zoology, University of Amsterdam, Netherlands

ABSTRACT

It is difficult to reconcile geological data on the age of Caribbean islands (for biogeographic purposes understood here as the time elapsed since their emergence above sea-level), with the supposed Mesozoic age of a vicariant origin, through fragmentation of a proto-Antillean archipelago. The Lesser Antilles in particular seem to be much younger. Several conflicting biogeographic examples are discussed. A possible method of dating Caribbean islands by means of species/area curves for stygobiont Malacostraca is pointed out.

1 INTRODUCTION

The present paper is mainly based on the study of stygobiont (= groundwater) Malacostraca of the West Indies. During a ten year program our team systematically sampled 43 islands, and a total of 1800 samples have been obtained (Stock 1983). Although the groundwater biota of the West Indies is not yet completely known, the present set of data form a sufficient foundation for biogeographic discussions.

The study of stygobionts for biogeographic purposes has certain advantages, the most important of which is the geological continuity of groundwater. In periods of drought, of low sea-levels and of glaciation, groundwaters have persisted. One of the best examples is the existence of a varied groundwater fauna in regions that are or have been glaciated. Up to a few years ago, it was assumed that such regions were devoid of stygobionts, but recent studies (Holsinger 1980, 1981) have demonstrated the presence of at least 12 species of Amphipoda (of which six are local endemics) and several aquatic Isopoda in the glaciated parts of North America and Canada.

Certainly, the physico-chemical properties of the groundwaters may have shown long term variation in geological time, but their stability has been much greater than that of surface waters. In addition, the level of the groundwater table may have changed, but presumably it has always been present, whereas surface water bodies often disappear over geological time.

The salinity of groundwaters no doubt increases during periods of drought. Since, however, in my view the majority of the stygobiont Malacostraca found nowadays in the Antilles are of marine origin, I assume that they have retained a capacity for coping with changes in salinity. The remarkable salt tolerance of certain epigean limnic Amphipoda [e.g. *Gammarus duebeni* (see Sutcliffe 1968a,b, 1971a-e), and *Echinogammarus berilloni* (see Vincent 1974)] confirm that this is not merely unsupported speculation.

2 VICARIANCE AND DISPERSAL

We owe Rosen (1976) acknowledgement for making an elegant attempt to apply a vicariant model to Caribbean biogeography. He postulated a Mesozoic proto-Antillean archipelago, the fragments of which drifted from their original position — at the place of the existing Central American land mass — to their present place — the Antillean chain. Unfortunately, his model suffers from a number of drawbacks enumerated here and discussed in detail below:

1. The existence of a proto-Antillean archipelago is supported by certain geological evidence (references in Hedges 1982), but the assumption that the present-day Lesser Antilles are derived from proto-Antillean fragments is contradicted by many geological data.

Rosen tried to compensate for pieces of geological evidence which were missing by using bigeographical data to corroborate his proposal that the Antillean islands, including those of the Lesser Antilles, were once (in the late Mesozoic) more or less contiguous. There are, however, several problems in animal groups other than fishes (which form Rosen's pièce de resistance), which will be enumerated in the following paragraphs.

2. Several limnic and terrestrial invertebrate species, widely distributed over the islands, have marine larval stages, and thus are more likely to owe their actual distribution not to fragmentation of a once continuous range, but to dispersal by sea currents.

3. Limnic animals without marine larval stages may nevertheless be unsuitable for biogeographic studies if they are capable of reproducing asexually. Single specimens, dispersed accidentally through some vector, can be founders of new populations in a more easy way than sexually reproducing animals.

4. Rosen mentions a number of cases (in Crustacea and Planaria) of identical species on different islands. During our survey of West Indian groundwaters, we have not encountered identical species that (a) reproduce sexually, (b) lack free-living larval stages or otherwise easily dispersed stages, and (c) are K-strategists (produce relatively small offspring and have long periods of brood-care). On the contrary, such groups have endemics on every suitable island.

Exceptions, that is the occurrence of identical species on different islands, are known, but only in those instances where the islands in question were interconnected during Pleistocene/Holocene sea level regression.

Of course, island endemism in itself is not in contradiction with the idea of vicariance, but the presence of such endemics on islands of different geological age is not in agreement with the idea of a fragmentation of one ancestral population (e.g. the various species of the amphipod *Metaniphargus* in Puerto Rico, Tintamarre, Marie-Galante, and Curaçao, all islands that merged in different epochs).

5. The supposed fragmentation of the proto-Antillean plate must have taken place in the late Mesozoic. If we could find nowadays identical species on different islands, as a result of the break-up (i.e. vicariance) of their original range, this would point to a lack of evolutionary differentiation in the long period elapsed since the fragmentation process caused isolation of such populations.

6. Certain biological processes (competition, predation) tend to obscure the biogeographic data.

7. In certain groups of stygobionts, vicariant processes other than fragmentation of the proto-Antilles, cannot be excluded.

I will briefly discuss points (1) to (7) in sequence.

1. The migration of the Lesser Antilles (away from the Mid Atlantic Ridge, Briden et al. 1979) does not necessarily contradict Rosen's vicariance model, although a movement toward the Atlantic Ocean would have provided more support for it.

Actually, almost all geologists believe that the islands of the Lesser Antilles have had no past continental connections (Hedges 1982) and arose independently, through combined or subsequent volcanic and positive geotectonic movements, from the deep-sea bottom.

In several islands even the untrained eye can witness evidence of the fact that they were once covered by sea water, and thus unfit for sustaining descendants of Mesozoic biotas. These islands (e.g. Curaçao, Bonaire, Barbados, Mona) are covered with uplifted coral limestones, not older than of Late Miocene age, showing several surf terraces above the actual sea level. The uplift rate has been considerable, the record being held by Barbados with 43 $cm/10^3$ year (Matthews 1973). In the Dutch Leeward group (Aruba, Bonaire, Curaçao), Focke (1978) distinguished three main and 12 substages of uplifting, with an uplift rate of 5 $cm/10^3$ year.

K-Ar dating and deep drilling in the lesser Antilles equally point to a post-Mesozoic (in fact Neogene to Pleistocene) emergence (see 3 below).

For almost every island in the Lesser Antilles for which the palaeohistory is known, similar scenarios come forward. See, for instance, Westercamp & Tazieff (1980) for Guadeloupe and Marie-Galante or Mendez (1978) for the Roques and Aves archipelagos.

At any rate, it is difficult to accept the survival on such islands of Mesozoic non-marine biotas, notwithstanding Ball's (1971) plea that in spite of catastrophic events, such as vulcanism or immersion, the non-marine biotas may not have been entirely destroyed. Though in itself not illogical, it does not seem to apply to the situation found in the Lesser Antilles.

2. One of Rosen's biogeographic examples illustrating phyletic relationships of West Indian islands once assumed to have been part of the proto-Antillean archipelago, is the distribution of fresh-water shrimps of the family Atyidae. However, Hunte (1979a) was able to rear the larvae of one of the shrimp species, *Micratya poeyi*, at marine salinities (32 ‰ *Jonga serrei* 20‰, *Potimirim mexicana* 14‰, see Hunte 1979b). Hunte (1976) also showed that the small (1.5 mm) zoea larvae of the atyids are carried to the estuaries of the fast flowing rivers in which the adults breed. The juveniles stay 53 to 74 days in the sea, long enough to be dispersed through the sea currents to an adjacent island.

Fragmentation of a once continuous range is thus not a necessary assumption to explain the present day identity of atyid faunas on different islands. In fact, older observations (e.g. Chace & Hobbs 1969) showed that in a single island, Dominica, all streams are inhabited by all species of shrimp. If Dominica had had refugia, like Ball (1971) supposes for Antillean *Dugesia*, it would have been more likely that the distribution patterns of the shrimps would reflect traces of these refugia.

The occurrence of one of the atyids, the hypogean *Typhlatya*, in the Galapagos and in various West Indian islands, seems to support Rosen's theory, but its occurrence on Ascension Island in the central Atlantic remains puzzling, unless we assume a marine ancestor or marine larval dispersal as the more likely cause. Ascension is geologically young (1.0 to 1.5 m.yr., based on K-Ar dating, Chace & Manning 1972), but of course certain scientists (Wilson 1963) believe that the island may be the last remnant of a once much greater and older archipelago, now sunken under sea level.

My personal observations show that *Typhlatya* apparently lives exclusively in anchialine habitats, i.e. habitats with underground connections to the sea. This supports the idea of larval dispersal through the marine environment.

It may also be mentioned in further support of the dispersal abilities noted above, that the terrestrial gecarcinid brachyuran crabs and coenobitid hermit crabs of the West Indies also have marine larval stages, and are also widely distributed over the islands.

3. As examples of asexually reproducing animals frequently found in West Indian hypo- and epigean fresh waters, we may cite the *Dugesia* group of planarian worms and certain genera of Ostracoda.

The *Dugesia* group was used by Rosen as one of the examples to support the proto-Antillean fragmentation idea. This example seems questionable in at least three respects: (a) The *Dugesia* group occurs in several West Indian islands, as well as on islands that are much younger than the supposed proto-Antillean fragments. Two such islands on which the planarians occur are St. Vincent and Curaçao. For St Vincent, K-Ar dating of volcanic rocks points to an age of <7.7 to 2.0-0.2 m.yr. (Briden et al. 1979). De Buisonjé's (1974) geological investigations show that Curaçao emerged in the Neogene. Deep drillings have shown that the underlying volcanic rocks are rather old (37-10 m.yr. Briden et al. 1979), but the structure and chemical composition of these larvas indicate that they have flown out in deeper waters. Moreover, in Curaçao, they are covered by Silica-rich pelagic sediments, which also indicate that they were situated far below sea level. At any rate, both islands do not seem to fit exactly in the image of a fragmented Mesozoic plate. (b) *Dugesia* are capable of asexual reproduction. In fact, most *Dugesia* found on certain islands are asexual (Benazzi 1982, De Vries pers. comm.). This leaves the possibility open that, starting from one or a very few individuals, larger isolated populations of *Dugesia* can be founded. (c) In his most recent paper, Ball (1983:414) states that 'each island (in the Caribbean) seems almost solely to harbour its own endemic form'. This more or less contradicts Ball's (1971) earlier assumptions, referred to by Rosen (1976), that identical species inhabit several islands. (d) In a number of West Indian ostracode genera, Broodbakker (1983a-d) has demonstrated the presence of parthenogenetic populations. The distribution of members of these genera over the islands is erratic (similar to the distribution of *Dugesia*) and is probably a result of chance dispersal rather than to vicariance. Of course, it should be borne in mind that long distance dispersal is more easy in ostracodes (through vectors like birds, or as resting stages) than in planarians (see also Ball 1983).

4. The great degree of endemism in stygobionts is striking. In the area under consideration, the peri-Caribbean and Caribbean genus *Metaniphargus* (Amphipoda) is the best example (Stock 1977a, Stock & Vermeulen in press, and unpublished data). The genus is differentiated into distinct taxa on each island. Those islands that have identical taxa, have been interconnected during the glacial period of sea level drops. This is the case in *Metaniphargus palpator* from St Martin, Tintamarre and Anquilla. A similar situation obtains in a species from another amphipod family, *Bogidiella virginalis* (Bogidiellidae) from various Virgin Islands.

5. As mentioned under (4), identical species of stygobiont Malacostraca occur only on islands that were connected in the glacial period. Of course, the high degree of endemism in stygobionts is not an indication per se that the fragmentation hypothesis is wrong. However, it cannot be accidental that several genera containing numerous West Indian

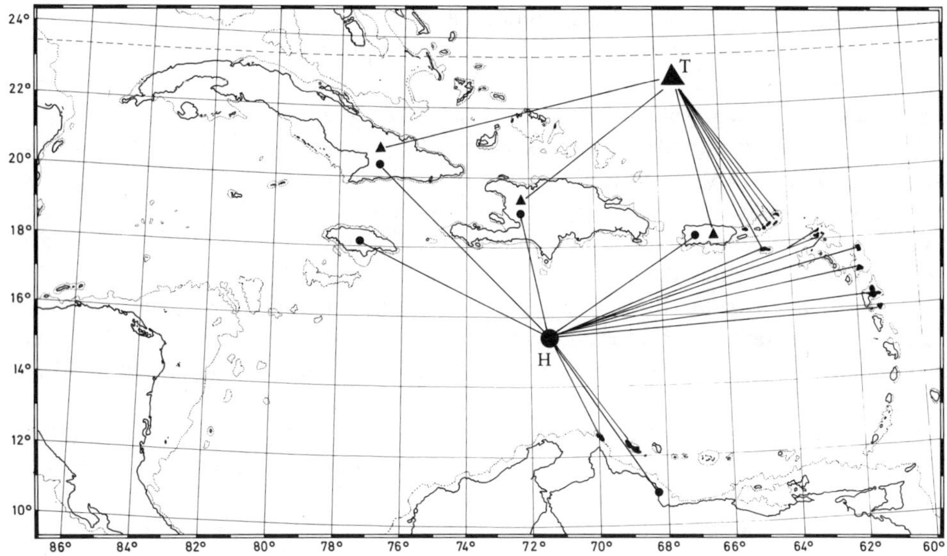

Figure 1. Distribution of hadziid Amphipoda (H) and Thermosbaenacea (T) over the Caribbean islands. Only inland (groundwater) taxa are recorded.

endemics of stygobiont Crustacea (hadziid Amphipoda, Thermosbaenacea, cyathurid Isopoda) lack close relatives on the Central and South American mainlands, but have closely allied taxa in the actual Antillean marine environment, or in Tethys derived biotas in the peri-Mediterranean belt. This seems to indicate, as stressed by Stock (1980) that they are descendants of Tethyan, marine ancestors that became adapted to groundwater biota during periods of relative low sea level. Such a scenario, which is essentially also vicariant, was called by Stock (1980) the Regression model.

6. Stock (1982a, 1983) has recently shown that the curious distribution patterns (Fig.1) of inland water (stybogiont) Thermosbaenacea and of hadziid Amphipoda over the Antillean islands can be explained in more than one way. The classical way is to assume that both groups represent North American tracks, but that the Thermosbaenacea came no further than the Greater Antilles and the islands of the Puerto Rico bank, whereas the hadziid Amphipoda were capable of invading the greater part of the Antillean chain.

A difficulty is that one finds islands with hadziids but without Thermosbaenacea and vice versa. Joint occurrence on one island is rare (observed only in Cuba, Hispaniola, Puerto Rico and St Croix), and even on islands of joint occurrence, co-existence in one locality occurs significantly less than would be expected if the two groups were distributed at random (details in Stock 1982a, 1983).

This observation leads to another way of explaining the observed distribution. It is not unlikely that Thermosbaenacea and hadziids represent two waves of 'invasion' (the word is used here in a passive sense, vide infra) of the groundwaters, beginning with marine ancestors. This 'invasion' may have taken place in an earlier (thermosbaenacean) or later (hadziid) geological epoch. The better competitive qualities (Stock 1983) of the more recent (hadziid) invaders have resulted in elimination of the older taxa (Thermosbaenacea) on almost all smaller islands. Only in the larger islands of the Greater Antilles, were sufficient refugial habitats available to permit survival of the Thermosbaenacea

In this example, the observed distribution patterns, both interinsular and intrainsular, appear to be real, and the avoidance of each other's presence is statistically significant. The explanation of these patterns remains speculative, but there is, in my view, no clear relationship with a fragmentation model of Rosen's conception.

7. The amphipod suborder Ingolfiellidea must be very old. This is not only evidenced by some morphological details (the presence of 'ocular lobes', presumed to be remnants of an extra cephalic somite), but also by the distribution of the taxa at the subgeneric level (i.e. identical subgenera in Europe and South America), which indicates a Mesozoic or Pangaean range before the formation of the Atlantic (Stock 1977b).

The distribution of the Ingolfiellidea over the West Indian islands is remarkable too: they do occur, in rather numerous different taxa, in the islands fringing the South American mainland (Aruba, Curaçao, Bonaire, Margarita, La Blanquilla, Los Testigos), but nowhere else, notwithstanding extensive sampling (Stock 1979, and unpublished data).

Here too, two explanations are possible: a South American track concept, or a colonization by an originally marine ancestor. Pleading against the idea of a South American track is the absence of Ingolfiellidea in northern South America, whereas standing in favor of a marine invasion idea is the presence of several marine ingolfiellids in the Caribbean.

Still, it remains disconcerting that ingolfiellids are lacking on the islands from Trinidad northward. If the proto-Antillean fragmentation was applicable to the ingolfiellids, one would expect members of so old a group to be present on at least some of the Greater Antilles. If the Regression model were true, one would expect them on at least some islands having a palaeohistory similar to that of the Leeward group.

In this respect, it may be useful to point to another old group of stygobionts, the Bathynellacea (Schminke 1974, see also Schminke 1979, for the northernmost record in the peri-Caribbean belt). For this group too, a Mesozoic distribution pattern is presumed, but its total absence from the islands in the Caribbean is hard to explain if these islands had past, Mesozoic, continental connections.

3 A BIOLOGICAL CALENDAR FOR GEOLOGICAL EVENTS?

In many of the previous arguments, a discrepancy came to light between island age (used here as the time elapsed since their emergence) according to geological data, or that presumed on the basis of the fragmentation model.

Elsewhere, I have proposed the idea (Stock 1982b) that species/area (S/A) curves might be useful in certain cases as a biological calendar for island age. A number of basic assumptions have been delineated in an earlier paper (Stock 1980). In short, I believe that the great majority of stygobionts in the West Indies have evolved from marine ancestors, by a vicariant process different from fragmentation of plates, according to a model that I called the Regression Model. Marine populations, mostly of meiofaunal elements, became 'stranded' by geotectonic positive movements (uplift) and gradually adapted to mixohaline coastal groundwaters, and in the end to fresh groundwaters as well. Because groundwaters (in contradistinction to continental surface waters) have had a greater stability in time, the extinction rate could well have been reduced, and many 'old' forms could survive in the groundwater where epigean forms could not.

Because every island, as it broke through the sea's surface, was surrounded by the sea and its marine biotas, the 'center of origin' for its fauna, so to speak, was right at the island doorstep. No long distance dispersal over a barrier was necessary for the ancestors of the stygofaunal elements to reach an emerging island.

In several cases, marine relatives (often belonging to the same genus as the Recent stygobionts) are still in existence, e.g. in the isopod genera *Microcharon* and *Microcerberus*, or in the amphipod genera *Bogidiella* and *Ingolfiella*. In other cases marine and stygofaunal groups belong to related genera (amphipods of the hadziid grouping and of the Paradaliscidae, isopods of the family Cirolanidae) and the Thermosbaenacea (Stock 1977a, Stock & Vermeulen 1982, Notenboom in press, Stock 1976).

The vicariant process — I prefer in point of fact the term 'passive process' — has been induced by the uplifting of the West Indian islands, mostly in Neogene/Pleistocene times, or has been aided by eustatic regressive sea level movements. The main difference between passive fragmentation processes (i.e. vicariance sensu stricto) and passive regression processes, is that in the former the geographic and genetic barriers become larger in time, whereas in the latter a constant contact with the ancestral populations remains intact. Notwithstanding the theoretical possibility of continuous exchange of genetic material, there is no disagreement between world stygobiologists that differences on at least the specific level have developed on a large scale.

The isolation process, it seems, has been helped to a considerable extent by the fact that the majority of the stybogionts have suppressed larval stages (which are passed inside the egg) and produce only a very small number of offspring (1 to 4 eggs per female per brood period). Such life history patterns lead to the formation of small, localized, inbreeding populations, and are supported by the distribution and biology of many stygobionts.

The importance of positive tectonic movements on the evoluion of stygobionts is shown best by a comparison of areas which have undergone uplift as opposed to sinking areas. Practically all islands of the Antillean chain are emerging; whereas, the islands of the Bahamas and the Turks and Caicos group are sinking.

In my view, the number of species of stygobionts per island is dependent on two factors: island size (the classical factor) and geological age of the island (the longer it has been exposed to invasions, or perhaps a better word is stranding, of marine animals, the more animals will succeed in penetrating its groundwaters).

Figure 2 shows a plot of stygobiont Malacostraca species against island size. The islands in the Lesser Antilles selected for this purpose are all situated at approximately the same latitude, are all of low altitude, and are all calcareous: viz. Aruba, Curaçao, Bonaire, Los Roques, La Blanquilla, La Tortuga, and Barbados.

Many characteristics of the regression curves are still under debate (slope, intersection with the abscissa, see Stock 1982b), but one feature looks very promising. It is not one regression curve that emerges, but two almost parallel lines, on the higher of which lie the older islands (emergence in the Late Miocene), and on the lower the younger islands (emergence in the Pleistocene).

Admittedly, the palaeohistory of the islands chosen in this example is simpler than that of the Greater Antilles. Nevertheless it is hoped that S/A curves of islands of debatable age might produce some biological evidence dating their emergence. In the case of the Lesser Antilles, the regression curves of Figure 2 show that one of the islands of questionble age, La Blanquilla, falls on the line of Aruba-Curaçao-Bonaire, all three of which are of Miocene age; it is presumed, therefore, that La Blanquilla emerged likewise in that epoch.

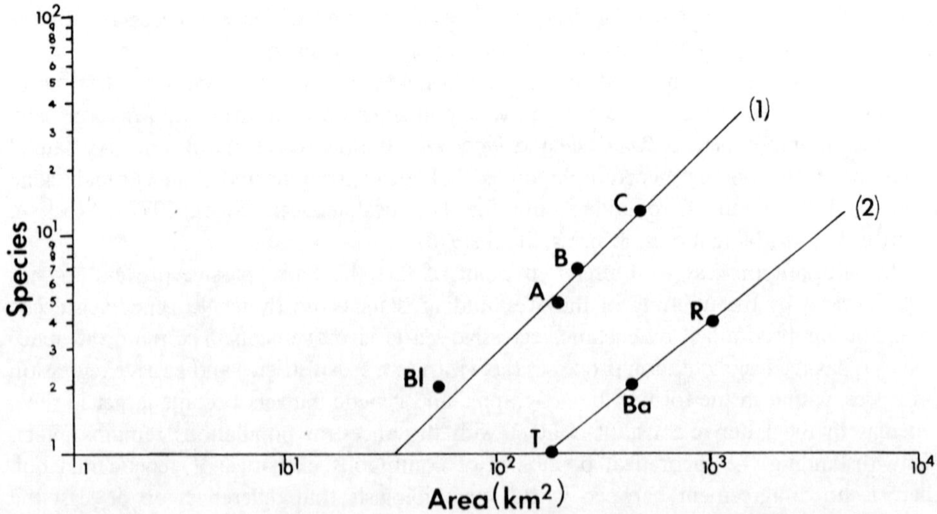

Figure 2. Power function regression of the number of stygofaunal Malacostraca species (S) against island area (A) in km^2; A – Aruba; B – Bonaire; Ba – Barbados; Bl – La Blaquilla; C – Curaçao; R – Los Roques; T – La Tortuga.

REFERENCES

Ball, I.R. 1971. Systematic and biogeographical relationships of some *Dugesia* species (Tricladida, Paludicola) from Central and South America. *Amer. Mus. Novit.* 2472: 1-25.

Ball, I.R. 1983. Planarians, plurality, and biogeographical explanations. In R.W.Sims et al. (eds.), *Evolution, time and space: the emergence of the biosphere.* pp 409-430. London: Syst. Ass.

Benazzi, M. 1982. Tricaldi cavernicoli italiani. *Lavori Soc. Ital. Biogeogr.* 8: 7-14.

Briden, J.C., D.C.Rex, A.M.Faller & J.F.Tomblin 1979. K-Ar geochrony and palaeomagnetism of volcanic rocks in the Lesser Antilles island arc. *Phil. Trans. Roy. Soc. Lond.* (A)291: 485-528.

Broodbakker, N.W. 1983a. The genus *Heterocypris* (Crustacea, Ostracoda) in the West Indies. *Bijdr. Dierk.* 53: 115-134.

Broodbakker, N.W. 1983b. The genus *Hemicypris* (Crustacea, Ostracoda) in the West Indies. *Bijdr. Dierk.* 53: 135-157.

Broodbakker, N.W. 1983c. The genus *Strandesia* and other Cypricercini (Crustacea, Ostracoda) in the West Indies. *Bijdr. Dierk.* 53: 327-368.

Broodbakker, N.W. 1983d. The subfamily Candoninae (Crustacea, Ostracoda) in the West Indies. *Bijdr. Dierk.* 53: 287-326.

Buisonjé, P.H.de 1974. Neogene and quaternary geology of Aruba, Curaçao and Bonaire. *Publs. Found. Sci. Res. Surinam Neth. Antilles* 78: 1-293.

Chace, F.A. Jr & H.H.Hobbs 1969. The freshwater and terrestrial decapod crustaceans of the West Indies with special reference to Dominica. *Bull. U.S. Natn. Mus.* 292: 1-258.

Chace, F.A. Jr & R.B. Manning 1972. Two new caridean shrimps, one representing a new family, from marine pools on Ascension Island (Crustacea: Decapoda: Natantia). *Smiths. Contr. Zool.* 131: 1-18.

Focke, J.W. 1978. Geologische aspecten van kwartaire koraal- en algenriffen op Curaçao, Bonaire, Aruba en Bermuda. Thesis, Univ. Leiden, ISBN 90-6231 958 3.

Hedges, S.B. 1982. Caribbean biogeography: implications of recent plate tectonic studies. *Syst. Zool.* 31: 518-522.

Holsinger, J.R. 1980. *Stygobromus canadensis*, a new subterranean amphipod crustacean (Crangonyctidae) from Canada, with remarks on Wisconsin refugia. *Can. J. Zool.* 58: 290-297.

Holsinger, J.R. 1981. *Stygobromus canadensis*, a troglobitic amphipod crustacean from Castleguard Cave, with remarks on the concept of cave glacial refugia. *Proc. 8th Int. Congr. Speleol.* 1.

Hunte, W. 1976. Biological studies of freshwater shrimps (Atyidae an Palaemonidae) in Jamaica. Ph.D. Thesis, Univ. of the West Indies, Kingston, Jamaica.

Hunte, W. 1979a. The complete larval development of the freshwater shrimp *Micratya poeyi* (Guérin-Méneville) reared in the laboratory (Decapoda, Atyidae). *Crustaceana* 5: 153-166.

Hunte, W. 1979b. The complete larval development of the freshwater shrimp *Atya innocous* (Herbst) reared in the laboratory (Decapoda, Atyidae). *Crustaceana* 5: 231-242.

Matthews, R.K. 1973. Relative elevation of Late Pleistocene high sea level stands: Barbados uplift rates and their implications. *J. Quat. Res.* 3: 147-153.

Mendez, J.Baamonde 1978. Archipiélago los Roques/islas de Aves. *Cuadernos Lagoven, Ecología*: 2-48.

Notenboom, J. 1984. *Arubolana parvioculata* n.sp. from the interstitial of an intermittent river in Jamaica with notes on *A.imula* Botosaneanu & Stock and *A.aruboides* (Bowman & Iliffe). *Bijdr. Dierk.* 54.

Rosen, D.E. 1976. A vicariance model of Caribbean biogeography. *Syst. Zool.* 24: 431-464.

Schminke, H.K. 1974. Mesozoic intercontinental relationships as evidenced by bathynellid Crustacea (Syncarida: Malacostraca). *Syst. Zool.* 23: 157-164.

Schminke, H.K. 1979. Eine neue Gattung der Familie Parabathynellidae (Bathynellacea, Syncarida) aus dem nördlichen Südamerika. *Bijdr. Dierk.* 49: 97-101.

Stock, J.H. 1976. A new genus and two new species of the crustacean order Thermosbaenacea from the West Indies *Bijdr. Dierk.* 46: 47-70.

Stock, J.H. 1977a. The taxonomy and zoogeography of the hadziid Amphipoda with emphasis on the West Indian taxa. *Stud. Fauna Curaçao* 55: 1-130.

Stock, J.H. 1977b. The zoogeography of the crustacean suborder Ingolfiellidea with descriptions of new West Indian taxa. *Stud. Fauna Curaçao* 55: 131-146.

Stock, J.H. 1979. New data on the taxonomy and zoogeography of ingolfiellid Crustacea. *Bijdr. Dierk.* 49: 81-96.

Stock, J.H. 1980. Regression model evolution as exemplified by the genus *Pseudoniphargus* (Amphipoda). *Bijdr. Dierk.* 50: 105-144.

Stock, J.H. 1982a. The influence of hadziid Amphipoda on the occurrence and distribution of Thermosbaenacea and cyclopoid Copepoda in the West Indies. *Polsk. Arch. Hydrobiol.* 29: 275-282.

Stock, J.H. 1982b. Stygobiont Crustacea Malacostraca from geologically older and younger Antillean islands: a biogeographic analysis. *Bijdr. Dierk.* 52: 191-199.

Stock, J.H. 1983. Predation as a factor influencing the occurrence and distribution of small Crustacea in West Indian groundwaters. *Bijdr. Dierk.* 53: 233-243.

Stock, J.H. & J.J.Vermeulen 1982. A representative of the mainly abyssal family Pardaliscidae (Crustacea, Amphipoda) in cave waters of the Caicos Islands. *Bijdr. Dierk.* 52: 3-12.

Stock, J.H. & J.J.Vermeulen (In press). Crustacés Amphipodes stygobies de la Guadeloupe en relation avec l'histoire géologique de l'île. *Mém. Biospéol.*

Sutcliffe, D.W. 1968a. Sodium regulation in the amphipod *Gammarus duebeni* Lilljeborg from freshwater localities in Ireland. *J. Exp. Biol.* 48: 339-358.

Sutcliffe, D.W. 1968b. Sodium regulation and adaptation to fresh water in gammarid crustaceans. *J. Exp. Biol.* 48: 359-380.

Sutcliffe, D.W. 1971a. Sodium influx and loss in freshwater and brackishwater populations of the amphipod *Gammarus duebeni* Lilljeborg. *J. Exp. Biol.* 54: 255-268.

Sutcliffe, D.W. 1971b. The effect of calcium on sodium influx and loss in *Gammarus* (Amphipoda). *J. Exp. Biol.* 54: 485-496.

Sutcliffe, D.W. 1971c. Regulation of water and some ions in gammarids (Amphipoda), 1. *J. Exp. Biol.* 55: 325-344.

Sutcliffe, D.W. 1971d. Regulation of water and some ions in gammarids (Amphipoda), 2. *J. Exp. Biol.* 55: 345-355.

Sutcliffe, D.W. 1971e. Regulation of water and some ions in gammarids (Amphipoda), 3. *J. Exp. Biol.* 55: 357-369.

Vincent, M. 1974. Préférendum ionique chez des Amphipodes epigés du Centre-Ouest. *Vie Milieu* (C)23: 65-79.

Westercamp, D. & H.Tazieff 1980. *Martinique, Guadeloupe, Saint-Martin, La Désirade. Guides géol. région Paris.* Masson.

Wilson, J.T. 1963. Continental drift. *Sci. Amer.* 208(4): 86-100.

3 Ecology and biogeography

Prologue

The third group of papers deals with the ecological and evolutionary significance of biogeographic patterns in body size among stomatopods, the implications that systematic revisions in chydorid cladocerans have for previous and future biogeographic studies, and recent studies on the biology of ostracodes that have great relevance for the way this important taxon can be used at present and past biogeographical studies. Thus, these three papers discuss new work on the ecological and evolutionary factors that shape morphology, an elucidate how this work allows new biogeographic inferences or suggests changes in previously held beliefs on crustacean biogeography.

In the first paper, Marjorie Reaka considers biogeographic patterns in stomatopod body size and their environmental correlates. She finds that stomatopods are significantly larger on soft than hard substrates, and concludes that the large size of temperate stomatopods is partly due to the greater abundance of soft sediments in high rather than in low latitudes. However, soft-bottom temperate species are still larger than tropical species from soft-bottoms. In addition, species from hard substrates clearly produce smaller post-larvae than those from soft substrates, while latitude and temperature are less clearly related to post-larval size. Reaka explains in detail the ecological and evolutionary consequences of observed patterns in adult and post-larval size in different environments. Some of these include the potential for consistent size ratios among sympatric species without the intervention of competition, less predictability in stomatopod species composition among small rather than large species, and the potential for higher rates of evolution among species with small post-larvae (that are less likely to travel long distances than large post-larvae, and to show more restricted distributions than large species).

David Frey describes his extensive systematic and ecological research on chydorid cladocerans. Based on detailed morphological comparisons and on breeding experiments, Frey concludes that the conventional wisdom that chydorids are cosmopolitan is largely incorrect. Instead, there are many more species than previously believed, with a great degree of latitudinal differentiation in North American lakes. Finally, he considers how climatic changes could have produced the newly discovered non-cosmopolitanism among chydorid cladocerans.

Kenneth McKenzie reviews recent biological studies on the morphology and distribution of ostracodes, and how this work changes our perspective on ostracode biogeography. This is particularly noteworthy because of the importance of ostracode fossils in paleo-biogeographic reconstructions. A particularly significant point is the degree to which the ornamentation of the ostracode shell appears to be a function of the availability of carbonate in the environment. This means that much of what was previously believed to be genetic variation in morphology is instead ecophenotypic variation. The significance of this for revisionary work in both living and fossil taxa is apparent and it is interesting that this runs counter to Frey's findings. Of course, it is likely that similar patterns may be seen in other organisms whose hard parts are of calcium carbonate, such as molluscs and other crustacean taxa. McKenzie also documents several instances of range changes in various taxa that are due to introduction by man. This means that ostracode distributions need to be interpreted cautiously, since the possibility for human intervention seems substantial.

Biogeographic patterns of body size in stomatopod Crustacea: Ecological and evolutionary consequences

MARJORIE L.REAKA
Department of Zoology, University of Maryland, College Park, USA

ABSTRACT

Species of stomatopods reach larger adult body sizes in soft (mud, sand) rather than hard (coral, rock) substrata, in deep rather than shallow habitats, in temperate rather than tropical latitudes, and in cold rather than warm environments. However, the proportion of species that inhabit soft substrata increases in deep habitats, at high latitudes, and in cold environments. Also, the proportion of species that occur in deep environments increases at high latitudes.

Due to these interactions, comparisons were restricted so that only one variable was considered, providing a more rigorous test of the effects of that factor upon body size. Body sizes for species of stomatopods that excavate their own burrows in mud or sand still exceed those for species that occupy pre-existing holes in rock or coral if comparisons are restricted to species in shallow tropical habitats. Taxa from cold environments still reach larger body sizes than those from warm environments when only species from soft bottoms are considered. However, depth is not significantly related to body size when only species from tropical soft bottoms are compared, and latitude is not significantly related to body size when only species from shallow soft bottoms are analyzed. The previously established overall correlations between large body size and high latitude and deep water therefore appear to be due to the disproportionate number of species that occupy soft bottoms in these cold environments.

Overall, patterns of body size and habitat use do not vary significantly among species of stomatopods in the East Atlantic, West Atlantic, and East Pacific oceans. However, the East Pacific stomatopod fauna from hard substrata in shallow tropical habitats is dwarfed compared to those in the West Atlantic. Body sizes of species from soft bottoms in shallow tropical habitats do not differ between the East Pacific and West Atlantic.

Of all the traits considered, postlarval size is most highly related to adult body size in species of stomatopods. Also, species inhabiting hard substrata produce significantly smaller postlarvae than those from soft bottoms. Although the relationships are more variable, latitude and temperature appear to be related to postlarval size in at least some cases.

In summary, for stomatopods in general, the factors that most influence adult body size are type of substratum occupied and temperature, while the factors that most influence postlarval size are adult body size, type of substratum inhabited by adults, and possibly temperature or latitude.

This study questions the factors that constrain body size in the active, aggressive species which occupy non-expandable holes in hard substrata compared to the larger species which burrow in mud or sand. I suggest that mechanical or metabolic constraints, predation, and possibly the nature of the substratum have limited body size in stomatopods from hard bottoms; however, these factors have not operated comparably in reef-dwelling decapods, and the question remains open. Similarly, the factors that have fostered dwarfism in East Pacific compared to West Atlantic species from hard bottoms remain to be determined.

Regardless of the causes of these shifts, the patterns of adult and postlarval size in different environments and geographical regions are ecologically and evolutionarily significant. A series of species arranged along an environmental or geographical cline could produce a community of sympatric species with contant ratios of body size without the intervention of competition. Also, species characterized by relatively large body size and long lived larvae should have high larval mortality, long lived iteroparous adults, less predictable recruitment at any one site, and less deterministic community structure than would be found in an assemblage of small species. Predation by stomatopods upon members of the cryptic reef community may be more severe in the West Atlantic than in the East Pacific due to the larger sizes of reef-dwelling stomatopods in the former rather than the latter region. The postlarvae of small species with reduced dispersal are more likely to settle near relatives and to have smaller geographic ranges than species with large long lived larvae. Most importantly, reduced adult body size and postlarval size will predispose species from hard substrata, from warm or tropical environments, and from certain geographic subregions (e.g. the East Pacific) toward higher rates of evolution.

1 INTRODUCTION

A variety of physical and ecological factors appear to cause shifts in body size in organisms. Whatever their causes, these shifts have strong consequences, since body size is a decisive factor determining the outcome of social interactions, including contests over territories or mates, and since competitive interactions among sympatric species are strongly mediated by body size relationships (e.g. Brooks & Dodson 1965, Morse 1974, E.O.Wilson 1975, see also Goulden et al. 1980). Additionally, vulnerability of prey and the number and type of prey available to predators frequently are determined by size (e.g. Dayton 1971, Paine 1976, Zipser & Vermeij 1978, Britton & Moser 1982). Adult body size often correlates with important life history characteristics such as longevity, fecundity, egg or offspring size, and larval type (Bonner 1968, Pianka 1970, Chia 1974, Fenchel 1974, Hendler 1975, Menge 1975, McCall 1977, Reaka 1979a, Barlow 1981, Strathmann & Strathmann 1982, Jablonski & Lutz 1983). In turn, the type of larval dispersal found in a species determines the predictability of recruitment into a local community (e.g. Thorson 1946, Sale 1977, 1980, Strathmann & Strathmann 1982). Metabolic rates and the size of territories, home ranges, and geographic ranges also are significantly related to body size (Van Valen 1973, Harested & Bunnel 1979, McNab 1980, Reaka 1980a). Both metabolic rates and the size of the geographic range have been shown to correlate with rates and patterns of evolution within lineages (McNab 1980, Jackson 1974, Hansen 1980, Reaka 1980a). Consequently, the factors that govern body size provide a critical foundation for the biotic interactions that structure many communities; many of the life history and geographic traits associated with body size have significant implications for both the ecology and evolution of the members of a community.

The purpose of this paper is, first, to examine some biogeographic patterns in body size in stomatopod crustaceans, including an analysis of new factors that can influence such patterns. Second, I hope to raise some new questions about the factors that constrain the upper limits of body size in organisms. Third, and most importantly, I wish to investigate the ecological and evolutionary consequences of these releases or constraints upon body size.

Before proceeding with the analysis, I will review briefly the major biogeographic patterns that are known for body size in organisms, and will tie together the disparate and heretofore unsynthesized wisps of literature that address the various factors thought to cause shifts in body size in contemporary species. Also, I will briefly identify here some of the known patterns and consequences of changes in body size among fossil lineages.

1.1 *Patterns of body size in contemporary species*

The pattern of body size that historically has drawn the greatest attention from biologists is a prevalent trend for increasing body sizes among populations and species at higher latitudes (Bergmann's Rule). This 'rule' was originally derived for endotherms (e.g. see Rensch 1960). However, subsequent analyses showed that the pattern held for many ectotherms as well. Lindsey (1966), for example, showed that body size increased with latitude among the majority of 12 500 species of vertebrate ectotherms. Among crustaceans, species of amphipods and myodocopid copepods are two to four times larger in high latitudes than are species from tropical environments (Barnard 1962, Poulsen 1965). Sizes of individuals in populations of an interstitial mystacocarid crustacean *Derocheilocaris typica* also increase with latitude (Hall & Hessler 1971). Populations of the decapods *Cyclograpsus cinereus* and *Sesarma reticulatum* increase in size with increasing latitude in the southern and northern hemispheres, respectively (Garth 1957, Abele 1973). Camp & Manning (1982) pointed out that a northern species of lysiosquilloid stomatopod, *Nannosquilla grayi* reaches considerably larger sizes (42 mm in length) than any of the other species of *Nannosquilla* in the West Atlantic, all of which have more southerly distributions and all of which are $\leqslant$ 28 mm in length. Reaka & Manning (1980) also noted that latitudinal gradients in body size occur within species of two East Pacific gonodactyloid stomatopods; *Gonodactylus stanschi* and *G. zacae* reach total body lengths of 41 mm and 59 mm in the Gulf of California, while populations of the same species reach only 15 mm and 37 mm in length, respectively, in Costa Rica. Thus, these latitudinal trends appear to be quite general both among and within species. However, few studies have concentrated upon taxa that may not conform to this pattern, or at least few have been published. Even more importantly, very few studies have critically analyzed other factors (e.g. depth distributions, habitat characteristics) that might interact with or bias the latitudinal correlations that have been established.

Body size also appears to increase with altitude and depth in some terrestrial and marine organisms. For example, body sizes of birds, frogs, and insects are larger in montane than lowland regions (James 1970, Berven et al. 1979, Levins 1969). Body size increases with depth in many crustaceans, but the pattern is not regular. Wolff (1956a,b) found that body size in isopods and tanaids increases with depth (>6000 m). However, no such pattern is present within the isopod genus *Storthyngura* (Menzies & George 1967), and no relationship between body size and depth is apparent for amphipods (Barnard 1962). Body size appears to increase with depth in some (but not all) myodocopid and podocopid ostracodes (Kornicker 1975, van Markhoven 1972). Thiel (1975) and Allen (1978), however, reported that most deep-sea molluscs are diminutive in size. Thus, the generality of the relationship between body size and depth, as well as the possibility that this pattern may be influenced by other variables (e.g. type of substratum, temperature, historical factors), remains unresolved.

A number of additional biogeographic patterns in body size have been documented for marine invertebrates. Closely related gastropods reach 20-50% larger adult sizes in East Pacific than in West Atlantic regions (Vermeij 1980). Following a similar trend, morphological defenses against predators are greater in tropical Indo-West Pacific and East Pacific than in Atlantic or temperate gastropods (Vermeij 1974, 1978, 1979), and the shells of similarly sized gastropods are stronger in the Indo-West Pacific than in the West Atlantic (Vermeij 1976).

Several biogeographic patterns of body size are known for geographic regions of more limited scale. In contrast to species from areas of high productivity (e.g. the west coast of Florida, western Venezuela, the coast of Chile, southwestern Africa), gastropods from the Florida Keys, Jamaica, and the coast of Brazil are relatively small (Stephenson & Stephenson 1950, Vermeij & Porter 1971, Vermeij 1978, 1980). Several workers have documented that the fauna (e.g. gastropod and bivalve molluscs, sipunculan and echiuran worms, stomatopod crustaceans) of atolls in isolated island groups are dwarfed compared to relatives in major island archipelagos or continental regions (Reaka & Manning 1986a,b and references therein, G.J.Vermeij pers. comm.). Although it seems likely that mid-ocean atoll environments are less productive than high island or mainland environments, the reasons for these cases of size reduction remain undetermined.

A number of correlations between body size and microhabitat are known, particularly for molluscs (see also Sebens 1983, for a discussion of ecological factors that contribute to local size structure in populations). In high intertidal species of gastropods, adult body size increases toward higher shore levels (Vermeij 1972), and in tree-associated gastropods from mangroves, littoral fringe species are smaller than intertidal species (Vermeij 1973a). However, in low intertidal species of gastropods, body size decreases toward higher shore levels (Vermeij 1972). On low wave-exposed shores, gastropods inhabiting open surfaces frequently reach larger sizes than those that inhabit similarly open but more sheltered environments; also, cryptic gastropods in low shore habitats are smaller than species from open surfaces (Vermeij 1973b).

1.2 *Environmental agents thought to cause shifts in body size*

The factors thought to influence body size in organisms include temperature, interspecific competition, primary productivity, size and availability of prey, predation pressure, and structure of the habitat. Competition probably has received the most attention and remains the most controversial.

Observational and experimental studies have shown that cell size, egg and larval size, size at metamorphosis, and adult body size are inversely related to temperature in a variety of invertebrate and vertebrate species (Ray 1960, Barnes & Barnes 1965, McLaren 1965, Lutz & Jablonski 1978, Berven et al. 1979, Grime & Mowforth 1982), suggesting the possibility that some biogeographical patterns of body size may be explained by temperature alone. However, several authors have explicitly doubted that ambient temperature explains latitudinal trends in body size. Scholander (1955) protested that trends for larger sized endotherms in colder regions could not be explained by surface to volume relationships, since other physiological mechanisms (insulation by fur, feathers, or fat; vascular mechanisms) are far more efficient in conserving heat than slight changes in surface to volume relationships (but see Mayr 1956). Rosensweig (1968) reported that body size correlates with latitude, but that primary productivity is a better predictor of body size in mammalian carnivores than latitude or temperature (see also discussion of productivity and body size in marine species above). Similarly, James (1970) found that the combined effects of temperature and humidity explain trends in the sizes of birds better than either latitude or temperature. McNab (1971) suggested that in mammals Bergmann's Rule applies only to small species of carnivores and granivores that encounter competitors in the lower latitudes of their range, and that latitudinal trends in body size may be correlated with prey size. Furthermore, Belyaev (1972, in Abele 1982:256) noted that relationship between temperature and body size is not straight-forward for many marine taxa, and Berven et al. (1979) showed that genetic components, as well as temperature, control body size in frogs. Shifts of body size in many ecological situations where temperature does not vary markedly suggests that, while temperature often may influence patterns of body size along latitudinal or perhaps depth and altitudinal gradients, the effects of temperature frequently are modified by other factors.

Following the initial discussion of character displacement by Brown & Wilson (1956), the phenomenon of divergence and convergence of body size and trophic structures as a result of competitive interactions among sympatric species has been the center of considerable interest (MacArthur & Wilson 1967, Schoener 1969a,b,c, 1970, 1972, Grant 1965, 1968, 1972, Cody 1973, D.S.Wilson 1975, Pearson & Mury 1979, Slatkin 1980).

Island populations are particularly prone to shifts in body size (Van Valen 1973), and lack of competitors was one of the early factors implicated. However, as discussed by Carlquist (1974) and Dunham et al. (1978), the causative agents in these shifts have been difficult to identify. In his review of the body sizes of insular populations compared to those of mainland relatives, Case (1978) concluded that more shifts in body size on islands are explained by the supply versus the demand of food than are explained by the presence or absence of predators. Drawing upon Johnston & Selander's (1973) observation that sexual dimorphism of house sparrows increases with latitude, and on his own results that Galapagos finches are larger on small dry islands than on larger islands with more varied habitat, Downhower (1976) provided a potentially general model which indicates that larger individuals can persist longer than small individuals in harsh or variably inclement environments because of greater stored energy reserves (see also Goulden et al. 1980). Also, historical factors or the structure of the habitat can constrain shifts in body size in island populations (Schoener 1970, Case 1978).

Competitive interactions also have been thought to be one of the factors responsible for constant ratios of adult body size in sympatric species assemblages (Hutchinson 1959, Schoener 1965, Grant 1968, Brown 1975, Diamond 1975, 1978, Abbott et al. 1977, Maiorana 1978, Oksanen et al. 1979). One of the conceptual problems with such a size structured community is how the juveniles of the larger species survive, since they must encounter intense competition from the adults of smaller species. Also, food type may not be sufficiently homogeneous and food size may not correlate with body size as clearly as would be necessary to cause widely spaced adult body sizes in sympatric species (see Slatkin 1980,

Strong & Simberloff 1981, Wiens & Rotenberry 1981). Some recent workers have suggested, additionally, that the observed sizes of species in some of these assemblages either are not structured by competition or are not different from those expected by chance (Horn & May 1977, Strong et al. 1979, Strong & Simberloff 1981, Simberloff & Boecklen 1981, Wiens & Rotenberry 1981, but see Grant & Abbott 1980, Hendrickson 1981).

Communities that are exposed to strong predation frequently become dominated by small individuals and species because predators selectively feed upon larger individuals in the prey populations (e.g. Brooks & Dodson 1965, Britton & Moser 1982). It is well known, for example, that body sizes of prey populations often decline as a result of overfishing by humans (e.g. lobsters in the north-eastern USA, snook in southwestern Florida, abalone on the west coast of the USA). Body sizes of coastal reef fishes are considerably smaller in heavily fished locations (e.g. Jamaica, Haiti, some parts of the Virgin Islands and Florida) compared to more lightly fished or protected areas (Belizan reefs, Buck Island National Park on St Croix, Looe Key and Key Largo Sanctuaries in Florida) (pers. obs., M.Hay pers. comm., Bohnsack pers. comm. and 1983). However, due to chance or other factors, some members of prey populations may obtain a size refuge from predators (Dayton 1971, Paine 1976, Zipser & Vermeij 1978), whereupon they attain unusually large sizes. Furthermore, these shifts in body size sometimes are associated with concomittant, genetically based shifts in life history characteristics. When adults of stream-dwelling guppies are exposed to severe predation by large fish predators, the guppies mature earlier at smaller sizes and produce more and smaller offspring at more frequent intervals than in situations where small fish predators inflict selective mortality on juvenile guppies (Reznick 1982, Reznick & Endler 1982). Therefore, it seems clear that in at least some ecological conditions strong predation can cause changes in body size toward either larger or, probably more commonly, toward smaller modal body sizes in local populations.

In conclusion, the extent to which temperature and particularly interspecific competition cause shifts in body size in natural populations remains a subject of active controversy. The arguments that productivity, the ability to store energy reserves, and predation influence patterns of body size are persuasive, but further documentation is needed to establish the frequency with which these factors operate in molding the size structure of natural communities.

1.3 *Evolutionary consequences of shifts in body size*

Numerous studies show that body size is an important factor affecting the evolutionary history as well as the ecology of a species. Many fossil lineages show consistent increases (and some show decreases) in body size over time (Newell 1949, Bonner 1968, Stanley 1973, 1979, Hallam 1975, Boucot 1976). Certain geological periods are characterized by gigantism, while others foster dwarf faunas (Marshall & Corruccini 1978). Particularly following rapid evolutionary increases in size, lineages of large structurally complex organisms appear to suffer higher rates of extinction than those of small body size (Stanley 1973, Hallam 1975). Wassersug et al. (1979) used a computer simulation model to determine that isolated faunas composed of large species are more likely to undergo catastrophic mass extinctions than faunas containing many small species. Furthermore, evolutionary changes in body size have been linked to radical evolutionary innovations such as neoteny, progenesis, or endothermy (Gould 1977, McNab 1978), and, as discussed below, Reaka

& Manning (1981, 1986c) showed that rates of evolution in stomatopod crustaceans are strongly influenced by body size.

Consequently, understanding the factors that control body size is a key element in understanding many aspects of the biology and evolution of a group of organisms. The objective of the present paper is to document the relationship between adult body size and environmental characteristics (latitude, depth, temperature, geographic subregion, substrate type) or biotic characteristics (postlarval size and dispersal ability) among species of stomatopod crustaceans. In addition, this report will examine the ecological and evolutionary significance of the patterns of body size found in stomapods from different environments.

2 METHODS

The stomatopods, or mantis shrimps, represent an old lineage that diverged from other malacostracan crustaceans as early as the Devonian (Holthuis & Manning 1969, Schram 1969, 1979, 1981, 1982, Reaka 1975a,b, Kunze 1983, Dahl 1983). Currently including about 350 species, all extant lineages of this largely tropical order are carnivorous burrow-dwellers with relatively homogeneous morphology. Compared to many other groups, the taxonomy, distribution, behavior, and ecology of the entire order are unusually well known. The stomatopod fauna of the Atlanto-East Pacific area is particularly well documented (see Manning 1969a, 1972, 1977, Reaka & Manning 1980, 1981, 1986c), and the data included in this paper are drawn primarily from these regions.

Total body length is used to represent size. Length is a comparable measure of body size among lineages, since the empirical relationship between body volume and body length does not vary significantly among different groups of stomatopods (Reaka unpubl. data). All measurements of body size are taken from my own collections, from specimens and records at the Smithsonian Institution, or from the published literature. Species in which the maximum size of adults is <50 mm are classified here as 'small', those from 50-150 mm as 'moderate', and those >150 mm as 'large' in body size. The number of individuals sampled and the use of specimens from museum collections could introduce error in the size categories assigned to species. However, stomatopods have indeterminate growth (Reaka 1979a,b), and immature individuals can be recognized by their morphology. Thus, it is unlikely that a large species would be misclassified as moderate or small because only juveniles had been collected. In addition, the size categories are broad (50-150 mm) and sample sizes are relatively large (see below), decreasing the likelihood that the size of a species would be misclassified due to sampling error. The overall mean and standard error for number of individuals collected per species is 63 (±15 SE). There is no significant correlation between maximum size of individuals recorded and the number of individuals collected per species ($r = 0.13$, $N = 150$, $p > 0.05$). Furthermore, as is documented below, the number of individuals collected does not vary significantly among different habitat types, or if sample sizes do vary among habitat types, they favor the null hypothesis. A bias due to the use of museum specimens would be most likely if sample sizes were small or unequal, or if collectors selected large or unusual specimens by hand. Because of their cryptic, burrow-dwelling habits, however, stomatopods rarely are collected by sighting and capturing individual specimens. Most museum specimens are collected 'en masse' at poison stations, which are nonselective for body size, or in trawls. Once killed at a poison station, all specimens usually are preserved. Trawls possibly could selectively retain large individuals, or only conspicuous specimens might be preserved, but statistical analyses showed that large stomatopods were not more or less numerous than small stomatopods in these analyses (see below). Consequently, the conclusions about different sizes of species found in different types of habitat do not result from the number of individuals included or the manner in which they were collected.

Substrata are categorized as 'hard bottom' (coral, rock, coralline algal matt) and 'soft bottom' (mud, sand, sometimes mixed with shell). Depths are defined as 'shallow' (where the species is recorded only from depths <100 m), 'moderate' (100-300 m), or 'deep' (>300 m). These depth categories differ from those used in Reaka (1980a), where finer subdivisions of depth distributions were analyzed for coral-dwelling stomatopods. Latitudinal distributions are classified as 'tropical' and 'temperate'. Tropical species are defined arbitrarily as those which occur at any depth within the latitudes where coral is known to grow (>20°C surface temperature, although note that this definition is delimited by latitude rather than temperature, since all depths are considered). Temperate species are those which occur at any depth outside of tropical latitudes (there are no boreal or polar stomatopods). 'Warm' environments are categorized here as those in tropical latitudes at shallow or moderate depths; species that live in temperate regions or deep tropical water are classified as inhabiting 'cold' environments. Postlarval sizes are termed 'small' (<10 mm total length), 'moderate' (10-20 mm), or 'large' (>20 mm). In a separate study with a different focus, Reaka & Manning (1986c) provide the range of known body sizes, the number of individuals recorded, and a summary of habitat and distribution for each species of stomatopod in the Atlanto-East Pacific region. Detailed references and sources of information on body size, distribution, and larval characteristics for each group of stomatopods are available in that publication.

In statistical tests, data were log transformed if necessary in order to meet assumptions of normal distributions and homogeneity of variance for t tests and one way analyses of variance; in the few cases where a significant Bartlett's test or small sample sizes suggested that the data still did not meet assumptions, a nonparametric test (e.g. Mann Whitney U test) was used. I have used $p < 0.05$ as the criterion for statistical significance, but have reported instances of $p = 0.06$ or $p = 0.07$ for the reader's information. Other details of statistical methos are explained in the text and tables.

3 RESULTS

3.1 *Adult body size and environmental factors*

3.1.1 *General comparisons*

Table 1 provides the incidence of species characterized as small, moderate, or large in body size for different habitats and geographic areas, as well as the relationship of postlarval size to adult body size. The number of taxa includes all Atlanto-East Pacific species for which adequate data on body size, habitat, distribution, or postlarval size are available.

3.1.1.1 *Substratum.* Clearly, species of stomatopods that inhabit soft bottoms reach significantly larger body sizes than those that occupy hard substrata (Table 1A). Larger sample sizes in a particular type of habitat possibly could increase the maximum size of individuals observed and thus increase the number of species that fall into larger size categories. However, more individuals have been collected in species from hard bottom ($\bar{X}$ = 134 ± 45 SE) than in those from soft bottom environments ($\bar{X}$ = 40 ± 13 SE)

Table 1. The relationship of adult body size to environment, faunal region, and postlarval size in Atlanto-East Pacific stomatopod taxa. For comparative purposes, the number of taxa in each size category are expressed as a proportion of the total taxa that are known to occur in a particular habitat or that have postlarvae of a particular size. All χ^2 tests were performed on raw data. As discussed in the text, one taxon was added to the data cells with no observations (*) to make statistical testing possible; χ^2 values are based on tests including the added datum. This artifact favors the null hypothesis.

	No. of taxa	Adult body size Small (<50 mm)	Moderate (50-150 mm)	Large (>150 mm)	Statistical test
A. Substratum					
Hard bottom (coral, rock)	36	64 %	36 %	0 %*	$\chi^2 = 12.74$
Soft bottom (mud, sand)	102	33 %	42 %	25 %	df = 2
					$p < 0.002$
B. Depth					
Shallow (<100 m)	102	57 %	30 %	13 %	$\chi^2 = 20.46$
Moderate (100-300 m)	20	25 %	60 %	15 %	df = 4
Deep (>300 m)	20	15 %	45 %	40 %	$p < 0.001$
C. Latitude					
Tropical	118	49 %	36 %	14 %	$\chi^2 = 6.21$
Temperate	28	25 %	46 %	29 %	df = 2
					$p < 0.050$
D. Temperature					
Warm	103	54 %	34 %	12 %	$\chi^2 = 12.41$
Cold	41	24 %	46 %	29 %	df = 2
					$p < 0.003$
E. Zoogeographic subregion					
East Atlantic	32	44 %	34 %	22 %	$\chi^2 = 4.67$
West Atlantic	69	39 %	46 %	14 %	df = 4
East Pacific	46	54 %	28 %	17 %	$p > 0.10$
F. Postlarval size					
Small (<10 mm)	38	82 %	18 %	0 %*	$\chi^2 = 35.41$
Moderate (10-20 mm)	71	42 %	39 %	18 %	df = 4
Large (>20 mm)	21	0 %*	57 %	43 %	$p < 0.001$

($t = 3.61$, df = 138, $p < 0.01$). Therefore, if sample size has any effect upon the conclusions, it is to bias the data toward the null hypothesis.

Species that excavate burrows in sand reach particularly large sizes. A West Atlantic sand-dweller, *Lysiosquilla scabricauda*, reaches 275 mm in length. The largest known stomatopod, *Lysiosquilla maculata*, also lives in sandy habitats and attains lengths of 385 mm in the Indo-West Pacific (Reaka & Manning 1986c). Many mud-dwelling squillids and harpiosquillids, as well as the bathysquillids, grow to very large sizes (200-300 mm) as well (Manning 1969a,b, 1980, Manning & Struhsaker 1976, Reaka & Manning 1986c). However, not one coral or rock-dwelling species achieves a body length greater than 150 mm, and most are smaller than 50 mm in length (Table 1A). In order to make statistical testing valid, one taxon was added to the category of large body size for taxa from hard substrata (asterisk in Table 1A). This artifact biases the statistical results toward the null hypothesis and lowers the χ^2 value. Nevertheless, the relationship between type of substratum and body size remains highly significant.

3.1.1.2 *Depth.* When all Atlanto-East Pacific stomatopod taxa are considered, the number of species that attain large body sizes increases significantly at greater depths (Table 1B). These data do not result from collecting artifacts, since the number of individuals recorded does not vary significantly among species collected from shallow, moderate, or deep waters (one way analysis of variance; $F = 0.75$; $df = 2,134$; $p > 0.05$).

The bathysquillids provide an example of the relationship between body size and depth. These stomatopods occupy deep, continental slope habitats (ca. 200-1300 m, Manning 1969a, Manning & Struhsaker 1976). Members of this ancient group of stomatopods (with affinities to the fossil sculdids) all reach large body sizes (maximum lengths 160-300 mm). On the other hand, some of the shallow-dwelling lysiosquilloids and squilloids attain similar or even larger sizes (e.g. *Lysiosquilla maculata*, above; *Harpiosquilla harpax* $\leqslant$ 335 mm, Manning 1969b, Reaka & Manning 1986c). In addition, the pattern of body sizes found at different depths for Atlanto-East Pacific stomatopods in the present study differs from the results obtained when only coral-dwelling gonodactyloids from all geographic regions are analyzed over a finer depth gradient (Reaka 1980a). In the latter, body size decreases in species that occur in deeper water compared to those that occupy exclusively shallow (0-10 m) habitats.

3.1.1.3 *Latitude.* Species of stomatopods that occur in temperate latitudes generally reach larger body sizes than taxa that are restricted to tropical latitudes (Table 1C). However, again exceptions occur, since some of the largest known stomatopods (e.g. *Lysiosquilla, Harpiosquilla*) are collected from tropical environments. There is no significant difference between the number of individuals collected in species from tropical or temperate habitats ($t = 0.76$, $df = 148$, $p > 0.05$).

3.1.1.4 *Temperature.* Large body sizes of stomatopods are also correlated significantly with cold environments (Table 1D). The number of individuals collected does not vary significantly among species from cold or warm habitats ($t = 0.59$, $df = 146$, $p > 0.05$).

3.1.2 *Interacting factors*

Some of the above trends may be interrelated. For example, stomatopods that burrow in soft bottoms attain larger body sizes than those restricted to holes in hard substrata, and stomatopods from deep water reach larger sizes than those from shallow water. Large body sizes in species from deep water may be related in part to cold temperatures; also, a significantly higher proportion of species inhabit soft bottoms in deep than in shallow habitats (Table 2A). The large body sizes observed at high latitudes also may be related to substratum affiliations. A higher proportion of species are found on soft bottoms in temperate than in tropical regions (Table 2C). Furthermore, temperate areas support more species than expected in moderate and deep habitats, while tropical waters support more species than expected in shallow environments (Table 2B). Consequently, it is desirable to examine in greater detail some of the general trends in body size given above.

3.1.3 *Restricted comparisons: The effects of a single variable*

3.1.3.1 *Substratum.* Taxa from soft bottoms still reach larger body sizes than those from hard substrata when only stomatopods occurring in shallow tropical habitats are considered (Table 3A, note that one datum was inserted into the empty category of large body size for stomatopods on hard bottoms, favoring the null hypothesis and lowering the χ^2 value

Table 2. The frequency of occurrence of stomatopod taxa in different environments. The number of taxa represents the total number of Atlanto-East Pacific species for which adequate data were available in both habitat types (e.g. the number of taxa for which both depth and substratum type are known). The percentages are provided to facilitate comparisons. They indicate, for example, the number of taxa found in a given depth range relative to the total number of taxa found on a given type of substratum. All χ^2 tests are based on raw data.

	No. of taxa	Depth Shallow	Moderate	Deep	Statistical text
A. Substratum					
Soft bottom	98	61 %	20 %	18 %	$\chi^2 = 12.09$
Hard bottom	37	92 %	3 %	5 %	df = 2
					p <0.003
B. Latitude					
Temperate	27	48 %	26 %	26 %	$\chi^2 = 9.42$
Tropical	116	78 %	11 %	11 %	df = 2
					p <0.010
		Substratum Soft bottom	Hard bottom		
C. Latitude					
Temperate	25	96 %	4 %		$\chi^2 = 7.26$
Tropical	114	70 %	30 %		df = 1
					p <0.008
D. Temperature					
Cold	38	92 %	8 %		$\chi^2 = 9.35$
Warm	98	66 %	34 %		df = 1
					p <0.003

below the actual but untestable value). This result is not biased by the number of individuals collected, since more individuals are reported for species from hard than soft bottoms in shallow tropical environments, favoring the null hypothesis ($t = 3.33$, df = 83, $p < 0.005$). Since few species from hard substrata occur at high latitudes or in deep waters, it is not possible to test the relationship of substratum type to body size within the temperate latitudes or at other depths.

3.1.3.2 *Depth.* Depth is *not* significantly related to body size when only taxa from soft bottom tropical habitats are considered (Table 3B). Numbers of individuals collected do not differ for taxa in shallow, moderate, or deep habitats in this analysis ($F = 0.30$; df = 2,72; $p > 0.05$). Unfortunately, taxa from temperate latitudes and hard bottoms provide too few data for valid statistical testing of the relationship between body size and depth in other restricted types of habitats.

When all soft bottom species (tropical and temperate) are included, body size possibly increases in deep water (Table 3C). This result, though, is likely to be biased by the preponderance of relatively large species living on soft bottoms in deep water in temperate zones (Tables 1, 2). Hence, this test represents a less rigorous comparison for the effects of depth than is found in Table 3B. The number of individuals collected for taxa included in Table 3C does not differ at different depths ($F = 0.79$; df = 2.94; $p > 0.05$).

Table 3. Restricted comparisons for the relationship between adult body size and environment or faunal region in all cases where sample sizes permitted analyses. Data are expressed and tested as in Table 1, and the asterisk again indicates that one taxon was added to the empty data cell in order to make testing possible. No species of large adult body size are observed in tropical shallow hard bottoms (dashes in H); consequently, statistical tests included only small and moderately sized species. The upper χ^2 test value in H includes all three zoogeographic subregions, while the lower test value represents only West Atlantic and East Pacific comparisons.

	No. of taxa	Adult body size: Small (<50 mm)	Moderate (50-150 mm)	Large (>150 mm)	Statistical test
A. Substratum (only shallow tropical species)					
Hard bottom (coral, rock)	34	62 %	38 %	0 %*	$\chi^2 = 5.47$
Soft bottom (mud, sand)	51	45 %	35 %	20 %	df = 2
					$p > 0.06$
B. Depth (only soft bottom tropical species)					
Shallow (<100 m)	52	44 %	39 %	17 %	$\chi^2 = 5.23$
Moderate (100-300 m)	13	38 %	46 %	15 %	df = 4
Deep (>300 m)	10	10 %	50 %	40 %	$p > 0.10$
C. Depth (only soft bottom species, tropical and temperate)					
Shallow (<100 m)	59	42 %	41 %	17 %	$\chi^2 = 8.58$
Moderate (100-300 m)	21	29 %	48 %	24 %	df = 4
Deep (>300 m)	17	6 %	59 %	35 %	$p = 0.07$
D. Latitude (only shallow soft bottom species)					
Tropical	50	46 %	34 %	20 %	$\chi^2 = 1.28$
Temperate	8	25 %	50 %	25 %	df = 2
					$p > 0.10$
E. Latitude (only soft bottom species, all depths)					
Tropical	79	37 %	43 %	20 %	$\chi^2 = 4.66$
Temperate	23	13 %	61 %	26 %	df = 2
					$p > 0.10$
F. Temperature (only soft bottom species)					
Warm	60	38 %	43 %	18 %	$\chi^2 = 7.59$
Cold	34	12 %	59 %	29 %	df = 2
					$p > 0.02$
G. Zoogeographic subregion (only tropical shallow soft bottom species)					
East Atlantic	13	46 %	31 %	23 %	$\chi^2 = 0.80$
West Atlantic	22	50 %	36 %	14 %	df = 4
East Pacific	17	41 %	41 %	18 %	$p > 0.10$
H. Zoogeographic subregion (only tropical shallow hard bottom species)					
East Atlantic	5	60 %	40 %	–	$\chi^2 = 5.59$
West Atlantic	14	43 %	57 %	–	df = 2
East Pacific	14	86 %	14 %	–	$p = 0.06$
					$\chi^2 = 5.60$
					df = 1
					$p = 0.02$

3.1.3.3 *Latitude.* Latitude is *not* significantly related to body size when only taxa from shallow soft bottom habitats are analyzed (Table 3D). The number of individuals collected is not different for the tropical and temperate species included in this analysis ($t = 0.84$, df = 56, $p > 0.05$).

Similarly, when taxa from soft bottoms at all depths are considered, species do not differ significantly in size at high and low latitudes (Table 3E). Numbers of individuals collected do not vary between taxa inhabiting soft bottoms (all depths) from temperate versus tropical latitudes ($t = 0.47$, $df = 100$, $p > 0.05$).

3.1.3.4 *Temperature.* Body size increases significantly in habitats characterized by low temperatures when only taxa from soft substrata are included (Table 3F). Numbers of individuals collected for taxa in these analyses do not vary between warm and cold environments ($t = 0.54$, $df = 92$, $p > 0.05$). Inadequate numbers of species occurring on hard substrata in cold environments prevent testing the relationship between body size and temperature in stomatopods from hard bottoms.

3.1.4 *Conclusions*

These results suggest that the environmental characteristics which are most directly correlated with adult body size in species of stomatopods are substratum type and temperature. The original overall correlations between large adult body size versus high latitudes and deep water appear to be spurious, and large body size in these environments appears to be influenced instead by the high proportions of species living on soft bottoms in these cold environments.

3.2 *Adult body size in different zoogeographic subregions*

3.2.1 *General comparisons*

When all species of stomatopods from the East Atlantic, West Atlantic, and East Pacific oceans are compared, there are no significant differences in body sizes among species in the three zoogeographic subregions (Table 1E). The number of species that occur in hard versus soft bottom habitats or in shallow versus moderate or deep water is not significantly different in the three oceans (Table 4A,B). Furthermore, the number of taxa that are found in temperate versus tropical, or in cold versus warm environments does not vary significantly among the three subregions (Table 4C,D).

3.2.2 *Restricted comparisons*

When only taxa from tropical shallow soft bottoms are considered, body size does not vary among the three geographic areas (Table 3G). The numbers of individuals collected in these taxa do not differ significantly among these oceans ($F = 0.50$, $df = 2,49$, $p > 0.05$).

On the other hand, when only taxa from tropical shallow hard bottoms are compared, body sizes probably differ ($p = 0.06$) among the three subregions (although only five species are recorded from the East Atlantic). Considering only West Atlantic and East Pacific species, the West Atlantic species attain significantly larger body sizes than their East Pacific counterparts ($p = 0.02$, Table 3H). The number of individuals collected in taxa from tropical shallow hard bottoms does not vary between the West Atlantic and East Pacific ($t = 1.79$, $df = 26$, $p > 0.05$).

3.3 *Postlarval size*

Independent evidence from rearing studies and field collections shows that egg size is positively correlated with size of larvae and postlarvae, and that size of postlarvae correlates

Table 4. The frequency of occurrence of stomatopod taxa in different types of environments in different zoogeographic subregions. As in Tables 1 and 2, data are represented in percentages for purposes of comparisons, and statistical tests are based on raw data.

Zoogeographic subregion	No. of taxa	Environment			Statistical test
A. Substratum		Soft bottom	Hard bottom		
East Atlantic	29	83 %	17 %		$\chi^2 = 2.24$
West Atlantic	64	75 %	25 %		df = 2
East Pacific	46	67 %	33 %		$p > 0.05$
B. Depth		Shallow	Moderate	Deep	
East Atlantic	28	71 %	14 %	14 %	$\chi^2 = 5.19$
West Atlantic	68	66 %	15 %	19 %	df = 4
East Pacific	44	82 %	14 %	5 %	$p > 0.05$
C. Latitude		Tropical	Temperate		
East Atlantic	32	72 %	28 %		$\chi^2 = 2.48$
West Atlantic	70	84 %	16 %		df = 2
East Pacific	49	84 %	16 %		$p > 0.05$
D. Temperature		Warm	Cold		
East Atlantic	31	68 %	32 %		$\chi^2 = 3.56$
West Atlantic	69	67 %	33 %		df = 2
East Pacific	45	82 %	18 %		$p > 0.05$

with length of larval life and dispersal capability in stomatopods (Reaka 1979a, 1980a, Reaka & Manning 1986c, see Steele 1977, and Strathman 1977, for similar trends in a number of crustaceans).

The size of settling postlarvae are known for 36 Atlanto-East Pacific species of stomatopods and their Indo-West Pacific cognates. Within genera of stomatopods, sizes of postlarvae are remarkably consistent (see Reaka & Manning 1986c). Consequently, approximate postlarval size can be estimated for 130 Atlanto-East Pacific taxa.

3.3.1 *Relationship to adult size*

When postlarvae are categorized as small (<10 mm), moderate (10-20 mm), or large (>20 mm) for all Atlanto-East Pacific genera in which postlarval size is known, the number of species characterized by larger postlarvae increases significantly with increased size of the adults (Table 1F). No small postlarvae are found in taxa characterized by large adults, and no large postlarvae are known in taxa with small adults. In order to test these data statistically, one taxon was inserted in each of the categories without representatives (asterisk in Table 1F). Although this artifact biases the results toward the null hypothesis and lowers the χ^2 value, the relationship between postlarval size and adult size is still highly significant.

These results are consistent with those drawn only from the 36 species in which postlarvae are definitely known. An analysis of variance shows that the mean size of known postlarvae increases significantly among taxa of small, moderate and large adult body size (Table 5A).

3.3.2 *Relationship to environmental factors*

Comparisons among the 36 species with known postlarvae show that postlarvae are

Table 5. The relationship of postlarval size (mean and standard error) to adult body size, environment, and faunal region. EA is East Atlantic, WA is West Atlantic, EP is East Pacific, and IWP is Indo-West Pacific. Data are analyzed, according to procedures discussed in the text, by t-tests and one way analyses of variance.

	Postlarval size (mm length)				Statistical test
A. Adult body size	Small species (<50 mm)	Moderate species (50-150 mm)		Large species (>150 mm)	
$\bar{X}$ (±SE)	12.6 (±1.4)	18.5 (±2.3)		21.5 (±1.7)	F = 4.36 df = 2,33
N	10	15		11	p < 0.025
B. Substratum	Hard bottom		Soft bottom		
$\bar{X}$ (±SE)	14.1 (±1.7)		18.7 (±1.5)		t = 2.06 df = 34
N	16		20		p < 0.05
C. Latitude	Tropical		Temperate		
$\bar{X}$ (±SE)	15.9 (±1.2)		22.2 (±2.9)		t = 1.96 df = 34
N	30		6		p = 0.06
D. Depth	Shallow		Moderate-deep		
$\bar{X}$ (±SE)	16.2 (±1.2)		16.4 (±3.7)		t = 0.06 df = 33
N	29		6		p > 0.10
E. Temperature	Warm		Cold		
$\bar{X}$ (±SE)	15.7 (±1.3)		18.6 (±2.3)		t = 1.14 df = 34
N	26		20		p > 0.10
F. Zoogeographic subregion	EA	WA	EP	IWP	
$\bar{X}$ (±SE)	17.3 (±1.9)	13.8 (±1.6)	26.0 (±2.6)	18.0 (±2.5)	F = 4.24 df = 3,32
N	11	15	4	6	p < 0.025

significantly smaller in stomatopods from hard than soft bottom habitats (Table 5B), and probably are smaller in species from tropical rather than temperate environments (Table 5C). No difference in postlarval size is detectable for species from different depths (Table 5D), or for species in warm versus cold environments (Table 5E). Because adult body size is highly correlated with postlarval size, it is necessary to control for interacting variables such as those observed in Table 2 when interpreting these data.

3.3.3 *Restricted comparisons of environmental factors*

Species that occupy soft bottoms produce significantly larger postlarvae than those from hard substrata when only species from cold environments are analyzed (Table 6A). However, there is no significant difference in the size of settling postlarvae for species that inhabit soft versus hard substrata when only species from shallow tropical (warm) environments are considered (Table 6B).

Tropical species probably (p = 0.06) produce smaller postlarvae than temperate species (Table 6C) when only soft bottom species are compared.

Table 6. Restricted comparisons for the relationship between postlarval size and isolated environmental factors in all cases where sample sizes permitted analyses. Data are analyzed by Mann Whitney U tests.

	Postlarval size (mm length)		Statistical test
A. Substratum	Hard bottom	Soft bottom	
(only cold environments)			
Median (Range)	11 (9-13)	21 (12-30)	U = 2
N	3	6	p = 0.025
B. Substratum	Hard bottom	Soft bottom	
(only shallow tropical = warm environments)			
Median (Range)	13 (6-29)	15 (9-25)	U = 73
N	13	13	p >0.10
C. Latitude	Tropical	Temperate	
(only soft bottom species, all depths)			
Median (Range)	15 (9-25)	21 (12-30)	U = 21
N	13	6	p = 0.06
D. Depth	Shallow	Moderate-deep	
(only tropical hard bottom species)			
Median (Range)	13 (6-29)	11 (9-13)	U = 16
N	16	3	p >0 10
E. Temperature	Warm	Cold	
(only soft bottom species)			
Median (Range)	15 (9-25)	21 (12-30)	U = 21
N	13	6	p = 0.06
F. Temperature	Warm	Cold	
(only hard bottom species)			
Median (Range)	13 (6-29)	11 (9-13)	U = 16
N	13	3	p >0.10

Because of sample sizes, it is possible to conduct restricted tests for the effect of depth upon postlarval size only for tropical hard bottom species; postlarval size does not vary with depth in these comparisons (Table 6D).

Species from warm environments probably (p = 0.06) produce smaller postlarvae than those from cold environments (Table 6E) when only soft bottom species are included in the analysis. Temperature does not influence size of postlarvae in species from hard substrata (Table 6F).

3.3.4 *Geographic comparisons*

The size of recorded postlarvae varies among geographic subregions (Table 5F), but these results must be discounted due to the strong influence of substratum. The large postlarvae known from the East Pacific all are from soft substrata, as are the majority of species with known postlarvae in the East Atlantic. In contrast, the majority of species from the West Atlantic with known postlarvae are from hard bottoms.

3.3.5 *Conclusions*

The factors that most closely correlate with postlarval size and dispersal ability are adult body size and substratum. Although the relationships are more variable, temperature and latitude may be related to postlarval size in some cases.

4 DISCUSSION

4.1 *Pattern and constraint in body size*

This study demonstrates strong correlations between adult body size of stomatopods and particular characteristics of the environment. The increased sizes of stomatopods found at high latitudes and in deep water are likely to result in part from low temperatures in these environments, since low temperatures impose slow growth rates and large terminal sizes in other organisms (Ray 1960, McLaren 1965, Berven et al. 1979, Grime & Mowforth 1982, but also see discussion in Clarke 1983). However, a high proportion of stomatopods occurring in deep environments or at high latitudes also occupy soft substrata, and one of the striking results of this study is the demonstration of a significant relationship between body size and type of substratum. Substratum undoubtedly exerts a major influence on the biology of most benthic marine organisms, but, oddly enough, this factor has not previously been considered as a component of shifts in body size along latitudinal or depth gradients.

The strong relationship between type of substratum and body size in mantis shrimps initially seems straight-forward, but closer examination raises enigmatic questions that require further research. Burrows provide a critical haven from fish predators and an important refuge in which to molt, mate, and brood young (Reaka 1975b, 1976, 1980b,c, Dingle & Caldwell 1972). Stomatopods that live in sand or mud excavate vertical or U-shaped burrows (Reaka & Manning 1981); these burrows can be enlarged easily as the stomatopod grows. Stomatopods from hard bottoms occupy pre-existing crevices or holes that have been bored by sponges, polychaetes, sipunculans, or bivalves. Except for minute chipping, these stomatopods cannot enlarge their burrows, and holes of larger size in the rubble become increasingly rare (Reaka & Moran in preparation). Considerable behavioral and morphological evidence suggests that stomatopods inhabiting coral and rock substrata are better adapted for interference competition than those which excavate their own burrows in soft substrata; most of this fighting is associated with defense of the burrow (Reaka & Manning 1981). Furthermore, manipulative field experiments demonstrate that competition for space occurs in coral-dwelling stomatopods (Reaka 1980c). These observations suggest that many stomatopods inhabiting hard substrata are limited by the availability of burrows, and that their body size cannot exceed the sizes of holes available within the coral and rock rubble.

However, numerous decapod crustaceans also inhabit holes in rock and coral substrata. Many of these show fighting behavior (e.g. xanthid and majid crabs, alpheid shrimps, thalassinids), and field experiments demonstrate that these populations also are increased by increasing the availability of burrows in local coral rubble habitats (Reaka 1980c and in preparation). Nevertheless, unlike stomatopods from hard substrata, some lineages of cavity-dwelling decapods reach very large sizes. Rather than being restricted to holes and crevices within coral rubble, members of these large lineages hide in shallow caves and under protective overhangs on the reef during the day and emerge to feed at night. Various spiny lobsters (*Panulirus, Justitia, Palinurellus*) and slipper lobsters (*Scyllarides*) reach large body sizes (often up to several hundred millimeters in length and many pounds in weight), and some crabs (*Carpilius, Daldorfia, Eriphea, Portunus, Mithrax*) attain very large body sizes on reefs. In one case, this transition to large size is shown dramatically even within one genus; individuals of small species of *Mithrax* occupy limited preformed holes in coral, while one species, *M. spinosissimus*, reaches carapace widths of almost 200 mm and inhabits caves or other shelters on reefs.

It is not clear why reef-dwelling stomatopods could not inhabit cavernous habitats on reefs as do these large decapods. Such habits would allow reef-dwelling stomatopods to attain body sizes similar to those of mantis shrimps that burrow in soft bottoms. Stomatopods from both types of habitat are raptorial predators and conform to a similar general body plan. In stomatopods from soft substrata, however, the raptorial maxilliped is slender, and spiked with grasping teeth. Stomatopods from hard bottoms generally bear a similar but inflated and heavily calcified hammering appendage (see Fig.1 and Manning 1969a, 1980, Reaka 1975b, Caldwell & Dingle 1975). This maxilliped operates by a click mechanism, where muscle tension is built up in the enlarged merus while the propodus and dactyl are held folded against the merus (Burrows 1969, Burrows & Hoyle 1972). Upon reaching a tension threshhold, the click mechanism suddenly is released, and the propodus and dactyl spring forward, usually with the dactyl still folded into the propodal sheath. The distal portion of the propodus and especially the proximal portion of the dactyl are inflated and heavily calcified, so that the distal portion of this folded appendage (the 'hammer') resembles a pendulum (Fig.1) as it swings forward at remarkable speeds (angular and linear velocities of 20000°/sec. and 10 m/sec. respectively, Burrows 1969). In some cases, the toothless dactyl also may be opened, and the needle-sharp tip is thrust forward as a spear. It is possible that in the coral-dwelling gonodactyloids the accumulation of muscle tension and sudden release of the pendulum-like 'hammer' would impose too much stress on the muscles and skeleton of the appendage if animals were larger than ca. 150 mm; similarly, the impact from such an appendage might fracture the skeletal structure of the 'hammer' itself. The possibility that mechanical constraints imposed by the heavy hammering appendage prevent coral-dwelling lineages of stomatopods from reaching larger body sizes needs to be investigated.

Additionally, coral-dwelling stomatopods, particularly the larger species, are more active and aggressive than species from soft bottoms (Reaka & Manning 1981). Growth rates, reproductive rates, egg number, egg size, and aggressive level all increase with larger adult body sizes in coral-dwelling gonodactyloids (Reaka 1978, 1979a,b, 1980a). By comparison, most of the very large reef-dwelling decapods (except portunids) are relatively sluggish (pers. obs.). Coral-dwelling stomatopods possibly cannot acquire sufficient food to meet high metabolic requirements if they are larger than the moderate sizes observed. Comparative studies of metabolic rates, such as those conducted by McNab (1980) for mammals, are badly needed in marine invertebrates. Such studies could provide a fruitful approach to understanding factors that constrain body sizes in some groups.

Although species inhabiting hard substrata are primarily tropical (in contrast to those from soft bottoms; Table 2D), the small size of taxa from hard bottoms is not explained by temperature, since soft bottom taxa still attain larger sizes than taxa from hard bottoms when only shallow tropical species are considered (Table 3A). There is no compelling reason to expect that productivity should be lower on hard substrata than on adjacent soft bottoms in shallow tropical oceans. Also, as shown in the introduction, it is not clear that increased competition for resources in hard versus soft bottom environments (Reaka & Manning 1981) would favor unilateral shifts toward smaller sized species. It

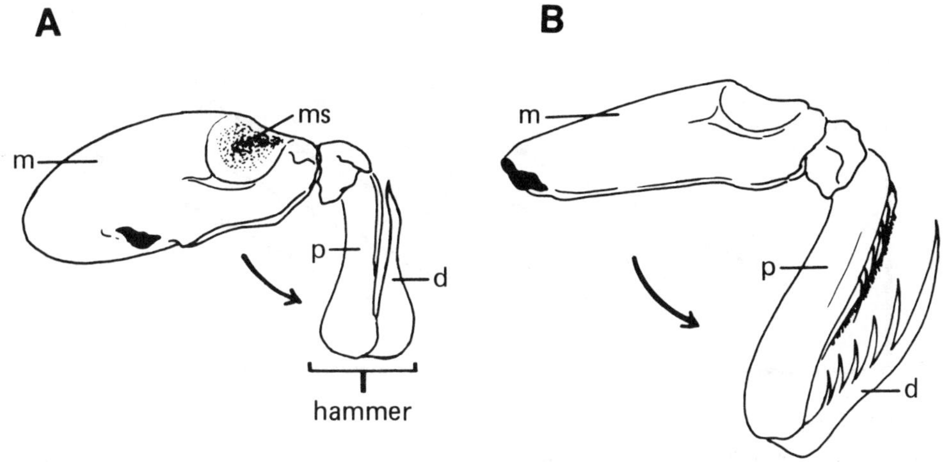

Figure 1. Raptorial second maxillipeds (mesial view) of (A) coral-dwelling species, *Gonodactylus aloha* Manning & Reaka 1981; and (B) mud-dwelling species, *Squilla empusa* Say 1818. Black area indicates the site of subterminal and terminal attachment, respectively, of the appendage to the thorax in the two stomatopods. m – merus; p – propodus; d – dactyl; ms – colored meral spot.

is possible that predation is more intense in reef than soft bottom environments, preventing the attainment of large size in species of stomatopods from hard bottoms. Hay (1981) has shown that herbivory by fishes is more intense on reefs than on adjacent sand flats where herbivores lack structural protection from other fish predators. However, predation has not prevented large decapods from occupying reef habitats. It may be significant that these large, relatively sluggish decapods are nocturnal, whereas reef-dwelling stomatopods are diurnal (pers. obs., and Dominguez & Reaka in review).

Although the amphi-American stomatopods inhabiting hard bottoms are derived from the same Miocene stock, present day taxa attain larger body sizes in the West Atlantic than in the East Pacific. Due to higher productivity in the East Pacific than the West Atlantic (Koblentz-Mishke et al. 1970), the dwarfism of East Pacific taxa probably is not explained by lower availability of prey for stomatopods in the latter region. Burrows in hard substrates are a known limiting resource for stomatopods (Reaka 1980c, Reaka & Manning 1981), and larger or more numerous burrows in the West Atlantic than in the East Pacific might account for the differences in stomatopod size between the two regions. However, the limited information available does not support this hypothesis. Boring bivalves provide an indication of the frequencies of boring organisms in the two regions, and the incidence of these organisms is higher in the East Pacific than in the West Atlantic (Highsmith 1980). More intense predation on East Pacific than on West Atlantic reefs may explain the relatively small sizes of the East Pacific stomatopod fauna, as discussed in the introduction. Vermeij (1976, 1978) found that in benthic molluscs morphological defenses against predators (particularly fishes and crabs) become less elaborate in a gradient from the Indo-West Pacific to the East Pacific and West Atlantic. Also, Bertness (1982) reported that gastropod shells glued to intertidal substrata are more frequently broken, apparently by predators, on the Pacific than the Atlantic coast of Panama. Experimental data for the impact of predation upon reef-dwelling stomatopods in the two areas are needed.

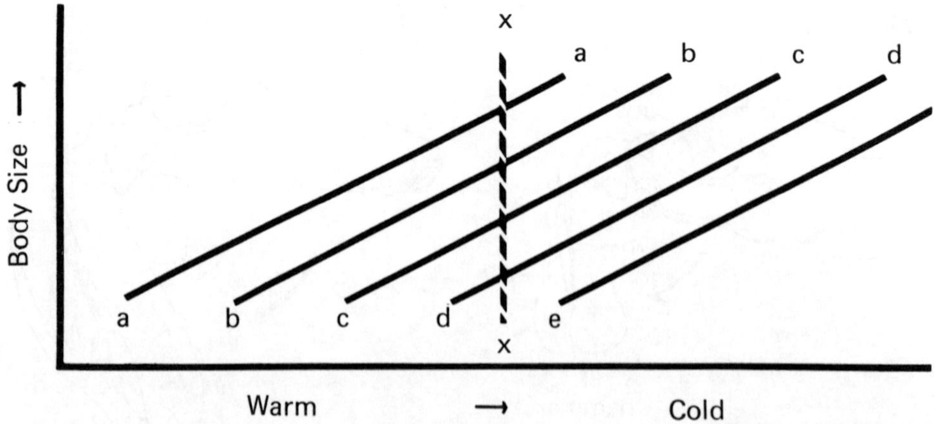

Figure 2. Clines of body size along an environmental gradient (e.g. warm to cold, low to high latitudes, or shallow to deep) for species with serially overlapping ranges. If the edges of the ranges are regularly spaced (a, b, c, d), then the sympatric species will show constant ratios of body sizes at a given geographical location (x).

Regardless of the mechanisms that mold body size, however, these patterns of body size can have powrful consequences for the ecology and evolution of these organisms.

4.2 *Ecological and evolutionary consequences of shifts in body size*

The observed shifts in adult and postlarval size with environmental characteristics have implications for the debates on character displacement and constant ratios of body size in communities, for life history patterns, and for patterns of evolution in different stomatopod taxa, habitats, and biogeographic regions.

The relationships between adult body size and environmental characteristics shown here demonstrate that one must use caution when examining possible instances of character displacement or ratios of body size among sympatric species. For example, if a predictable cline in body size occurs along a latitudinal or depth gradient, and if species distributions form regular overlapping patterns, then the four species at locality 'x' in Figure 2 will show constant ratios of body size. This pattern of body sizes would result not from competition but from regular spacing of the edges of the distributions (Fig.2a,b,c,d). Although no one has applied this argument to the controversy over constant ratios of body size among species, Grant (1972) has noted the necessity for examining environmentally imposed clines before interpreting cases of character displacement in pairs of species.

Stomatopods are the top predators in an abundant cryptic reef community, where they smash gastropods, bivalves, and crustaceans, as well as soft bodied prey (Reaka 1980c). They are very active, presumably have high metabolic rates, and can be expected to have a significant effect upon the benthic community. Due to the larger sizes attained, stomatopod predators are likely to exert a stronger ecological and evolutionary impact upon their cryptic prey in the West Atlantic than in the East Pacific.

The relationship between body size and environment in stomatopods also has strong consequences for dispersal in different habitats. Larvae which settle at large sizes spend long periods (up to several months) in the plankton, while those which settle at smaller sizes spend only a few weeks in the plankton (Manning & Provenzano 1963, Pyne 1972, Morgan & Provenzano 1978, Provenzano & Manning 1978). Even within the coral-dwelling gonodactyloids, larger species with larger postlarvae have wider geographic ranges than smaller species; also, species size is more highly correlated with longitudinal than latitudinal distributions, suggesting that larval dispersal by currents is greater in larger than in smaller species (Reaka 1980a). It is reasonable to hypothesize that longer planktonic lives may be associated with greater larval mortality (predation, starvation) (Thorson 1950, Vance 1973), with less predictable recruitment into any one site (Thorson 1946), and with less deterministic community structure (Sale 1977, 1980) in large rather than in small species of stomatopods. In turn, high mortality of juveniles relative to adults favors longevity and iteroparity of adults (Murphy 1968, Schaffer 1974, Stearns 1976, Barclay & Gregory 1981), well exemplified by stomatopods (Reaka 1976, 1979a).

Additionally, in species of larger body size, larvae are increasingly likely to be swept away from parental populations because of their longer planktonic phases, and settling juveniles are less likely to be related to adults than in species of small size. Like many other planktonic larvae (Alldredge & King 1977, 1980, Porter & Porter 1977, Robichaux et al. 1981), stomatopod larvae show vertical migrations, where they hide on the substratum by day and emerge into the water column to feed at night (Reaka in preparation). This behavior reduces dispersal, so that the larvae of smaller species with relatively short planktonic phases may settle in the same bay as the parental population in many cases. Consequently, the opportunities for kin selection or rapid genetic shifts in local populations are greater in small than in large species of stomatopods (see also Shapiro 1983, for a discussion of mechanisms promoting kin selection in reef fishes).

Reaka & Manning (1981, 1986c) examined evolutionary rates in Atlanto-East Pacific stomatopods by quantifying the incidence of conspecific, cognate, and endemic taxa, as well as the frequency of species multiplication and apparent extinctions, among zoogeographic subregions. These measures of evolutionary change were compared among stomatopods of different adult body sizes, dispersal potential, habitat characteristics, and competitive and behavioral regimes. Phyletic divergence among subregions and species multiplication within subregions are most dramatically correlated with adult body size and dispersal potential ($p < 0.01$). Species with the largest adult body size and the greatest dispersal potential evolve the most slowly. Furthermore, evolutionary innovations (such as major habitat shifts) occur only in the smallest members of different lineages (gonodactylids, pseudosquillids, lysiosquilloids, squilloids) of stomatopods (Reaka 1980a, Reaka & Manning 1981, 1986c). The strong relationship between adult body size, postlarval size, and environmental characteristics found in the present study, then, will predispose members of particular communities (e.g. those on reefs versus soft bottoms, species from warm or tropical versus cold or temprate environments) toward lower dispersal and more rapid evolutionary change. Indeed, the trends for higher rates of evolution in stomatopods from hard substrata than in those from soft bottoms and in those from tropical versus temperate environments (Reaka & Manning 1986c) are fully explained by the difference in their body sizes and dispersal abilities. Examining an array of fish, molluscs, crustaceans, and polychaetes from the West Atlantic and East Pacific, Rosenblatt (1967) and Vermeij (1978) also have found that the degree of taxonomic divergence is greater in species from hard rather than soft bottoms. Rosenblatt (1963) noted that fishes from sandy habitats have greater powers of dispersal than those from rocky shores.

The small body size of stomatopods from hard substrata in the East Pacific relative to the West Atlantic also has strong implications for patterns of evolution in these two subregions. One would expect high evolutionary rates in these East Pacific stomatopods due to their small size. Some limited data support this suggestion. All of the 15 species of stomatopods that inhabit hard substrata in the East Pacific are endemic, while only 81% of the 16 species that inhabit hard substrata in the West Atlantic are endemic. Also, two lineages of stomatopods that inhabit hard bottoms apparently have become extinct in the East Pacific, in contrast to one lineage of stomatopods that occupies such substrata in the West Atlantic (Reaka & Manning 1986a). Various factors, especially geographic isolation, have been implicated in the high rates of evolution noted for many groups of organisms in the East Pacific (Briggs 1974, Vermeij 1978). The significance of dwarfism for high rates of evolutionary change in the East Pacific fauna, however, has not previously been considered. Particularly in combination with other factors, reduced body size is likely to accelerate rates of evolution in many benthic organisms because of the strong correlation between body size and dispersal characteristics in marine invertebrates (Chia 1974, Menge 1975, Reaka 1979a, 1980a, Strathmann & Strathmann 1982, Jablonski & Lutz 1983, Reaka & Manning 1986a). The extent of dwarfism in other East Pacific marine organisms has not been examined with respect to this question.

ACKNOWLEDGEMENTS

I thank Drs. Raymond Manning, David Reznick, Geerat Vermeij, and the editors for helpful discussions or comments on the manuscript. I am grateful to the National Science Foundation (Biological Oceanography) for support.

REFERENCES

Abbott, I., L.K.Abbott & P.R.Grant 1977. Comparative ecology of Galapagos Ground Finches (*Geospiza* Gould): evaluation of the importance of floristic diversity and interspecific competition. *Ecol. Monogr.* 47: 151-184.

Abele, L.G. 1973. Taxonomy, distribution and ecology of the genus *Sesarma* (Crustacea Decapoda, Grapsidae) in eastern North America, with special reference to Florida. *Amer. Midl. Nat.* 90: 375-386.

Abele, L.G. 1982. Biogeography. In L.B.Abele (ed.), *The Biology of Crustacea*, Vol.1. New York: Academic Press.

Alldredge, A.L. & J.M.King 1977. Distribution, abundance, and substrate preferences of demersal reef zooplankton at Lizard Island Lagoon, Great Barrier Reef. *Mar. Biol.* 41: 317-333.

Alldredge, A.L. & J.M.King 1980. Effects of moonlight on the vertical migration patterns of demersal zooplankton. *J. Exp. Mar. Biol. Ecol.* 44: 133-156.

Allen, J.A. 1978. Evolution of the deep sea protobranch bivalves. *Phil. Trans. Roy. Soc. Lond.* (B)284: 387-401.

Barclay, H.J. & P.T.Gregory 1981. An experimental test of models predicting life-history traits. *Amer. Nat.* 117: 944-961.

Barlow, G.W. 1981. Patterns of parental investment, dispersal, and size among coral-reef fishes. *Envir. Biol. Fishes* 6: 65-85.

Barnard, J.L. 1962. South Atlantic abyssal amphipods collected by R.V. Vema. *Abyssal Crustacea, Vema Res. Ser.*1: 1-78.

Barnes, H. & M.Barnes 1965. Egg size, nauplius size, and their variation with local, geographical, and specific factors in some common cirripedes. *J. Anim. Ecol.* 34: 391-402.

Belyaev, G.M. 1972. Hadal bottom fauna of the world ocean. *Inst. Oceanogr. Acad. Sci. USSR* (Israel Progr. Sci. Transl. Jerusalem).

Bertness, M.D. 1982. Shell utilization, predation pressure, and thermal stress in Panamanian hermit crabs: an interoceanic comparison. *J. Exp. Mar. Biol. Ecol.* 64: 159-187.

Berven, K.A., D.E.Gill & S.J.Smith-Gill 1979. Countergradient selection in the green frog, *Rana clamitans. Evolution* 33: 609-623.

Bohnsack, J.A. 1983. Changes in reef fish populations at Looe Key National Marine Sanctuary. *Amer. Zool.* 23: 986.

Bonner, J.T.B. 1968. Size change in development and evolution. *J. Paleontol.* 42: 1-15.

Boucot, A.J. 1976. Rates of size increase and of phyletic evolution. *Nature* 261: 694-696.

Briggs, J.C. 1974. *Marine Zoogeography.* New York: McGraw Hill.

Britton, R.H. & M.E.Moser 1982. Size specific predation by herons and its effects on the sex-ratio of natural populations of the mosquito fish *Gambusia affinis* Baird and Girard. *Oecologia* 53: 146-151.

Brooks, J.L. & S.I.Dodson 1965. Predation, body size and composition of plankton. *Science* 150: 28-35.

Brown, J.H. 1975. Geographical ecology of desert rodents. In M.L.Cody & J.M.Diamond (eds.), *Ecology and Evolution of Communities.* Cambridge: Harvard Univ. Press.

Brown, W.L. & E.O.Wilson 1956. Character displacement. *Syst. Zool.* 5: 49-64.

Burrows, M. 1969. The mechanics and neural control of the prey capture strike in the mantid shrimps *Squilla* and *Hemisquilla. Z. vergl. Physiol.* 62: 361-381.

Burrows, M. & G.Hoyle 1972. Neuromuscular physiology of the strike mechanism of the mantis shrimp, *Hemisquilla. J. Exp. Zool.* 179: 379-394.

Caldwell, R.L. & H.Dingle 1975. Ecology and evolution of agonistic behavior in stomatopods. *Naturwiss.* 62: 214-222.

Camp, D.K. & R.B.Manning. 1982. Five new species of *Nannosquilla* from the Northwestern Atlantic (Crustacea: Stomatopoda). *Smith. Contr. Zool.* 368: 1-15.

Carlquist, S. 1974. *Island Biology.* New York: Columbia Univ. Press.

Case, T.J. 1978. A general explanation for insular body size trends in terrestrial vertebrates. *Ecology* 59: 1-18.

Chia, F.S. 1974. Classification and adaptive significance of developmental patterns in marine invertebrates. *Thal. Jugosl.* 10: 121-130.

Clarke, A. 1983. Life in cold water: the physiological ecology of polar marine ectotherms. *Oceanogr. Mar. Biol. Ann. Rev.* 21: 341-453.

Cody, M.L. 1973. Character convergence. *Ann. Rev. Ecol. Syst.* 4: 189-212.

Dahl, E. 1983. Malacostracan phylogeny and evolution. In F.R.Schram (ed.), *Crustacean Phylogeny.* Rotterdam: Balkema.

Dayton, P.K. 1971. Competition, disturbance, and community organization: the provision and subsequent utilization of space in a rocky intertidal community. *Ecol. Monogr.* 41: 351-389.

Diamond, J.S. 1975. Assembly of species communities. In M.L.Cody & J.S.Diamond (eds.), *Ecology and Evolution of Communities.* Cambridge: Harvard Univ. Press.

Diamond, J.S. 1978. Niche shifts and the rediscovery of interspecific competition. *Amer. Sci.* 66: 322-331.

Dingle, H. & R.L.Caldwell 1972. Reproductive and maternal behavior of the mantis shrimp *Gonodactylus bredini* Manning (Crustacea: Stomatopoda). *Biol. Bull.* 142: 417-426.

Dominguez, J.H. & M.L.Reaka 1984. Temporal activity patterns of reef-dwelling stomatopods: a test of alternative hypotheses. Ms. in review.

Downhower, J.F. 1976. Darwin's finches and the evolution of sexual dimorphism in body size. *Nature* 263: 558-563.

Dunham, A.F., D.W.Tinkle & J.W.Gibbons 1978. Body size in island lizards: a cautionary tale. *Ecology* 59: 1230-1238.

Fenchel, T. 1974. Intrinsic rate of natural increase: the relationship with body size. *Oecologia* 14: 317-326.

Garth, J.S. 1957. The Crustacea Decapoda Brachyura of Chile. *Repts. Lund Univ. Chile Exped. 1948-49* 29: 3-131.

Gould, S.J. 1977. *Ontogeny and Phylogeny.* Cambridge: Harvard Univ. Press.

Goulden, C.E., L.L.Henry & A.J.Tessier 1980. Body size, energy reserves, and competitive ability in three species of Cladocera. *Ecology* 63: 1780-1789.

Grant, P.R. 1965. The adaptive significance of some size trends in island birds. *Evolution* 19: 355-367.

Grant, P.R. 1968. Bill size, body size, and the ecological adaptations of bird species to competitive situations on islands. *Syst. Zool.* 17: 319-333.

Grant, P.R. 1972. Convergent and divergent character displacement. *Biol. J. Linn. Soc.* 4: 39-68.

Grant, P.R. & I.Abbott 1980. Interspecific competition, island biogeography and null hypotheses. *Evolution* 34: 332-341.

Grime, J.P. & M.A.Mowforth 1982. Variation in genome size — an ecological interpretation. *Nature* 299: 151-153.

Hall, J.R. & R.R.Hessler 1971. Aspects in the population dynamics of *Derocheilocaris typica*(Mystacocarida, Crustacea). *Vie et Milieu*, Ser.A, 22: 305-326.

Hallam, A. 1975. Evolutionary size increase and longevity in Jurassic bivalves and ammonites. *Nature* 258: 439-446.

Hansen, T.A. 1980. Influence of larval dispersal and geographic distribution on species longevity in neogastropods. *Paleobiology* 6: 193-207.

Harested, A.S. & F.L.Brunnell 1979. Home range and body weight — a re-evaluation. Ecology 60: 389-402.

Hay, M.E. 1981. Herbivory, algal distribution, and the maintenance of between-habitat diversity on a tropical fringing reef. *Amer. Nat.* 118: 520-540.

Hendler, G. 1975. Adaptation significance of the patterns of ophiuroid development. *Amer. Zool.* 15: 691-715.

Hendrickson, J.A. 1981. Community-wide character displacement re-examined. *Evolution* 35: 794-809.

Highsmith, R.C. 1980. Geographic patterns of coral bioerosion: a productivity hypothesis. *J. Exp. Mar. Biol. Ecol.* 46: 177-196.

Holthuis, L.B. & R.B.Manning 1969. Stomatopoda. In R.C.Moore (ed.), *Treatis on Invertebrate Paleontology, Part R, Arthropoda 4*, Vol.2. Lawrence: Geol. Soc. Amer. & Univ. Kansas.

Horn, H.S. & R.M.May 1977. Limits to similarity among co-existing competitors. *Nature* 270: 660-661.

Hutchinson, G.E. 1959. Homage to Santa Rosalia, or why are there so many kinds of animals? *Amer. Nat.* 93: 145-158.

Jablonski, D. & R.A.Lutz 1983. Larval ecology of marine benthic invertebrates: paleobiological implications. *Biol. Rev.* 58: 21-89.

Jackson, J.B.C. 1974. Biogeographic consequences of eurytopy and stenotopy among marine bivalves and their evolutionary significance. *Amer. Nat.* 108: 541-560.

James, F.C. 1970. Geographic size variation in birds and its relationship to climate. *Ecology* 51: 365-390.

Johnston, R.F. & R.K.Selander 1973. Evolution in the house sparrow: III. Variation in size and sexual dimorphism in Europe and North and South America. *Amer. Nat.* 107: 373-390.

Koblentz-Mishke, I.J., V.V.Volkovinsky & J.B.Kabanova 1970. Plankton primary production of the world ocean. In W.S.Wooster (ed.), *Scientific Exploration of the South Pacific.* Washington: Nat. Acad. Sci.

Kornicker, L.S. 1975. Antarctic Ostracoda (Myodocopina). *Smith. Contr. Zool.* 163: 1-720.

Kunze, J.C. 1983. Stomatopoda and the evolution of the Hoplocarida. In F.R.Schram (ed.), *Crustacean Phylogeny.* Rotterdam: Balkema.

Levins, R. 1969. Thermal acclimation and heat resistance in *Drosophila* species. *Amer. Nat.* 103: 483-499.

Lindsey, C.C. 1966. Body sizes of poikilotherm vertebrates at different latitudes. *Evolution* 2: 456-465.

Lutz, R.A. & D.Jablonski 1978. Larval shell morphometry: a new paleoclimatic tool? *Science* 202: 51-53.

MacArthur, R.H. & E.O.Wilson 1967. *The Theory of Island Biogeography.* Princeton: Princeton Univ. Press.

Maiorana, V.C. 1978. An explanation of ecological and developmental constants. *Nature* 273: 375-377.

Manning, R.B. 1969a. *Stomatopod Crustacea of the Western Atlantic.* Coral Gables: Univ. Miami Press.

Manning, R.B. 1969b. A review of the genus *Harpiosquilla* with descriptions of three new species. *Smith. Contr. Zool.* 36: 1-41.

Manning, R.B. 1972. Stomatopoda. Eastern Pacific Expeditions of the New York Zoological Society. *Zoologica*56: 95-113.

Manning, R.B. 1977. *A Monograph of the West African Stomatopod Crustacea.* Atlantide Rept. No.12. Copenhagen: Scandinavian Sci. Press.

Manning, R.B. 1980. A list of the families and genera of recent stomatopod Crustacea, with diagnosis of six new families. *Proc. Biol. Soc. Wash.* 93: 362-372.

Manning, R.B. & A.J.Provenzano 1963. Studies on development of stomatopod Crustacea. I. Early larval stages of *Gonodactylus oerstedii* Hansen. *Bull. Mar. Sci. Gulf Caribb.* 13: 467-487.

Manning, R.B. & P.Struhsaker 1976. Occurrence of the Caribbean stomatopod, *Bathysquilla microps*, off Hawaii, with additional records for *B.microps*and *B.crassispinosa. Proc. Biol. Soc. Wash.* 89: 439-450.

Marshall, L.G. & R.S.Corruccini 1978. Variability, evolutionary rates, and allometry in dwarfing lineages. *Paleobiology* 4: 101-119.

Mayr, E. 1956. Geographic character gradients and climatic adaptation. *Evolution* 10: 105-108.

McCall, P.L. 1977. Community patterns and adaptive strategies of the infaunal benthos of Long Island Sound. *J. Mar. Res.* 35: 221-266.

McLaren, I. 1965. Some relationships between temperature and egg size, body size, development rate, and fecundity of the copepod *Pseudocalanus*. *Limnol. Oceanogr.* 10: 528-538.

McNab, B.K. 1971. On the ecological significance of Bergmann's Rule. *Ecology* 52: 845-854.

McNab, B.K. 1978. The evolution of endothermy in the phylogeny of mammals. *Amer. Nat.* 112: 1-21.

McNab, B.K. 1980. Food habits, energetics, and the population biology of mammals. *Amer. Nat.* 116: 106-124.

Menge, B.A. 1975. Brood or broadcast? The adaptive significance of different reproductive strategies in the two intertidal sea stars *Leptasterias hexactis* and *Pisaster ochraceus*. *Mar. Biol.* 31: 87-100.

Menzies, R.J. & R.Y.George 1967. A re-evaluation of the concept of hadal or ultra-abyssal fauna. *Deep Sea Res.* 14: 703-723.

Morgan, S.G. & A.J.Provenzano 1978. Development of pelagic larvae and postlarva of *Squilla empusa*(Crustacea, Stomatopoda), with an assessment of larval characters within the Squillidae. *Fish. Bull.* 77: 61-90.

Morse, D.H. 1974. Niche breadth as a function of social dominance. *Amer. Nat.* 108: 818-830.

Murphy, G.I. 1968. Pattern in life history and environment. *Amer. Nat.* 102: 390-404.

Newell, N.D. 1949. Phyletic size increase — an important trend illustrated by fossil invertebrates. *Evolution* 3: 103-124.

Oksanen, L.S., S.D.Fretwell & O.Jarvinen 1979. Interspecific competition and the limiting similarity of close competitors: the problem of size gaps in some community arrays. *Amer. Nat.* 114: 117-129.

Paine, R.T. 1976. Size-limited predation: an observational and experimental approach with the *Mytilus-Pisaster* interaction. *Ecology* 57: 858-873.

Pearson, D.L. & E.J.Mury 1979. Character divergence and convergence among tiger bettles (Coleoptera: Cicindellidae). *Ecology* 60: 557-566.

Pianka. E.R. 1970. On 'r' and 'K' selection. *Amer. Nat.* 104: 592-597.

Porter, J.W. & K.G.Porter 1977. Quantitative sampling of demersal plankton migrating from different coral reef substrates. *Limnol. Oceanogr.* 22: 553-556.

Poulsen, E.M. 1965. Ostracoda-Myodocopa. Part II. Rutidermatidae, Sarsiellidae and Asteropidae. *Dana Rept.* 65: 1-484.

Provenzano, A.J. Jr & R.B.Manning 1978. Studies on development of stomatopod Crustacea. II. The later stages of *Gonodactylus oerstedii* Hansen reared in the laboratory. *Bull. Mar. Sci.* 28: 297-315.

Pyne, R.R. 1972. Larval development and behaviour of the mantis shrimp, *Squilla armata* Milne Edwards (Crustacea: Stomatopoda). *J. Roy. Soc. N.Zeal.* 2: 121-146.

Ray, C. 1960. The application of Bergmann's and Allen's Rules to the poikilotherms. *J. Morphol.* 106: 85-108.

Reaka, M.L. 1975a. Phylogeny of malacostracan Crustacea. Amer. Zool. 15: 816.

Reaka, M.L. 1975b. Molting in stomatopod crustaceans: I. Stages of the molt cycle, setagenesis, and morphology. *J. Morphol.* 146: 55-80.

Reaka, M.L. 1976. Lunar and tidal periodicity of molting and reproduction in stomatopod Crustacea: A selfish herd hypothesis. *Biol. Bull.* 150: 468-490.

Reaka, M.L. 1978. The effects of an ectoparasitic gastropod, *Caledoniella montrouzieri*, upon molting and reproduction of a stomatopod crustacean, *Gonodactylus viridis*. *Veliger* 21: 251-254.

Reaka, M.L. 1979a. The evolutionary ecology of life history patterns in stomatopod Crustacea. In S.Stancyk (ed.), *Reproductive Ecology of Marine Invertebrates*. Columbia: Baruch Library in Marine Science, Univ. S.Carolina Press.

Reaka, M.L. 1979b. Patterns of molting frequencies in coral-dwelling stomatopod Crustacea. *Biol. Bull.* 156: 328-342.

Reaka, M.L. 1980a. Geographic range, life history patterns, and body size in a guild of coral-dwelling mantis shrimps. *Evolution* 34: 1019-1030.

Reaka, M.L. 1980b. On learning and living in holes by mantis shrimp. *Anim. Behav.* 28: 111-115.

Reaka, M.L. 1980c. Resource limitation in mobile cryptic species: food or space? *Amer. Zool.* 20: 885.

Reaka, M.L. & R.B.Manning 1980. The distributional ecology and zoogeographical relationships of stomatopod Crustacea from Pacific Costa Rica. *Smith. Contr. Mar. Sci.* 7: 1-29.

Reaka, M.L. & R.B.Manning 1981. The behavior of stomatopod Crustacea, and its relationship to rates of evolution. *J. Crust. Biol.* 1: 309-327.

Reaka, M.L. & R.B.Manning 1986a. The stomatopod Crustacea of Enewetak Atoll. In P.Colin, D.M.Deveney & E.S.Reese (eds.), *The Natural History of Enewetak Atoll.* Washington: U.S. Dept. Energy.

Reaka, M.L. & R.B.Manning 1986b. The ecology of the stomatopod Crustacea of Enewetak Atoll: An analysis of their distribution, behavior and color patterns. *Smith. Contr. Mar. Sci.*

Reaka, M.L. & R.B.Manning 1986c. The significance of body size, larvae, and habitat for rates of morphological evolution in stomatopod Crustacea. *Smith. Contr. Zool.*

Rensch, B. 1960. *Evolution above the species level.* New York: Columbia Univ. Press.

Reznick, D. 1982. The impact of predation on life history evolution in Trinidadian guppies: genetic basis of observed life history patterns. *Evolution* 36: 1236-1250.

Reznick, D.N. & J.A.Endler 1982. The impact of predation on life history evolution in Trinidadian guppies. *Evolution* 36: 160-177.

Robichaux, D.M., A.C.Cohen, M.L.Reaka & D.Allan 1981. Experiments with zooplankton on coral reefs, or, will the real demersal plankton please come up? *Mar. Ecol.* 2: 77-94.

Rosenblatt, R.H. 1963. Some aspects of speciation in marine shore fishes. In J.P.Harding & N.Tebble (eds.), *Speciation in the Sea.* London: Syst. Assn. Pub. No.5.

Rosenblatt, R.H. 1967. The zoogeographic relationships of the marine shore fishes of tropical America. *Stud. Trop. Oceanogr. Univ. Miami* 5: 579-587.

Rosensweig, M.L. 1968. The strategy of body size in mammalian carnivores. *Amer. Midl. Nat.* 80: 299-315.

Sale, P.F. 1977. Maintenance of high diversity in coral reef fish communities. *Amer. Nat.* 111: 337-359.

Sale, P.F. 1980. The ecology of fishes on coral reefs. *Oceanogr. Mar. Biol. Ann. Rev.* 18: 367-421.

Schaffer, W.M. 1974. Optimal reproductive effort in fluctuating environments. *Amer. Nat.* 108: 783-790.

Schoener, T.W. 1965. The evolution of bill size differences among sympatric congeneric species of birds. *Evolution* 19: 189-213.

Schoener, T.W. 1969a. Models of optimal size for solitary predators. *Amer. Nat.* 103: 277-313.

Schoener, T.W. 1969b. Size patterns in West Indian *Anolis* lizards: I. Size and species diversity. *Syst. Zool.* 18: 386-401.

Schoener, T.W. 1969c. Optimal size and specialization in constant and fluctuating environments. Brookhaven Symposia in Biology, Diversity and Stability in Ecological Systems. Upton, N.Y.: Brookhaven National Laboratory.

Schoener, T.W. 1970. Size patterns in West Indian *Anolis*lizards. II. Correlations with the sizes of particular sympatric species — displacement and convergence. *Amer. Nat.* 104: 155-174.

Schoener, T.W. 1982. The controversy over interspecific competition. *Amer. Sci.* 70: 586-595.

Scholander, P.F. 1955. Evolution of climatic adaptation in homeotherms. *Evolution* 9: 15-26.

Schram, F.R. 1969. Some middle Pennsylvanian Hoplocarida (Crustacea) and their phylogenetic significance. *Fieldiana: Geol.* 12: 235-289.

Schram, F.R. 1979. The genus *Archaeocaris* and a general review of the Paleostomatopoda (Hoplocarida: Malacostraca). *Trans. San Diego Nat. Hist.* 19: 57-66.

Schram, F.R. 1981. On the classification of the Eumalacostraca. *J. Crust. Biol.* 1: 1-10.

Schram, F.R. 1982. The fossil record and evolution of Crustacea. In L.G.Abele (ed.), *The Biology of Crustacea*, Vol.1. New York: Academic Press.

Sebens, K.P. 1983. Size structure and growth rates in populations of colonial and solitary invertebrates. In M.L.Reaka (ed.), *The Ecology of Deep and Shallow Coral Reefs.* Rockville: Office of Undersea Research, NOAA.

Shapiro, Y. 1983. On the possibility of kin groups in coral reef fishes. In M.L.Reaka (ed.), *The Ecology of Deep and Shallow Coral Reefs.* Rockville: Office of Undersea Research, NOAA.

Simberloff, D. & W.Boecklen 1981. Santa Rosalia reconsidered: size ratios and competition. *Evolution* 35: 1206-1228.

Slatkin, M. 1980. Ecological character displacement. *Ecology* 61: 163-177.

Stanley, S.M. 1973. An explanation for Cope's Rule. *Evolution* 27: 1-26.

Stanley, S.M. 1979. *Macroevolution: Pattern and Process.* San Francisco: W.H.Freeman.

Stearns, S.C. 1976. Life history tactics: A review of the ideas. *Q. Rev. Biol.* 51: 3-47.

Steele, D.H. 1977. Correlation between egg size and developmental period. *Amer. Nat.* 111: 371-372.

Stephenson, T.A. & A.Stephenson 1950. Life between the tidemarks in North America. I. Florida Keys. *J. Ecol.* 38: 354-402.

Strathmann, R.R. 1977. Egg size, larval development, and juvenile size in benthic marine invertebrates. *Amer. Nat.* 111: 373-376.

Strathmann, R.R. & M.F.Strathmann 1982. The relationship between adult size and brooding in marine invertebrates. *Amer. Nat.* 119: 91-101.

Strong, D.R. & D.S.Simberloff 1981. Straining at gnats and swallowing ratios: character displacement. *Evolution* 35: 810-812.

Strong, D.R. Jr, L.A.Szyska & D.S.Simberloff 1979. Tests of community-wide character displacement against null hypothesis. *Evolution* 33: 897-913.

Thiel, H. 1975. The size structure of the deep-sea benthos. *Int. Rev. ges. Hydrobiol.* 60: 575-606.

Thorson, G. 1946. Reproduction and larval development of Danish marine bottom invertebrates. *Meddel. Komm. Danmarks Fisk. Havundersog. Ser. Plankton* 4: 1-523.

Thorson, G. 1950. Reproductive and larval ecology of marine invertebrates. *Biol. Rev.* 25: 1-45.

Van Markhoven, F.P.C.M. 1972. Bathymetry of recent marine Ostracoda in the northwest Gulf of Mexico. *Trans. Gulf Coast Assoc. Zool. Soc.* 12: 241-252.

Van Valen, L. 1973. Body size and numbers of plants and animals. *Evolution* 27: 27-35.

Vance, R.R. 1973. On reproductive strategies in marine benthic invertebrates. *Amer. Nat.* 107: 339-352.

Vermeij, G.J. 1972. Intraspecific shore-level size gradients in intertidal molluscs. *Ecology* 53: 693-700.

Vermeij, G.J. 1973a. Molluscs in mangrove swamps: physiognomy, diversity, and regional differences. *Syst. Zool.* 22: 609-624.

Vermeij, G.J. 1973b. West Indian molluscan communities in the rocky intertidal zone: a morphological approach. *Bull. Mar. Sci.* 23: 351-386.

Vermeij, G.J. 1974. Regional variations in tropical high intertidal assemblages. *J. Mar. Res.* 32: 343-357.

Vermeij, G.J. 1976. Interoceanic differences in vulnerability of shelled prey to crab predation. *Nature* 260: 135-136.

Vermeij, G.J. 1978. *Biogeography and Adaptation: Patterns of Marine Life.* Cambridge: Harvard Univ. Press.

Vermeij, G.J. 1979. The architectural geography of some gastropods. In J.Gray & A.J.Boucot (eds.), *Historical Biogeography, Plate Tectonics, and the Changing Environment.* Corvallis: Oregon State Univ. Press.

Vermeij, G.J. 1980. Gastropod shell growth rate, allometry, and adult size. Environmental implications. In D.C.Rhoads & R.A.Lutz (eds.), *Skeletal Growth of Aquatic Organisms.* New York: Plenum Press.

Vermeij, G.J. & J.W.Porter 1971. Some characteristics of the dominant intertidal molluscs from rocky shores in Pernambuco, Brazil. *Bull. Mar. Sci.* 21: 440-454.

Wassersug, R.J., H.Yang, J.M.Sepkoski & D.M.Raup 1979. The evolution of body sizes on islands: A computer simulation. *Amer. Nat.* 114: 287-295.

Wiens, J.A. & J.T.Rotenberry 1981. Morphological size ratios and competition in ecological communities. *Amer. Nat.* 117: 592-599.

Wilson, D.S. 1975. The adequacy of body size as a niche difference. *Amer. Nat.* 109: 769-784.

Wilson, E.O. 1975. *Sociobiology: The New Synthesis.* Cambridge: Harvard Univ. Press.

Wolff, T. 1956a. Isopoda from depths exceeding 6000 meters. *Galathea Rept.* 2: 85-157.

Wolff, T. 1956b. Crustacea Tanaidacea from depths exceeding 6000 meters. *Galathea Rept.* 2: 187-241.

Zipser, E. & G.J.Vermeij 1978. Crushing behavior of tropical and temperate crabs. *J. Exp. Mar. Biol. Ecol.* 31: 155-172.

The non-cosmopolitanism of chydorid Cladocera: Implications for biogeography and evolution

DAVID G.FREY
Department of Biology, Indiana University, Bloomington, USA

ABSTRACT

Cladocera were long thought to be a group in which a few species had widespread, sometimes almost ubiquitous distributions. Recent taxonomic revisions in the group reveal that the contrary is probably the case, i.e. this is a group in which discretely defined members of closely related species have rather restricted distributions. Cladocerans are probably a very old group whose origins go back to an originally ubiquitous fauna rather uniformly distributed on the supercontinent of Pangaea. Consideration of ability to hybridize and time of gamogenesis can offer some insights into sorting out the biogeographic and ecologic histories of specific cladoceran taxa.

1 INTRODUCTION: STATEMENT OF THE PROBLEM

Cladocera are small crustaceans that occur mainly in standing fresh waters. A few species occur in the oceans near continents, and an interesting small suite of species in such inland brackish waters as the Caspian Sea and the Aral Sea. They also occur in flowing waters, but very few studies have been made on them there. They form two contrasting associations in basins that are virtually allopatric. One group of species (e.g. the genera *Daphnia, Bosmina, Diaphanosoma*) forms a major component of the offshore zooplankton, which spend their entire lives in the water away from any substrate. A second much larger group occurs in the littoral zone associated with macrophytes, sessile algae, rocks and stones, or sediments. These littoral animals crawl over the objects present or conceal themselves in the sediments, obtaining their sustenance from the algae-plus-detritus that forms an extensive covering. Swimming is done mainly to move from one temporary resting place to another, not to keep the animals suspended in the water away from the bottom. Also present occasionally in this littoral community are some swimming members from the same families that furnish the zooplankton. Sometimes one can find planktonic forms in the littoral zone, but seldom, except mainly for the *Chydorus sphaericus* complex, do any littoral species occur abundantly offshore. Thus, a person studying the offshore zooplankton often has little appreciation of the richness of the littoral cladoceran community.

The number of species of Cladocera that occur in the world is unknown, although an estimate of 400 has been made (Frey 1966a). Lists of species of Cladocera from the north-temperate portions of Eurasia and North America contain mostly the same names, and added from the southern portions of these continents are species that are widely distributed throughout the tropics and subtropics. South America, Africa, and Australia/ New Zealand likewise contain many of these pan-tropical species as well as many of those present in the northern continents. It seems obvious from these listings that many species occur on more than one continent and that many are so widespread as to be considered cosmopolitan. Cosmopolitanism, in fact, has become the guiding theme or 'Leitmotiv' of cladoceran distribution. For example, in Birge's chapter on the Cladocera of the United States (1918) there are 57 species in the family Chydoridae, of which 35 (= 61%) were described from Europe, nine from North America, six from South America, three each from Australia and Sri Lanka, and one from Asia Minor. For the other ten families the pattern is similar, with most of the species having been described from Europe. As the two subsequent treatments of North America (Brooks 1959, Pennak 1953, 1978) are based closely on Birge, the situation is unchanged today except for *Daphnia* and *Bosmina*.

Study of the Cladocera began in Denmark (Müller 1776, 1785) and then spread to other countries in northern Europe (Frey 1982c). By 1878 most of the genera and species presently known from Europe had been described, but only about 20 papers in all were concerned with Cladocera from other continents. In 1878 Birge obtained his Ph.D. from Harvard with a thesis on the taxonomy of the Cladocera he had collected in Massachusetts and Wisconsin (Frey 1982d). In his first paper (Birge 1879), and in three others published through 1910, he described a total of 33 new species, only 14 of which survived to his monograph (Birge 1918). Quite certainly Birge was not familiar with all the work going on in Europe, which began with a few scattered papers in the early 1800s and then blossomed out vigorously in the 1860s, 1870s, and later. He purchased Richard's specimen collection containing many European samples in 1903 but evidently did not study the specimens closely. Seeing species from North America that resembled those in Europe, he decided they were the same, and hence he gave them the same names.

Persons from Europe about this time also were starting to look at collections from other continents, using a similar methodology. Being familiar with the European species and seeing species from elsewhere that were grossly similar in morphology, they gave them the European names. This led to the use inappropriately of monographs for identification. For example, Lilljeborg (1901) produced a magnificent volume on the Cladocera of Sweden, containing many detailed illustrations of all the species included. Persons working on Cladocera from other continents often used Lilljeborg's illustrations for their taxa, rather than looking at the specimens closely and making their own illustrations (Frey 1982d). Also, because Birge's chapter is well written and well illustrated, it has been used widely to identify Cladocera from virtually everywhere. Even today many workers are content to place names on Cladocera from any part of the world on the basis of superficial resemblances without conducting any detailed study or comparisons. After all, the reasoning goes, if the Cladocera are cosmopolitan, there is little need to document their morphology.

Thus, the present view held by most persons working with the Cladocera is that the individual species are often widely distributed, even cosmopolitan, but that they can be quite highly variable within a given species. This widespread acceptance of the concept of cosmopolitanism has obstructed creative progress in studying the Cladocera as species and as members of communities. The concept has developed at least partly, perhaps largely, through a complacency generated by too loose a definition of species.

2 EVENTS LEADING TO QUESTIONING AND EVENTUAL REFUTATION OF CLADOCERAN COSMOPOLITANISM

During the last century, scientists largely restricted their studies of cladoceran taxonomy to the European fauna. They had little or no knowledge of taxa from other parts of the world. When additional species were occasionally described from Europe, enough characters were added to the descriptions to be able to distinguish the old taxa from the new ones, but little more than this. When specimens from other continents were examined, frequently the European names were used for them without any further expansions of the descriptions. Later decisions on names were usually easier, because the names were already established in the literature. However, cosmopolitanism has never been defined nor substantiated by any good objective evidence.

One process necessary for cosmopolitanism to occur is the production of diapause or resting eggs. These normally result from bisexual mating, although some species of *Daphnia* that are obligate parthenogens can produce them as well (e.g. Olofsson 1918, Herbert 1983). There are suggestions in the literature, such as the recorded presence in collections of abundant ephippial females but no males, that indicate various species in the Arctic may also be parthenogenetic. But where the ice-free season lasts more than a month or two, it is quite certain that in non-*Daphnia* species males are involved in the production of resting eggs.

Whether they are produced unisexually (parthenogenetically) or bisexually (gamogenetically), these resting eggs allow passive distribution of the Cladocera from one waterbody to another by wind, birds, insects, mammals, or other means. A single resting egg can start a new population, which is a clone of the parent until gamogenesis provides for genetic recombination.

Resting eggs are important. They provide the means of reestablishing populations each spring in northern lakes where the active populations die out over winter, and in seasonal waterbodies that are completely dry for variable periods of time. They facilitated the widespread northward migration of many species into areas of North America that had been covered by ice during the last glacial age. They provide the means of moving about locally into newly established waterbodies (Frey 1976). They also probably provide for the migration northward of southern species in North America and their occasional bloom in a lake quite far north of their normal occurrence. However, the limited areas of distribution of many species of chydorids suggest that passive distribution of resting eggs is not the dominant factor controlling distribution. Physiological and ecological requirements of the species must be the major factors.

Different families have different ways of taking care of resting eggs. The daphniids have the most elaborate ephippia, containing a foamy or sponge-like development that causes the ephippia to float and hence be scattered easily by various means. The chydorids and macrothricids, on the other hand, frequently glue or cement their ephippia to the substrate, and as a result they are not so easily dispersed. The sidids do not produce any ephippia at all but release their resting eggs directly into the water, where they sink to the bottom. Thus, there are three completely different methods of providing for the resting eggs, possibly all equally effective in terms of the biology, survival, and distribution of the species, because members from the four families mentioned contribute to the Cladocera populations in nearly all waterbodies, regardless of their origin and location. It is quite certain from the work of Sars (e.g. 1901) and others that all lakes contain in their sediments viable resting eggs of the species occurring in the lakes. Sars' studies on the Cladocera from outside Europe are based mainly on animals that hatched from samples of sediments placed in aquaria.

Different species in the Cladocera have quite precise ecological and physiological requirements that determine where they can exist and thus control their distribution. So little is known about the nuances involved that there is difficulty, for example, in using the cladoceran remains found in lake sediments to estimate the conditions in the lake and its watershed at the time the animals were alive. North-temperate lakes of glacial origin were invaded in very early postglacial time by nearly all the species presently occurring in them (Frey 1955, 1961a, Goulden 1964, Harmsworth 1968, Whiteside 1970). During postglacial time there have been fluctuations in relative abundance of the various species and the occasional invasion and establishment of a new species.

My own research has been focused almost exclusively on the Chydoridae. It began with the collection of exoskeletal components in lake sediments from Austria (Frey 1955) and Germany (Frey 1958, 1961a), struggling with their identifications, and then finding a pattern of pores on the heads (Frey 1959) that helped to establish the four subfamilies presently recognized (Frey 1966b). This pattern, for example, nicely separates the genera *Alona* and *Alonella* into different subfamlies, whereas previously their species had been shifted from one genus to the other by various investigators.

Recently we have begun looking more closely at the North American fauna. It became apparent only gradually (as this was contrary to all accepted belief for the Cladocera) that the American species which had the same names as their European counterparts nevertheless differed morphologically from them. Further study and detailed morphological comparisons, involving intensive use of the scanning electron microscope, have convinced workers that most, perhaps all, species of chydorids previously claimed to be widely distributed represent, instead, species-groups or species-complexes. No member of these groups by itself encompasses the full geographic range of the group or complex. The guiding principle now is to regard no species as cosmopolitan until this condition has been substantiated by good evidence. To date there is little such evidence for any of the rather numerous species studied. One cannot use literature references by themselves, because both the written and illustrated descriptions tend to be completely inadequate and sometimes even erroneous. Needed are large instantaneous populations of specimens that include parthenogenetic females over their full size range plus ephippial females and the three instars of males. Many of the useful characters change as a function of size, or else are confined to a particular reproductive stage. Hence, the unit for comparison is not a type specimen but rather a population, because frequently species cannot be characterized adequately from the morphology of a single specimen.

This great change in our conceptual understanding of the chydorids has developed over the past ten years. Ideas about non-cosmopolitanism received their first expression in the scientific literature very recently (Frey 1982d), but as these concepts have advanced considerably since then, it is appropriate to restate them and to speculate as to where they are leading. To accomplish this, I shall give a brief summary of the major events that led to our present disregard for cosmopolitanism in the chydorids.

2.1 *Eurycercus*

Species of this genus are common and widely distributed in the Northern Hemisphere and are the largest chydorids, reaching a maximum length of 6 mm in *E.glacialis*. Hence, aquatic biologists have been attracted to them and have considered them closely. Until 1971 just two species were recognized in the world, both confined to the Northern Hemisphere except for disjunct populations in Argentina and South Africa (Frey 1971). *Eurycercus lamellatus* was regarded as widespread and abundant throughout the Holarctic region, occurring southward to Mexico City, Algeria, and northern India. The other species, *E.glacialis*, occurs mostly in the Far North at the edges of the land masses in North America, Greenland, and Eurasia, and southward to Newfoundland, Denmark-Netherlands, and possibly to one locality in Hokkaido, although the identification of the latter animals as *E.glacialis* is not certain (Frey 1971). Both species, particularly *E.lamellatus*, were considered to be good examples of a cosmopolitan distribution.

Studying the ecology of these two species in Denmark led to an appreciation of the fine details of their morphology. A third species was described (Frey 1973), from samples collected in eastern Siberia in 1971, a fourth from northern Sweden (Frey 1975), and a fifth species from the Southeastern United States (Frey 1978). Hann (1980, 1982) separated the common *Eurycercus* in North America into a northern species and a southern species, for a total of seven species. All the new species described thus far are included in what was formerly considered *E.lamellatus*. However, this species in the strict sense does not occur in North America at all, and only the diminutive species in the southeastern states is in the same subgenus. Moreover, the taxa in Argentina and South Africa are probably different from their corresponding species in North America and Europe, although this matter has not yet been resolved. Likewise in North America a number of other undescribed species occur including two in what is currently called *E.glacialis*. How many species exist cannot even be guessed at the moment, although the genus is certainly very diverse. A suite of 16 meristic and continuous-range characters was used (Frey 1973, Hann 1982) to help define the species, but for a full description the ephippial females and the three instars of males are also needed. These, unfortunately, are seldom available. Previous investigators were so impressed by the size of the animals and the uniqueness of their broad, markedly flattened postabdomen provided with a continuous fringe of saw-like teeth that they failed to appreciate differences in other characters. Instead, to separate *E.lamellatus* from *E.glacialis* they counted the teeth on the postabdomen and determined whether the antennal seta arose proximally or distally. Because of the strong overlap in tooth counts on the postabdomen, these characters are not adequate to separate *E.lamellatus* sensu stricto from *E.glacialis* much less to define the other species now known to exist (Frey 1973).

2.2 *Chydorus sphaericus*

This, the type species of the family, has been considered to be the most widespread and abundant of all the Cladocera in the world. Unfortunately, many identifications of it have been based on a process of elimination — if specimens could not be identified readily as one of the other species, then they were called *C.sphaericus*.

Several samples of littoral Cladocera from Salmon Lake in western Montana contained ephippial females of what was first called *Chydorus sphaericus*. However, some of these specimens were carrying two ephippial eggs, whereas other females in the same samples, supposedly of the same species, were carrying only one egg. The normal condition expected in all chydorids, except species in the genera *Eurycercus* and *Saycia*, is to have just a single ephippial egg. The observation of two resting eggs, although interesting, probably would have remained as a note on the inventory card, except that a coincidental inquiry from Fryer (Fryer & Frey 1981) stimulated a detailed study of these *Chydorus.*

Two distinct species were found (Frey 1980). A detailed comparison of American forms with a population from Denmark allowed a neotype to be established in the latter. The comparison also revealed that neither of the Montanan species was *C.sphaericus* sensu stricto. The somewhat smaller species with a single ephippial egg and a short labrum was established as the first member of a new species group, *Chydorus brevilabris.* Examination of many populations from low latitudes around the world that had been labelled *Chydorus sphaericus* revealed that the morphological type present was *C.brevilabris*, not *C.sphaericus.* Furthermore, the second species with the two ephippial eggs, although a member of the *C.sphaericus* group, was not *C.sphaericus* sensu stricto because of a number of significant characters, especially in reproductive individuals (Frey 1985). Examination of many other collections from North America has shown undescribed species of the *C.brevilabris* group frequently to be the only species present, especially in southerly populations. Species of the *C.sphaericus* group tend to occur toward the north and at higher elevations, but nothing has been found yet that is *C.sphaericus* sensu stricto. Thus, it is quite possible that *C.sphaericus* sensu stricto does not occur in North America at all, although some potentially positive samples from Southampton Island are not yet analyzed. The number of species in the two species groups that may occur in North America is completely unknown, although there are certainly several in each group. They will be difficult to sort out.

Thus, the chydorid that had been considered to be the most widespread and abundant of all Cladocera, and certainly the best possible example of cosmopolitanism, is now known to include a yet to be determined number of species in at least two species groups. Even in Europe there is more than one species in the *C.sphaericus* group, but no one has attempted to describe them. Any studies to do this must be based on large point-time populations containing the stages involved in gamogenetic reproduction. Single or a few specimens from a specific locality will not be adequate.

The problem in working with these minute animals is the virtual lack of any detailed descriptions and accurate drawings in the literature and the often complete absence of specimens in museums. Lilljeborg's 1901 monograph (Rodhe & Frey 1982) is the best taxonomic study, even though it is limited to the species that occur in Sweden. The transfer of names from one continent to another was considered acceptable, and still is by many persons, but the resulting confusion is horrendous. Redescriptions of the primary taxa are needed, as is the placement of suitable specimens in museums. Only then can morphological comparisons be made between widely distributed taxa that presently have the same names.

2.3 *Ephemeroporus barroisi*

Described as *Pleuroxus barroisi*, this species is small and has one tooth on the shell and four teeth on an expanded plate-like labrum. A taxon with these characteristics has been recorded many times from South Asia, where it is one of the most frequent and abundant chydorids. It also occurs in the Mediterranean Region, Africa, the southern United States, northern South America, Australia, and New Zealand (Smirnov 1971), indicating clearly that it is cosmopolitan in distribution. It had been included in the genus *Chydorus*, but because of the arrangement of the ventral setae on the shell, the shape of the posterior portion of the headshield, and the presence of just one median headpore, which occurs only in intar I, Frey (1982e) transferred it and a number of closely related species to the new genus *Ephemeroporus.*

In this genus there are many species, mostly undescribed and most of which are known only from Florida or which extend for short distances into adjacent southeastern states. Three species in Florida with one tooth on the shell and four on the labrum would have to be called *E.barroisi* at the present state of our knowledge. *E.barroisi* remains a vague species, because no type specimens are available, the type locality in Syria has been destroyed, and the original description, although reasonably detailed and reasonably well illustrated, is inadequate for present purposes. I have specimens of a species with this general morphology from Iran, but as they do not match up closely with the published description of *E.barroisi*, they probably represent a different species.

Hence, it is certain that quite a number of species in the *E.barroisi* group occur at low latitudes around the world. All are presently called *E.barroisi*, because descriptions of the various species have focused on obvious and easily observed characters. However, these characters, we are learning, define mostly the species groups, not the individual species.

In addition to the *E.barroisi* species group there are other species in the genus *Ephemeroporus* (Frey 1982e). One is *E.hybridus* described from Paraguay, which has one tooth on the shell and only one tooth on the labrum. Again, in Florida and neighboring states there are a number of undescribed species with these same tooth characters, but they differ among themselves in body size, size of teeth, and various other characters that help define species. Because the morphology remains constant, each species can be recognized confidently wherever it is found in lakes of Florida and neighboring states. They constitute another part of the rapidly accumulating evidence that the chydorids are much more diverse and much more limited in distribution than generally believed. The concept of cosmopolitanism does not apply to them at all.

In summary, the genus *Ephemeroporus* occurs widely in the Tropics and Subtropics, with particular species existing as far north as Isle Royale in Lake Superior (Birge 1918) and as far south as New Zealand (Sars 1904). At present 16 species are known, fewer than half of which have been described (Frey 1982e). Most of the undescribed species occur in Florida, which may be a reflection of the intensive and extensive studies made there, but which also must reflect the large number of natural, low-electrolyte, acidic waterbodies in this region. Such habitats are particularly favorable for chydorids, and so contain many species. Florida is possibly a unique area for chydorids, and may have provided particularly good conditions for radiation in the genus *Ephemeroporus.*

2.4 *Species in North America with European names*

As mentioned earlier, 35 out of 57 chydorids in Birge's monograph for the United States were described from Europe. Are the populations on the two sides of the Atlantic Ocean the same species, as claimed? The instances already considered indicate that *Eurycercus lamellatus* does not occur in North America at all, and that *Chydorus sphaericus* very likely does not occur in the United States either, possibly not even in Canada, although it may be present in the far North. Likewise, *Ephemeroporus barroisi* from Syria quite certainly does not occur in North America either, although this matter is not yet resolved.

Comparing taxa from Europe and North America is not a simple matter, because one needs from both continents large populations that contain all the states and stages needed to define the species fully. The descriptions of most of the European species are very incomplete and are based mainly on parthenogenetic females. For the three taxa reported thus far in this review, the conclusion is inescapable that different species are involved on the two sides of the Atlantic. In addition, two full scale comparisons of taxa in Europe and North America that had been given the same names show that the taxa are different at the species level. This is just the beginning, as some other instances of implied cosmopolitanism are quite certainly erroneous also.

2.4.1 *Oxyurella tenuicaudis in Europe versus O.brevicaudis in North America*

O.tenuicaudis (Sars 1862) is a name that has been used over much of the world for a very distinctive species of chydorid. However, detailed comparison of North American with European populations shows that although the taxa look grossly similar, they are very strikingly different when the important details are examined. Consequently, the North American taxon has been described as *O.brevicaudis* (Michael & Frey 1983).

Oxyurella tenuicaudis is one of the 16 species of chydorids Sars described from Norway and one of 52 species of all Cladocera he described from that country (Frey 1982b). All Sars' descriptions of these species are short, in Latin, and unaccompanied by any illustrations. Moreover, there are almost no specimens in the Sars Museum in Oslo that Sars unequivocally had used in preparing his descriptions, and in fact very few of any specimens come from Norway. Sars made colored portraits of the living Cladocera and did little to save the actual specimens. He had a long unpublished manuscript from 1861 that contains many camera lucida drawings of details of the species he described, as well as colored portraits, but for only one of his 52 species (*Limnosida frontosa*) did he ever publish illustrations (Sars 1865). His projected monograph of the Cladocera of Norway ended abruptly after this first part, because the next genus in the sequence was *Daphnia*, which Sars never understood to his full satisfaction. The main result is that these 51 taxa have very meager published original descriptions and no published original illustrations. Yet, most of them are claimed to occur on other continents.

2.4.2 *Disparalona rostrata in Europe versus D.leei in North America*

Disparalona rostrata had been claimed to occur in North America as well as in Europe, but a detailed comparison shows that the taxa on the two continents are completely distinct from one another. *D.rostrata* definitely does not occur in North America at all.

The situation is actually more complicated than this simple statement suggests. Birge (1879) described the new species *Alonella acutirostris* from the United States, and later, after becoming acquainted with the species present in Europe, he decided his taxon was *A.rostrata* instead, and so used this name in his monograph (Birge 1918). However, both species occur in North America (Frey 1961b). The illustration Birge used is clearly of *A.acutirostris*, but the verbal description contains elements of both species. Brooks (1959) changed the name to *A.acutirostris* in his revision of Birge's treatment without giving any indication that *A.rostrata* also exists. He used Birge's illustration and the same description. Having only this treatise available, Chien (1970) could not identify specimens he collected in Mississippi, and so he described them as the new species *Alonella leei.* However, *A.leei* was shown later to be the species then being called *D.rostrata* in North America (Frey 1974). Finally, when *D.rostrata* was eventually shown not to occur in North America, the taxon present reassumed Chien's name (Michael & Frey 1984).

2.4.3 *Alonopsis elongata in Europe versus A.americana in North America*

Doolittle (1912), studying the food habits of fishes in New England, found specimens of *A.elongata* in smallmouth bass fingerlings, and a new species *A.aureola* in smallmouth bass and Sunapee golden trout. Because of the very restricted distribution of *Alonopsis* in New England, Kubersky hypothesized that possibly *A.elongata* had been introduced from Europe during early colonial times and that a very slow rate of dispersal had kept it localized. However, careful comparison showed that the North American specimens are easily distinguishable from *A.elongata*, specimen by specimen, over the entire size range. The postabdomen and claw of the mature male also are strikingly different from *A.elonata*, so that the North American taxon was described as the new species *Alonopsis americana*. *Alonopsis aureola* quite obviously represents immature specimens of *A.americana.*

In Europe *A.elongata* occurs mainly west of the Ural Mountains, with very few positive identifications from localities to the east. The only other species with a similar morphology is *A.americana* from North America. Presently there are 11 positive records of this taxon in New England, mostly in New Hampshire in the restricted region between Lake Sebago and Vermont (Kubersky 1973, 1977), plus two records in Newfoundland and seven in Nova Scotia (Chengalath 1980). This highly restricted distribution of *A.americana* in the northeast is very interesting, and as yet without explanation. Some other chydorids, such as *Alona bicolor* (Frey 1965, 1969), have almost equally restricted distributions in New England and adjacent parts of Canada, as does the macrothricid *Parophryoxus tubulatus* (see Doolittle 1909, Chengalath 1980, 1982).

2.4.4 *Alona costata and A.rustica in Europe and North America*

Sars described *A.costata* from Norway in 1862, and Scott described *A.rustica* from the Shetland Islands in 1895. Although the two species are quite similar in their morphology, they can be distinguished readily. The taxon Mäemets (1958) described from Estonia as *A.estonica* was later shown by Frey (1965) to be *A.rustica.* Since separation of the two taxa, *A.rustica* has been found widely in Europe (Hrbáček et al. 1978), in North America (Frey 1982a), and on other continents (e.g. *A.iheringi* in Brazil, Sars 1901). In Europe the two taxa are allopatric (Flössner 1967), whereas in North America they commonly occur together in the same waterbodies (Frey 1969). This suggests that at least one of the species (*A.rustica*) is markedly different in its ecology on the two sides of the Atlantic Ocean.

The first morphological comparison of any taxa that were claimed to be the same in Europe and North America was carried out on these two species (Flössner & Frey 1970). At this 'early' time workers still were not lookin closely enough at the morphological details. Characters used were size, length/height ratio, size of ocellus versus size of eye, a headpore index (interpore distance/postpore distance), shape of labrum, and, most importantly, details of the postabdomen and postabdominal claw. Using these characters the taxa called *A.costata* on the two continents were judged to be the same, but those called *A.rustica* were considered to be different at the level of subspecies, the one in North America being described as *A.rustica americana* (Flössner & Frey 1970).

Since 1970 studies on other species of chydorids have been carried out in greater detail, using characters of the three instars of males and of ephippial females and ephippia in addition to those of parthenogenetic females. These have greatly changed our entire regard for species and species-limits in the Cladocera. A recomparison of *A.costata* and *A.rustica* on the two continents would be valuable, because I suspect that attention to these reproductive instars and to other significant body parts would probably show that the members of each pair of taxa are quite different at the species level.

2.5 *Pleuroxus denticulatus in North America and Europe*

This comparison reverses the usual sequence of a species first being described from Europe and then reported from North America, in that *P.denticulatus* was described initially from North America (Birge 1879) and then later reported from Europe and a number of localities on other continents, chiefly Asia. In North America the taxon is common, whereas on other continents it is erratic in occurrence and generally less common. Flössner & Kraus (1977) wished to compare morphologically specimens from Europe with those from North America to determine the distinctness or degree of overlap of various characters. They found some differences, but they also found so much overlap they couldn't even define any geographic subspecies.

Again, the study of Flössner & Kraus was based on a rather selected suite of characters, not the entire morphotype. However, more importantly, it assumed that only a single species occurred on each continent. As for Europe I cannot say, but for North America there are at last two distinct species which can occur simultaneously in the same waterbody. No studies have been completed on them yet, and hence we cannot say how they might modify the conclusions Flössner & Kraus reached. The species called *P.denticulatus* in North America has a very broad distribution from coast-to-coast and from northernmost Alaska southward to Guatemala and possibly even to Panama (Shan & Frey 1983). It was long doubted that only a single species occupied this entire region. It is now certain there are more, and after these have been sorted out, the relationship between the *P.denticulatus* species in North America and those occurring elsewhere must be considered again.

2.6 *Chydorus faviformis*

Chydorus faviformis is a striking species with a deep honeycomb-like pattern on its shell and head, which Birge (1893) described from northern Wisconsin. He was so struck by the honeycomb pattern that he could say little else about the morphology of the species. His illustrations are likewise undiagnostic.

More recently, honeycomb-patterned *Chydorus* have been reported from India, Malaysia, China, Australia, Argentina, and Brazil — all called *C.faviformis* (China — Yeh 1956, Chiang & Du 1979, northern India — Khan et al. 1978, Malaysia — Idris 1979, 1983, Idris & Fernando 1980, Argentina — de Ferrato 1966). Quite large collections exist from all these areas except Argentina and Brazil. Preliminary study indicates that two major species groups are represented, one with large meshes on the shell (about 120 on each side) and the other with small meshes (more than 150 and up to more than 200). It is certain that none of these other taxa from various parts of the world is the same as the North American *C.faviformis*. Hence, *C.faviformis*, regarded as representing another cosmopolitan species, obviously is not. The male of the North American *C.faviformis* loses its honeycomb pattern on molting to maturity, and it also has a postabdomen resembling that of *C.sphaericus* (Frey 1982c). None of the males of these other taxa have these characters.

2.7 *Chydorus linguilabris in North America*

In the southeastern states and extending erratically along the Atlantic coast into Nova Scotia and Newfoundland, a small *Chydorus* occurs with a striking pattern of reticulation completely covering the head and shell (Frey 1982f). Because Daday (1898) had described *Chydorus reticulatus* from Sri Lanka, it had to be ascertained whether the North American population was the same as his. However, field samples from Sri Lanka contained two reticulated species, and it was impossible to tell from Daday's description which was his species. By best judgment I attributed one as his and described the other as *Chydorus invaginatus.*

These two species from South and Southeastern Asia are rather closely related to one another, but scarcely related at all to the species in North America. The males are completely different in many respects, which together with various significant differences in the females, indicate that *C.invaginatus* cannot even be included in the same species group with *C.reticulatus.* This demonstrates, as does the ongoing study of *C.faviformis*, that various outstanding characters (such as honeycombing or strong reticulations) are not necessarily conservative, but that they can have arisen independently a number of times.

3 GEOLOGIC AGE OF THE CLADOCERA

Except for ephippia, the Cladocera leave almost no remains large enough to be recovered by the ordinary techniques used by geologists. Consequently, the stratigraphic record of Cladocera is very poor, and hence the time of origin of the Cladocera is completely uncertain. Nevertheless, for reasons given below, the Cladocera are considered to be very old.

Remains are abundant in sediments of present day lakes. Studies from sediments older than about 15 000 years are rare. Only a study by Frey (1962) from the Eemian Interglacial of Denmark and a study by Hann & Karrow (1984) from the Don Valley Brickyard deal with fossil cladocerans. Two features of the Frey study are significant: 1) no morphological differences were found between the species in the Eemian and the same species in Denmark today; and 2) the relative abundance by species was about the same in the Eemian as in Denmark today, indicting that the species have continued essentially unchanged in ecology and physiology over this time period.

From the Tertiary there are records of the ephippia of *Daphnia* and *Moina* from the western United States and from Western Europe (Frey 1964). The ephippia of *Daphnia florissantensis* from the Oligocene of Colorado are the oldest. They are from the subgenus *Ctenodaphnia*, the species of which typically occur in marginal waterbodies. This indicates that not only were Cladocera present in the early Tertiary, but also that they had evolved the production of ephippia to protect their resting eggs (*Daphnia* had already separated by that time into the two present subgenera characterized in part by different kinds of ephippia). Moreover, Goulden (1968) illustrated a *Moina* ephippium from the Late Miocene of California, which he considered either the species *Moina micrura* existing today or else one very close to it. This might suggest a long period of evolution prior to the Tertiary, and a great stability since that time.

Earlier occurrences of cladocerans are known, but are very scanty and uncertain. Smirnov (1970) described four species of *Archedaphnia* and one of *Propleuroxus* from the Permian of eastern Kazakhstan. He presented photomicrographs of several holotypes and simple pen sketches of various other specimens. These show chiefly outlines, and in addition some details that may represent a considerable amount of subjective interpretation by the author. On the other hand, Fryer (pers. comm.) has found an object from the Lower Cretaceous of Australia that he believes is an ephippium closely related to *Simocephalus.* I have examined an excellent photograph of the object and agree it could well be a daphniid ephippium. Thus, the Cladocera are old, and they must have been well diversified by the time Pangaea was breaking up and drifting apart, possibly leaving about the same suite of primordial genera and species on each continent.

4 HYBRIDIZATION

4.1 *Hybridization in chydorids*

The great age of the Cladocera is also supported by the results of attempts to cross closely related species. Hybridization requires the simultaneous availability of two populations with gamogenetic individuals; each success at crossing yields just a single resting egg. Then the diapause of the egg must be broken, which is frequently the most difficult part of all. We have been able to initiate gamogenesis in laboratory populations by the control of photoperoid and temperature, occasionally also by crowding and food supply.

The first attempts at experimental crossings (Shan & Frey 1968, Shan 1974) used *Pleuroxus denticulatus* from widely scattered locations across North America. These studies are particularly important for they indicate that populations in the South are genetically different from populations in the North, because the one successful cross between the two regions yielded offspring that were often deformed, and none them achieved reproductive maturity. The complication here, realized just very recently, is that there are different species of '*P.denticulatus*' even in the North, and that the crosses attempted probably did not always involve the same species.

Later came attempted crosses by Kubersky (1973, 1977) between populations of *Alonopsis americana* from two localities in New England and *A.elongata* from England. Production of ephippia containing resting eggs was scored as a success, because no attempts at hatching the resting eggs were successful — except for a few that had been exposed briefly to dilute KOH to provide a necessary shock. Crosses within the individual populations were successful, as also were crosses between the New England populations. However, crosses between the New England populations and the one from England were not successful at all in any combination of males and females. Thus, the populations on the two continents are completely isolated from one another genetically.

Hann (1980) achieved more than 100 crosses among various populations of *Eurycercus longirostris* from northern Indiana, *E.vernalis* from North Carolina, and between these two species. About 95% of these crosses produced resting eggs, but as none of the eggs could be induced to hatch, success had to be judged merely by their production. Success was as great between Indiana and North Carolina as within populations from either locality. The northern populations became gamogenetic in autumn at declining photoperiod and the southern populations in spring at increasing photoperiod. To make hybridization possible, the northern populations were maintained in gamogenetic condition in the laboratory by manipulating the photoperiod and then crossed with gamogenetic individuals collected in the field in North Carolina. The northern and southern populations are not distinct from one another on the basis of the morphology of individual specimens, but they can be separated easily by regression statistics and by frequencies of various allozymes. They are regarded as distinct species (Hann 1980, 1982). The interesting feature here is that the two species seem to be capable of interbreeding successfully, but they are kept from doing so in nature by becoming gamogenetic at different times of the year, besides which they are geographically separated.

In summary, these crossing attempts indicate that morphologically distinct species (*Alonopsis*) are incapable of producing resting eggs when crossed, that northern and southern populations (species?) of *Pleuroxus denticulatus* can produce resting eggs that hatch, but are incapable of yielding reproductive individuals; and that northern and southern species of *Eurycercus* when crossed produce resting eggs easily, although nothing is known about their hatchability and survival. Shan (1971) and Shan & Frey (1983) succeeded in producing a hybrid population of *P.denticulatus x P.procurvus*, both parents coming from Lake Itasca, Minnesota. The population thrived parthenogenetically for a couple of years and developed gamogenetic individuals, which, in attempts at backcrossing them in the laboratory, could not produce resting eggs among themselves nor with their parental populations. Again, this rather meager evidence merely suggests that most species of Cladocera are genetically isolated from one another. Thus, in northern lakes in which under favorable condition 20 or more species of chydorids can be gamogenetic at the same time, there has never been any indication of interspecific crossing, even though these occasionally can be induced in the laboratory. When this occurs the hybrids are capable only of parthenogenetic reproduction (Shan & Frey 1983).

4.2 *Hybridization in other Cladocera*

An attempt to summarize the concepts of speciation and evolution in other Cladocera is difficult. Even today persons are attempting to explain the spectacular diversity among *Bosmina* in the Baltic region as the result of introgression among several parent species that invaded the region in late glacial-early postglacial time. Lieder (1983), for example, suggested this complex diversity was produced genetically over the past 10 000 years. Hofmann (1978, 1984), on the other hand, from a study of Holocene *Bosmina* microfossils in two north German lakes concluded that there has been no introgression in these localities. Others (see e.g. Brooks 1957) think that introgression is continually occurring in *Daphnia*, based on what they consider to be individuals with intermediate characters among wild populations. These claims may be true, because there is no necessity that the planktonic species should behave in the same way as the benthic species. Most chydorids certainly are well isolated from one another, the planktonic species are possibly less well isolated.

Certainly the allozyme studies being done on *Daphnia, Simocephalus, Eurycercus*, and *Bosmina* are pertinent, because they are developing insight into the genetic makeup of these Cladocera and the ways in which the species function in short-term adaptation and long-term evolution. Hebert (1983), for example, is concerned with evolution among obligate parthenogens (*Daphnia pulex*), as well as how many clones of these animals exist in a particular region and how many in different lakes of that region. Such studies are extremely valuable for showing how the Cladocera, especially those that reproduce only by parthenogenesis, function genetically.

5 SOME IMPLICATIONS OF NON-COSMOPOLITANISM

Working with the Cladocera is frustrating because of the large amount of morphological variation in the group from region to region, even on the same continent. One has to attempt to distinguish between the morphological characteristics of the species and those features of the morphology of populations that may represent adaptation to specific environmental features of the lake or region. Sorting out these features within these small animals can be accomplished only by studying and comparing entire populations, mostly based on their morphology, and also by recording their time of gamogenesis and other features. Regrettably, most chydorids are too small in size for individual allozyme analyses utilizing present techniques.

5.1 *Gamogenesis: Northern species in the South*

The work recently initiated, especially in Florida in 1979 and also in Louisiana-Mississippi in 1977 and 1978, has generated most of our present ideas concerning gamogenesis timing (Frey 1982a). Gamogenesis occurs in anticipation of the season that provides the major stresses for the species. In our northern lakes the various species become gamogenetic in autumn in preparation for winter. Under extreme conditions, all active individuals die, leaving the species to be maintained only by their resting eggs, which hatch next spring as a new favorable season begins. In the South, winter is not a major stress for most species, but apparently summer is. The southern animals remain vigorously parthenogenetic over winter and become weakly and variably gamogenetic in spring.

The above study also showed that many species present in the Southeast seem to be the same as species in the North. These are the Cladocera that become gamogenetic in spring or early summer in the South. They generally tend to be replaced in these lakes over summer by 'southern' species, which become gamogenetic in autumn and then disappear entirely or at least suffer a major decline in their population density over winter. However, there is a tendency toward a seasonal replacement of 'northern' species that occur in the lakes from late autumn into spring or early summer by 'southern' species that occur in summer and autumn.

The studies to date on *Pleuroxus denticulatus* (Shan & Frey 1968, 1983) and the two species of *Eurycercus* (Hann 1980, 1982) suggest strongly that these northern species in the South differ at the level of full species from their counterparts in the North. I and others believe that the members of quite a few other species pairs are genetically differentiated, although up to now we have no evidence for this, just general or intuitive impressions from studying them with a microscope.

One hypothesis to explain this distribution is that during the ice ages, northern species necessarily were displaced from the northern lakes by the development of the glaciers. Some species entered the waterbodies of the southeastern states, where either immediately or else more gradually in response to interglacial warming, changed their time of gamogenesis to springtime, thus allowing survival in the inclement weather of summer. With amelioration of climate upon the melting of the glaciers, those populations that had remained toward the north and had retained their autumnal gamogenesis reinvaded the new waterbodies being formed. The populations in the South, now having become adjusted there to the severity of the summers, remained in the South. The differences in timing of gamogenesis between the northern and southern populations of these species isolated them genetically and permitted speciation to occur.

5.2 *North-south displacement along the Atlantic Coast*

If species in the North were displaced southward during the glacial ages, then it is equally likely that southern species advanced northward during interglacial climatic ameliorations. The Atlantic coastal strip is particularly favorable for such movements. A number of southern species, such as *Chydorus bicollaris, Disparalona dadayi,* and *Eurycercus microdontus* have advanced northward to the New Jersey Pine Barrens. Similarly, a number of northern species, such as *Chydorus bicornutus*, reach their southernmost limits in New Jersey, where it is possible for northern and southern species to occur in the same waterbodies.

Chydorus linguilabris, a southern species because of the frequency of its occurrence and abundance in Florida and North Carolina, is found in the Pine Barrens. However, unlike the other species mentioned, it also occurs northward into Nova Scotia and even Newfoundland. There are a few scattered records from Massachusetts and Maine, all within 15 km of the coast, but nothing at all from farther inland from numerous samples. The records from farther north are also isolated occurrences, all from softwater lakes that are naturally acidic. This is the kind of habitat that seems particularly favorable for chydorids, since they contain large numbers of species.

The corollary hypothesis, therefore, is that species from the South pushed northward during the interglacial thermal maxima, some of them getting to Newfoundland. During the subsequent climatic regression, any southern species in the North generally tended to become extinct. Only *C.linguilabris*, so far as we know, has persisted here in isolated localities. Future investigations should attempt to determine when *C.linguilabris* and other species might have invaded these lakes by analysis of their exoskeletal remains in cores. There undoubtedly has been continuing latitudinal adjustment of cladoceran species, particularly along the Atlantic Coast, in response to climatic changes.

5.3 *East-west discontinuities along the Gulf Coast*

Various groups of organisms, both plant and animal, exhibit distributional discontinuities along the Gulf coast. The species tend to occur in Texas or Florida, or in both, but without any occurrences in between.

For example, in the genus *Ephemeroporus* only the species *E.acanthodes* and one member of the *E.hybridus* group occur in Louisiana-Texas-Oklahoma. Nearly all the other North American species of the genus are confined to Florida and nearby areas in Georgia and South Carolina. None of these species has been collected yet from as far north as the Pine Barrens. Just one species (presently called *E.hybridus*) occurs erratically in northern Michigan and on Isle Royale in Lake Superior (perhaps another example of the occasional persistence of relict populations of southern species that advanced north during interglacial periods?). The discontinuous distribution of *Ephemeroporus* along the Gulf means either that the nature of the waterbodies excludes certain species of the genus because of their individual species requirements, or else that the species occurring east or west have not been able to move to the other region and establish populations. Lowered sea level during the glacial advances must also be considered as a distributional factor.

5.4 *Endemic species*

Endemism can have different meanings. There are a few species of chydorids confined

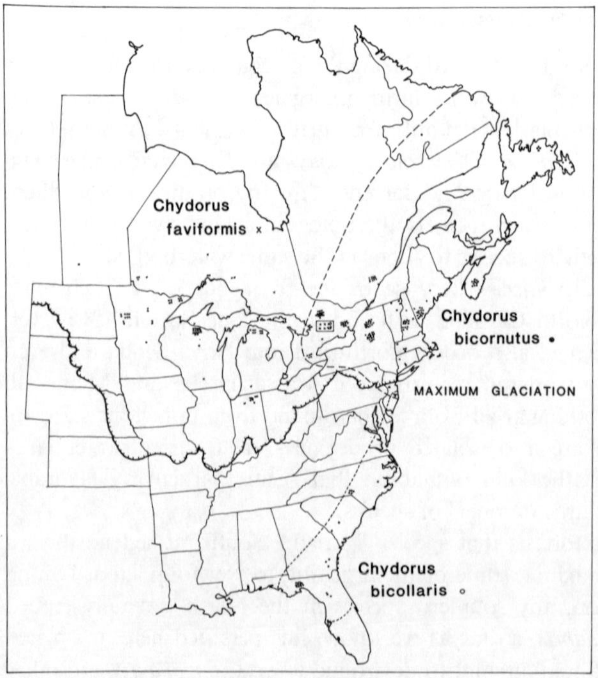

Figure 1. Known occurrences of *Chydorus faviformis, C.bicornutus*, and *C.bicollaris* in Eastern North America. Most of the records for *C.bicornutus* and *C.bicollaris* are from Frey (1982g). Those for *C.faviformis*, including both literature and collection records, have not been published previously. Many of the Canadian records are from Chengalath (1980), although we have extensive additional records from Ontario. Hann (1975) has 10 records of *C.bicornutus* and 45 of *C.faviformis* from Ontario, but she did not specify their localities.

to particular lakes, such as various species in the deepwater meiobenthos of Lake Baikal (Vasil'yeva & Smirnov 1969, 1975) and *Alona smirnovi* in Lake Ohrid (Petkovski & Flössner 1972), all of which seem to be true endemics. However, if the chydorids (and Cladocera in general?) are as restricted in distribution as they seem to be, confined to relatively small areas of the earth's surface, then these also should be considered endemic for the particular region.

Endemism obviously has a strong time dimension to it. Where species now occur represents only a moment in the time continuum of their distribution. These distributions, at least on a large land mass like North America, have fluctuated over geologic time, even as recently as the last few thousand years. During the last few million years that constitute the Quaternary, conditions have changed drastically and precipitously a number of times between interglacial and glacial periods. Such fluctuations forced migrations and speciation. Although the chydorid Cladocera seem to be inherently stable and conservative, they, too, continued to migrate during the Quaternary, producing the populations of northern species in the South, which are differentiating from their parent populations that have retreated back to the North.

5.5 *Distribution in North America*

Birge's treatment (1918) of distributions of Cladocera is poor, based as it is on a rather small number of collections with large distributional gaps, and because Birge unjustifiably extrapolated some of these records into larger distributions when in reality the species are much more narrowly restricted. Although very little work on the systematics of Cladocera in North America, except for *Daphnia* (Brooks 1957) and *Moina* (Goulden 1968), has been done since Birge's time, a number of summaries of the Cladocera for particular states or regions exist, to which can be added my own very extensive collections from many parts of North America and those of Chengalath (1980, 1982) from Canada.

Preliminary analysis of this material is showing broad patterns of distribution, many of which are probably related to the events of glaciation-deglaciation. For example, of the three honeycombed species of *Chydorus* in North America, *C.faviformis* and *C.bicornutus* occur only in the glaciated region of the Northeast and *C.bicollaris* strictly in non-glaciated areas of the Southeast, with *C.bicornutus* and *C.bicollaris* barely overlapping in the New Jersey Pine Barrens (Fig.1). *Pleuroxus procurvus* likewise is confined to the glaciated region, although it occurs across the entire continent and extends southward only at high elevations in the western mountains into New Mexico (Shan & Frey 1983). Quite a few species seem to have a very restricted distribution in New England and the Maritime Provinces, sometimes extending as far west as Lake Huron. *Chydorus bicornutus* is among these, as also are species with much more restricted distributions, such as *Chydorus canadensis, Alonopsis americana,* and *Alona bicolor.*

Our knowledge of the Cladocera in western North America is very meager. R.W.Kiser recently deposited his extensive collections, mainly from Washington and Alaska, with the US National Museum, where they are now accessible. Many of the species in the West have the same names as those in the East, although differences in fine details suggest genetic differentiation and careful study may well show some of them to be separate species.

In North America an assemblage of cladocerans exists which superficially resembles what occurs in Europe and other continents. However, I believe this assemblage has largely developed wholely within North America from taxa that were on the continent very early, perhaps near the beginning of the breakup of Pangaea. These taxa have dispersed via resting eggs, and occasionally have been displaced from areas in the North, chiefly in response to major disruptions and climatic changes during the Pleistocene. The fact that many North American species have the same names as Cladocera on other continents should not delude us into thinking that they necessarily are the same taxa. Strong positive evidence is needed to support such a belief, especially in light of the fact that a number of cladoceran species suspected of being cosmopolitan have already been shown to be a series of narrowly distributed species.

REFERENCES

Birge, E.A. 1878. Crustacea Cladocera collected in Fish Pond and Glacialis, Cambridge, Mass. 1876, and at Madison, Wis. 1877. Ph.D. thesis, Harvard Univ.

Birge, E.A. 1879. Notes on Cladocera. *Trans. Wis. Acad. Sci. Arts Lett.* 4: 77-109.

Birge, E.A. 1892. List of Crustacea Cladocera from Madison, Wisconsin. *Trans. Wis. Acad. Sci. Arts Lett.* 8: 379-398.

Birge, E.A. 1893. Notes on Cladocera, III. *Trans. Wis. Acad. Sci. Arts Lett.* 9: 275-317.

Birge, E.A. 1910. Notes on Cladocera, IV. *Trans. Wis. Acad. Sci. Arts Lett.* 16: 1017-1066.

Birge, E.A. 1918. The water fleas (Cladocera). In H.B.Ward & G.C.Whipple (eds.), *Fresh-water Biology.* New York: Wiley.

Brooks, J.L. 1957. The systematics of North American *Daphnia. Mem. Conn. Acad. Arts Sci.* 13: 1-180.

Brooks, J.L. 1959. Cladocera. In W.T.Edmondson (ed.), *Fresh-water Biology*, 2nd ed. New York: Wiley.

Chengalath, R. 1980. Survey of the littoral Cladocera of Canada. Ottawa, Ont. Mimeogr. iv, 205 pp.

Chengalath, R. 1982. A faunistic and ecological survey of the littoral Cladocera of Canada. *Can. J. Zool.* 60: 2668-2682.

Chiang, Sieh-chih & Nan-shan Du 1979. *Fauna Sinica, Crustacea, Freshwater Cladocera.* Peking: Science Press, Academica Sinica.

Chien, S.M. 1970. *Alonella fitzpatricki* sp.n. and *A.leei* sp.n.: new Cladocera from Mississippi. *Trans. Amer. Micros. Soc.* 89: 532-538.

Daday, E. 1898. Mikroskopische Süsswasserthiere aus Ceylon. *Termés. Füzetek, Anhangsheft* 21: 1-123.

Doolittle, A.A. 1909. New Cladocera from New England. *Proc. Biol. Soc. Wash.* 22: 153-156.

Doolittle, A.A. 1912. Notes on the occurrence of the crustacean *Alonopsis* in America, with description of a new species. *Proc. U.S. Nat. Mus.* 43: 561-565.

De Ferrato, A.M. 1966. Nuevos cladoceros para las aguas Argentinas. *Physis, Buenos Aires* 26: 397-403.

Flössner, D. 1967. *Alona rustica* Scott und *Alona costata* Sars (Cladocera, Chydoridae). Zur Kenntnis der Morphologie und Ökologie einer Zwillingsart. *Limnologica (Berlin)* 5: 417-434.

Flössner, D.& D.G.Frey 1970. A morphological comparison of European and North American populations of *Alona rustica* Scott and *Alona costata* Sars (Cladocera, Chydoridae). *Limnologica (Berlin)* 7: 325-337.

Flössner, D. & K.Kraus 1977. On the variability and taxonomy of *Pleuroxus denticulatus* Birge (Cladocera: Chydoridae). *J. Fish. Res. Bd. Can.* 34: 463-476.

Frey, D.G. 1955. Längsee: a history of meromixis. *Mem. Ist. Ital. Idrobiol.,* suppl.8: 141-164.

Frey, D.G. 1958. The late-glacial cladoceran fauna of a small lake. *Arch. Hydrobiol.* 54: 209-275.

Frey, D.G. 1959. The taxonomic and phylogenetic significance of the head pores of the Chydoridae (Cladocera). *Int. Rev. ges. Hydrobiol.* 44: 27-50.

Frey, D.G. 1961a. Developmental history of Schleinsee. *Verh. Internat. Verein. Limnol.* 14: 271-278.

Frey, D.G. 1961b. Differentiation of *Alonella acutirostris* (Birge, 1879) and *Alonella rostrata* (Koch, 1841) (Cladocera, Chydoridae). *Trans. Amer. Micros. Soc.* 80: 129-140.

Frey, D.G. 1962. Cladocera from the Eemian Interglacial of Denmark. *J. Paleontol.* 36: 1133-1154.

Frey, D.G. 1964. Remains of animals in Quaternary lake and bog sediments and their interpretation. *Arch. Hydrobiol. Ergebn. Limnol.* 2: 1-114.

Frey, D.G. 1965. Differentiation of *Alona costata* Sars from two related species (Cladocera, Chydoridae). *Crustaceana* 8: 159-173.

Frey, D.G. 1966a. Cladocera in space and time. *Mar. Biol. Assoc. India, Proc. Symposium Crustacea* 1: 1-9.

Frey, D.G. 1966b. Phylogenetic relationships in the family Chydoridae. *Mar. Biol. Assoc. India, Proc. Symposium Crustacea* 1: 29-37.

Frey, D.G. 1969. Further notes on *Alona bicolor* Frey, 1965 (Cladocera, Chydoridae), with a description of the male. *Trans. Amer. Micros. Soc.* 88: 380-386.

Frey, D.G. 1971. Worldwide distribution and ecology of *Eurycercus* and *Saycia* (Cladocera). *Limnol. Oceanogr.* 16: 254-308.

Frey, D.G. 1973. Comparative morphology and biology of three species of *Eurycercus* (Cladocera, Chydoridae) with a description of *Eurycercus macracanthus* sp.nov. *Int. Rev. ges. Hydrobiol.* 58: 221-267.

Frey, D.G. 1974. Reassignment of *Alonella fitzpatricki* and *A.leei* Chien, 1970 (Cladocera, Chydoridae). *Trans. Amer. Micros. Soc.* 93: 162-170.

Frey, D.G. 1975. Subgeneric differentiation within *Eurycercus* (Cladocera, Chydoridae) and a new species from northern Sweden. *Hydrobiologia* 46: 263-300.

Frey, D.G. 1976. The distribution and ecology of *Eurycercus glacialis* (Cladocera, Chydoridae) in Western Europe. *Verh. Internat. Verein. Limnol.* 19: 2922-2934.

Frey, D.G. 1978. A new species of *Eurycercus* (Cladocera, Chydoridae) from the southern United States. *Tulane Stud. Zool. Bot.* 20: 1-26.

Frey, D.G. 1980. On the plurality of *Chydorus sphaericus* (O.F.Müller) (Cladocera, Chydoridae), and designation of a neotype from Sjælsø, Denmark. *Hydrobiologia* 69: 83-123.

Frey, D.G. 1982a. Contrasting strategies of gamogenesis in northern and southern populations of Cladocera. *Ecology* 63: 223-241.

Frey, D.G. 1982b. G.O.Sars and the Norwegian Cladocera: a continuing frustration. *Hydrobiologia* 96: 267-293.

Frey, D.G. 1982c. Honeycombing of the carapace in the chydorid Cladocera: the elusive male of *Chydorus faviformis*. *J. Crust. Biol.* 2: 469-476.

Frey, D.G. 1982d. Questions concerning cosmopolitanism in Cladocera. *Arch. Hydrobiol.* 93: 484-502.

Frey, D.G. 1982e. Relocation of *Chydorus barroisi* and related species (Cladocera, Chydoridae) to a new genus and descriptions of two new species. *Hydrobiologia* 86: 231-269.

Frey, D.G. 1982f. The reticulated species of *Chydorus* (Cladocera, Chydoridae): two new species with suggestions of convergence. *Hydrobiologia* 93: 255-279.

Frey, D.G. 1982g. The honeycombed species of *Chydorus* (Cladocera, Chydoridae): comparison of *C.bicornutus* and *C.bicollaris* n.sp., with some preliminary comments on *faviformis*. *Can. J. Zool.* 60: 1892-1916.

Frey, D.G. 1985. A new species of the *Chydorus sphaericus* group (Cladocera: Chydoridae) from western Montana. *Int. Rev. ges. Hydrobiol.* 69: 3-20.

Fryer, G. & D.G.Frey 1981. Two-egged ephippia in the chydorid Cladocera. *Freshwater Biol.* 11: 391-394.

Goulden, C.E. 1964. The history of the cladoceran fauna of Esthwaite Water (England) and its limnological significance. *Arch. Hydrobiol.* 60: 1-52.

Goulden, C.E. 1968. The systematics and evolution of the Moinidae. *Trans. Amer. Phil. Soc.* N.S. 58(6): 1-101.

Hann, B.J. 1975. Taxonomy of Chydoridae in Ontario and genus *Chydorus* (Worldwide). M.S. thesis, Department of Biology, University of Windsor, Ontario.

Hann, B.J. 1980. Population differentiation in the *Eurycercus (Bullatifrons)* species complex (Chydoridae, Cladocera) in eastern North America. Ph.D. thesis, Indiana University.

Hann, B.J. 1982. Two new species of *Eurycercus (Bullatifrons)* from Eastern North America (Chydoridae, Cladocera). Taxonomy, ontogeny, and biology. *Int. Rev. ges. Hydrobiol.* 67: 585-610.

Hann, B.J. & P.F.Karrow 1984. Pleistocene paleoecology of the Don and Scarborough Formations, Toronto, Ontario based on cladoceran microfossils. I. The Don Valley Brickyard. Boreas.

Harmsworth, R.V. 1968. The developmental history of Blelham Tarn (England) as shown by animal microfossils, with special reference to Cladocera. *Ecol. Monogr.* 38: 223-241.

Hebert, P.D.N. 1983. Clonal diversity in cladoceran populations. In C.E.King & P.S.Dawson (eds.), *Population Biology: Retrospect and Prospect.* New York: Columbia Univ.

Hofmann, W. 1978. *Bosmina (Eubosmina)* populations of Grosser Plöner See and Schöhsee lakes during late-glacial and postglacial times. *Polsk. Arch. Hydrobiol.* 25(1/2): 167-176.

Hofmann, W. 1984. Postglacial morphological variation in *Bosmina longispina* Leydig (Crustacea, Cladocera) from the Grosser Plöner-See (north Germany) and its taxonomic implications. Submitted.

Hrbáček, J., V.Kořínek & D.Frey 1978. Cladocera. In J.Illies (ed.), *Limnofauna Europaea. Eine Zusammenstellung aller die europäischen Binnengewässer bewohnenden mehrzellingen Tierarten mit Angaben über ihre Verbreitung und Ökologie*, 2nd ed. Stuttgart: Fischer.

Idris, B.A.G. 1979. The taxonomy and distribution of Malaysian freshwater Cladocera (Crustacea). M.S. thesis, Univ. Waterloo.

Idris, B.A.G. 1983. *Freshwater zooplankton of Malaysia (Crustacea: Cladocera).* Penerbit Universiti Pertanian Malaysia.

Idris, B.A.G. & C.H.Fernando 1980. Cladocera of Malaysia and Singapore with new records, redescriptions and remarks on some species. *Hydrobiologia* 77: 233-257.

Khan, M.A., B.A.Subla & D.P.Zutshi 1978. A new crustacean for India. *Geobios* 5: 81.

Kubersky, E.S. 1973. Worldwide distribution and ecology of *Alonopsis* (Cladocera: Chydoridae). Ph.D. thesis, Indiana Univ.

Kubersky, E.S. 1977. Worldwide distribution and ecology of *Alonopsis* (Cladocera: Chydoridae) with a description of *Alonopsis americana* sp.nov. *Int. Rev. ges. Hydrobiol.* 62: 649-685.

Lieder, U. 1983. Introgression as a factor in the evolution of polytypical plankton Cladocera. *Int. Rev. ges. Hydrobiol.* 68: 269-284.

Lilljeborg, W. 1901. Cladocera Sueciae, oder Beiträge zur Kenntniss der in Schweden lebenden Krebsthiere von der Ordnung der Branchiopoden und der unterordnung der Cladoceren. *Nova Acta reg. Soc. Sci. Upps.* Ser.3: i-vi; 1-701.

Mäemets, A. 1958. Systematisch-ökologische Bemerkungen über einige interessante Wasserflöhe (Cladocera) Estlands. *Eesti, NSV Teaduste Akad. Toim., VII Köide, Biol. Ser.* 4: 260-271.

Michael, R.G. & D.G.Frey 1983. Assumed Amphi-Atlantic distribution of *Oxyurella tenuicaudis* (Cladocera, Chydoridae) denied by a new species from North America. *Hydrobiologia* 106: 3-35.

Michael, R.G. & D.G.Frey 1984. Separation of *Disparalona leei* (Chien, 1970) in North America from *D.rostrata* (Koch, 1841) in Europe (Cladocera, Chydoridae). *Hydrobiologia* 114: 81-108.

Müller, O.F. 1776. *Zoologiae Danicae Prodromus seu Animalium Daniae et Norvegiae indigenarum characteres, nomina, et synonyma imprimis popularium.* Havniae.

Müller, O.F. 1785 *Entomostraca seu Insecta Testacea, quae in aquis Daniae et Norvegiae reperit, descripsit et iconibus illustravit.* Lipsiae et Havniae.

Olofsson, G. 1918. Studien über die Süsswasserfauna Spitzbergens. *Zool. Bidrag Uppsala* 6: 183-648.

Pennak, R.W. 1953. *Fresh-water invertebrates of the United States.* New York: Ronald.

Pennak, R.W. 1978. *Fresh-water invertebrates of the United States*, 2nd ed. New York: Wiley.

Petkovski, T. & D.Flössner 1972. Eine neue *Alona*-Art (Crustacea: Cladocera) aus dem Ohridsee. *Fragm. Balcanica* 9: 97-106.

Rodhe, W. & D.G.Frey (eds.) 1982. *Wilhelm Lilljeborg, Cladocera Sueciae. Facsimile Reissue of the Original Edition with a Prologue.* Stockholm: Almqvist & Wiksell.

Sars, G.O. 1861 Om de i Christiania's Omegn forekommende Ferskvandskrebsdyr. Unpublished MS, Dept. Manucripts, Univ. Library of Oslo.

Sars, G.O. 1862. Hr. Studios medic. G.O.Sars fortsatte sit Foredrag over de af ham i Omegnen af Christiania iagttagne Crustacea Cladocera. *Forh. Vidensk.-Selskab. Christiania*, Aar 1861: 250-302.

Sars, G.O. 1865. *Norges Ferskvandskrebsdyr. Første Afsnit Branchiopoda. I. Cladocera Ctenopoda (Fam. Sididae & Holopedidae).* Christiania.

Sars, G.O. 1901. Contributions to the knowledge of the freshwater Entomostraca of South America, as shown by artificial hatching from dried material. Part I. Cladocera. *Arch. Math. Naturvidensk.* 23: 1-101.

Sars, G.O. 1904. Pacifische Plankton-Crustaceen. (Ergebnisse einer Reise nach dem Pacific. Schauinsland 1896/97). *Zool. Jahrb. Syst.* 19: 629-646.

Scott, T. 1895. The inland waters of the Shetland Islands. Notes on the species represented by the drawings on Plate V. *Ann. Rep. Fish. Bd. Scotland* 13: 189-191.

Shan, R.K. 1971. Genetic stability in the Cladocera. *Trans. Amer. Micros. Soc.* 90: 116.

Shan, R.K. 1974. Reproduction in laboratory stocks of *Pleuroxus* (Chydoridae, Cladocera) under the influence of photoperiod and light intensity. *Int. Rev. ges. Hydrobiol.* 59: 643-666.

Shan, R.K. & D.G.Frey 1968. Induced interbreeding between two stocks of a chydorid cladoceran. *BioScience* 18: 203-205.

Shan, R.K. & D.G.Frey 1983. *Pleuroxus denticulatus* and *P.procurvus* (Cladocera, Chydoridae) in North America: distribution, experimental hybridization, and the possibility of natural hybridization. *Can. J. Zool.* 61: 1605-1617.

Smirnov, N.N. 1970. Cladocera (Crustacea) iz permskikh otlozheniy Vostochnogo Kazakhstana. *Paleontol. Zhur.* 3: 95-100.

Smirnov, N.N. 1971. *Chydoridae Fauny Mira.* Fauna SSSR, Nov. Ser. No.101. Rakoobrasnyye, T.1, Vyp.2.

Vasil'yeva, G.L. & N.N.Smirnov 1969. Chydoridae (Cladocera) Baykala. *Zool. Zhurnal* 48: 184-196.

Vasil'yeva, G.L. & N.N.Smirnov 1975. Ekologiya Chydoridae (Cladocera) Baykala. *Zool. Zhurnal* 54: 1293-1305.

Whiteside, M.C. 1970. Danish chydorid Cladocera: modern ecology and core studies. *Ecol. Monogr.* 40: 79-118.

Yeh, C.C. 1956. Cladocera from the lake Tung-Tsien-Hu. *Acta Hydrobiol. Sinica* 1: 43-60.

Ostracoda: New aspects of their biogeography

KENNETH G.McKENZIE
Riverina College of Advanced Education, Wagga Wagga, NSW, Australia

ABSTRACT

The approach to biogeography by modern ostracodologists, while continuing to rely on uniformitarian principles, takes account of a large amount of new basic data, much of it from hitherto little known regions and little studied geological periods. Environmental factors are now examined more precisely. In particular, recent research has concentrated upon carbonate equilibrium, oxygen content, osmoregulation, food supply, trace elements, and solute composition, in addition to more familiar factors such as salinity, temperature, depth and substrate. In the context of the space-time continuum, recent work tends increasingly to postulate the existence of mosaic evolution. Ostracode geneticists have recorded a diversity of karyotype patterns which are consistent with the mosaic evolution hypothesis, and which can be interpreted meaningfully for the better elucidation of persistent biogeographic problems, such as the implications of parthenogenesis for distributions of continental taxa. Among the many dispersal mechanisms recently studied, greater emphasis is now given to the role of man, citing the historically well-documented example of the introduction of ostracodes into the Italian rice fields ecosystem. There is also a greater reliance on Ostracoda in biostratigraphy, particularly in relation to changes in sea level, climate and a paleogeography disposed in accordance with Expanding Earth as well as Plate Tectonic theories. Economic applications continue to provide solid incentives for much ostracode-based biogeographic analysis.

1 INTRODUCTION

The last review of ostracode biogeography by this author (McKenzie 1973) appeared during a period of stasis following earlier pioneer studies of marine and freshwater ostracode distributions during the Mesozoic and Cenozoic (Krommelbein 1966, McKenzie 1967, 1971). Since then there has been a remarkable increase of effort and several changes of emphasis.

Much of the new data comes from China where over 2000 new species have been described since 1978 (Table 1). Much also comes from the Paleozoic (Jones & McKenzie 1980) and Mesozoic (Al-Furaih 1980, Dingle 1981, Tambareau 1982a,b), although Cenozoic studies continue to multiply apace (Kornicker 1981, Gou et al. 1983, Deevey 1983).

Table 1. New taxa from China described and figured over the past six years. Data only from references provided to the author during and since a visit to China in July 1983; numbers thus likely to be conservative.

Geological age	Number of species	New genera
Cambrian	170	30
Devonian	150	10
Permo-Triassic	150	6
Jurassic	40	1
Cretaceous	375	3
Paleogene	650	25
Neogene	225	4
Quaternary	240	14
Totals	2 000	93

No less remarkable have been the changes in approach. Previously, studies were less detailed. A typical summary of the usual environmental factors that were discussed appears in McKenzie (1973). The modal concepts of Simpson (1940) concerning corridors, filters and sweepstakes routes were not applied to Ostracoda until a generation later (McKenzie 1967), as were the ideas of Sewell (1940) regarding ocean currents as dispersion agents (Neale 1976). Both papers were concerned with interpretations of Tethyan ostracode distributions. Numerous researchers have investigated the dispersal of continental taxa. Their results will be cited below where appropriate, but major emphasis will be given to the as yet little appreciated role of man. Evidently, pre-human distributions represent a set of data which differ significantly from the man produced distributions.

The latest studies of marine Ostracoda emphasize the stratigraphic equivalence (correlation) or relations of widely distributed taxa; and there is recognition of the fact that this new stratigraphic information can be linked to changes in climate, paleogeography, and sea level. Whereas the earlier work analyzed distributions in terms of Continental Drift and Plate Tectonics, recent papers deal with interpretations of continental movements in terms of Expanding Earth theory as well.

Environmental factors which are now stressed include particularly carbonate equilibrium, oxygen content, available light, osmoregulation, food supply (nutrient), trace elements, and solute composition. The effects of these factors are discussed either in terms of physiological response as expressed by the ostracode carapace, or as they affect ostracode distributions.

Among the speciation and distribution patterns now being evaluated, the idea of mosaic evolution is gaining wider currency (Colin & Danielopol 1980), and will be shown to be anchored in the results of recent genetic studies (Tetart 1978).

The advantages of Ostracoda for biogeographic research is related to their excellent fossil record and well-understood Recent distributions. Also important is the reliance that ostracode specialists place on uniformitarian principles (Neale 1983) which is bolstered by the evidence for the continuity of ostracode assemblages over long epochs, especially in post-Paleozoic environments (McKenzie 1976). The basic biogeographic data are now interpreted in the light of more detailed knowledge of the biology of these fascinating crustaceans, with the linkage from fossil data and uniformitarianism indicating greater maturity and confidence among ostracodologists.

2 ENVIRONMENTAL CONSIDERATIONS

2.1 *The carbonate equilibrium*

Pora (1969) established the rhopic factor, i.e. the ionic equilibrium of aquatic environments, by utilizing the ratios Mg/Ca and/or Mg + Ca/total ions. Subsequently, Martin (1977) showed that magnesium accumulation in the crustacean exoskeleton could be as much a result of physico-chemical phenomena as physiological and biochemical ones. The significance of these results for ostracodes, as for other crustaceans, lies in the fact, demonstrated by Turpen & Angell (1971), that when crustaceans molt there is little reabsorption of carbonate from the exuviae, nor is carbonate stored in the living tissue — hemolymph or gut wall. Lately, Rosenfeld (1982) has discovered a calcitic/apatitic granular layer, secreted by the epidermis and attached to the internal face of the carapace/valve. He speculates that it recrystallizes to help form the carbonate of the succeeding molt stage. This recrystallization depends on input from carbonate in the surrounding aquatic environment.

The research group at Bordeaux (Rousselle et al. 1980, Peypouquet et al. 1981) has developed a model which correlates the rhopic factor and architectural variations of ostracode shells. According to this model, Ostracoda aggrade (add carbonate to) their genotypic ornament in environments where temperature and pH are high, CO_2 is being given off, and there is an excess of carbonate. They degrade (reduce) their genotypic ornament in environments where temperature and pH are lower, CO_2 is retained, and there is undersaturation with respect to carbonate. Clearly, their response is ecophenotypic. In terms of such a 'dialectic of the coevolution of organisms and environments' (Rousselle et al. 1980) many ornamented Ostracoda exhibit polymorphisms which former workers attributed to speciation, i.e. to genotypic rather than phenotypic variation. The new approach has obvious significance for interpreting the dispersal/vicariance of taxa which are widely distributed in time and space.

Ducasse & Rousselle (1978, 1979) completed a pioneering study (Keen 1976) of polymorphism in the genus *Hammatocythere*, a typical Eocene-Oligocene ostracode of the Aquitaine Basin, and Neale (1983: Fig.5) subsequently summarized its distribution pattern. The progressive (in a temporal sense) establishment of this environmentally-cued poymorphism means that *Hammatocythere* is a reliable biostratigraphic marker in the Aquitaine Eocene. In the Oligocene, fluctuating environmental conditions controlled further polymorphisms and the genus migrated from Aquitaine towards Paris, Brussels and the Isle of Wight (Neale 1983).

Because there was no subsequent recapitulation of the ancestral morphotype, two processes are at work in this case. Firstly, there is the immediate phenotypic response (polymorphism); and secondly, a longer term genotypic response representing evolution of the ancestral stock over the interval Middle Eocene to Oligocene.

McKenzie & Peypouquet (1984) isolated the phenotypic response in a new species of *Bradleya praemackenziei* Whatley and Downing, 1983 from the Miocene near Mornington, Victoria, Australia. In their study, 10 cm thickness of sediment (representing about 10 000 years) was sampled and prepared, then *Bradleya* individuals were picked from the washings. Figures 1 and 2 indicate clearly the influence of rhopic factor variations in determining phenotypic variations.

Rhopic factor architectural variations, of course, are not independent of other factors such as temperature and turbidity which control the CO_2 buffer and pH. However, the emphasis is quite different from earlier studies that linked temperature to ostracode distributions, such as that by Hazel (1971) for the Yorktown Formation (Late Miocene-Early Pliocene) of Virginia and North Carolina in which the paleotemperatures of fossil assemblages were inferred on the basis of known temperature regimes of Recent species, viz. a strictly uniformitarian analysis.

2.2 *Oxygen content*

The oxygen content of seawater is known to vary as a function of depth and amount of organic matter. Peypouquet (1979) used this fluctuation to propose a novel resolution of a problem in interpreting the typical deepwater genera *Krithe* and *Parakrithe.* Most specialists were aware that similar-looking *Krithe* and *Parakrithe*, often from the same sample, varied considerably in their carapace marginal areas. After a numerical study, Kaesler & Lohmann (1976) decided that these were environmentally responsive phenotypic variations. Peypouquet (1979) went further and linked the marginal (vestibule) variations to dissolved oxygen content in an inverse relationship. He considers that narrow vestibules indicate well-oxygenated water (about 4.5 ml/l or greater) and that large vestibules indicate oxygen deficient water (about 2 ml/l or even less). The hypothesis was tested on numerous samples ranging from the Norwegian Sea to the Equator in the Atlantic Ocean and found to be consistent with observed data.

Peypouquet (1977) proposed a physiological explanation for his hypothesis, suggesting that the respiratory functions of Ostracoda are exercised via subdermal blood cells in the inner region of the inner lamella. Bearing in mind that the respiratory pigment of entomostracans (which include Ostracoda) is hemoglobin (HB), experimental data on the fluctuation of HB as a function of the dissolved oxygen (O_2) content of the ambient medium indicate that the observed relationship between external dissolved O_2 and the pigment is an inverse one. This is a general property of Crustacea having hemoglobin as their respiratory pigment. A medium deficient in O_2 induces a marked increase in the HB concentration of an entomostracan placed in it and when such experimental animals are returned to well-oxygenated water the superfluous pigment cells degenerate. Oxyhemoglobin cannot be detected in the pigment at external (= ambient medium) dissolved O_2 tensions below 28 mm. Through the inverse relationship just described, however, entomostracans by a proportionate increase in hemoglobin can maintain the total number of oxygen molecules transported in their blood to the tissues even when ambient medium dissolved O_2 is below the level needed to saturate the in vivo pigment fully (Lockwood 1968).

Figure 1. *Bradleya praemackenziei* Whatley and Downing, 1983: (1) Juvenile female, left valve, lateral view, × 100; (2) Same specimen as (1), detail of valve ornament, × 750; (3) Adult male (genotypic), right valve, lateral view, × 100; (4) Adult female (genotypic), left valve, lateral view, × 100; (5) Same specimen as (4), detail of valve ornament, × 500; (6) Adult female (phenotypic, aggraded), left valve, lateral view, × 100; (7) Same specimen as (6), detail of aggraded ornament, × 750; (8) Adult female (phenotypic, degraded), left valve, lateral view, × 100; (9) Same specimen as (8), detail of degraded ornament, × 200; 'e' indicates the eye tubercle.

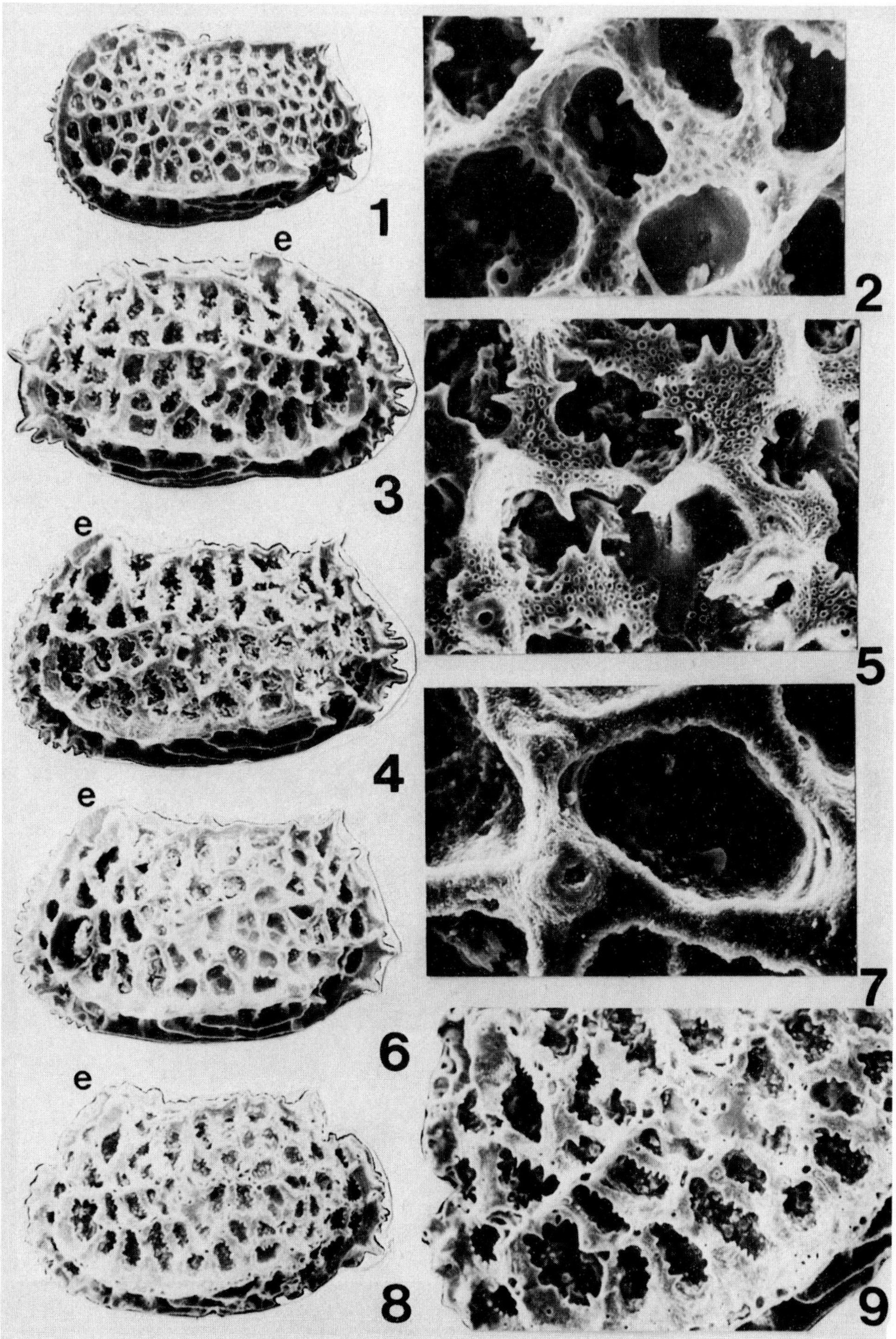
1
2
e
3
5
e
4
7
e
6
e
8
9

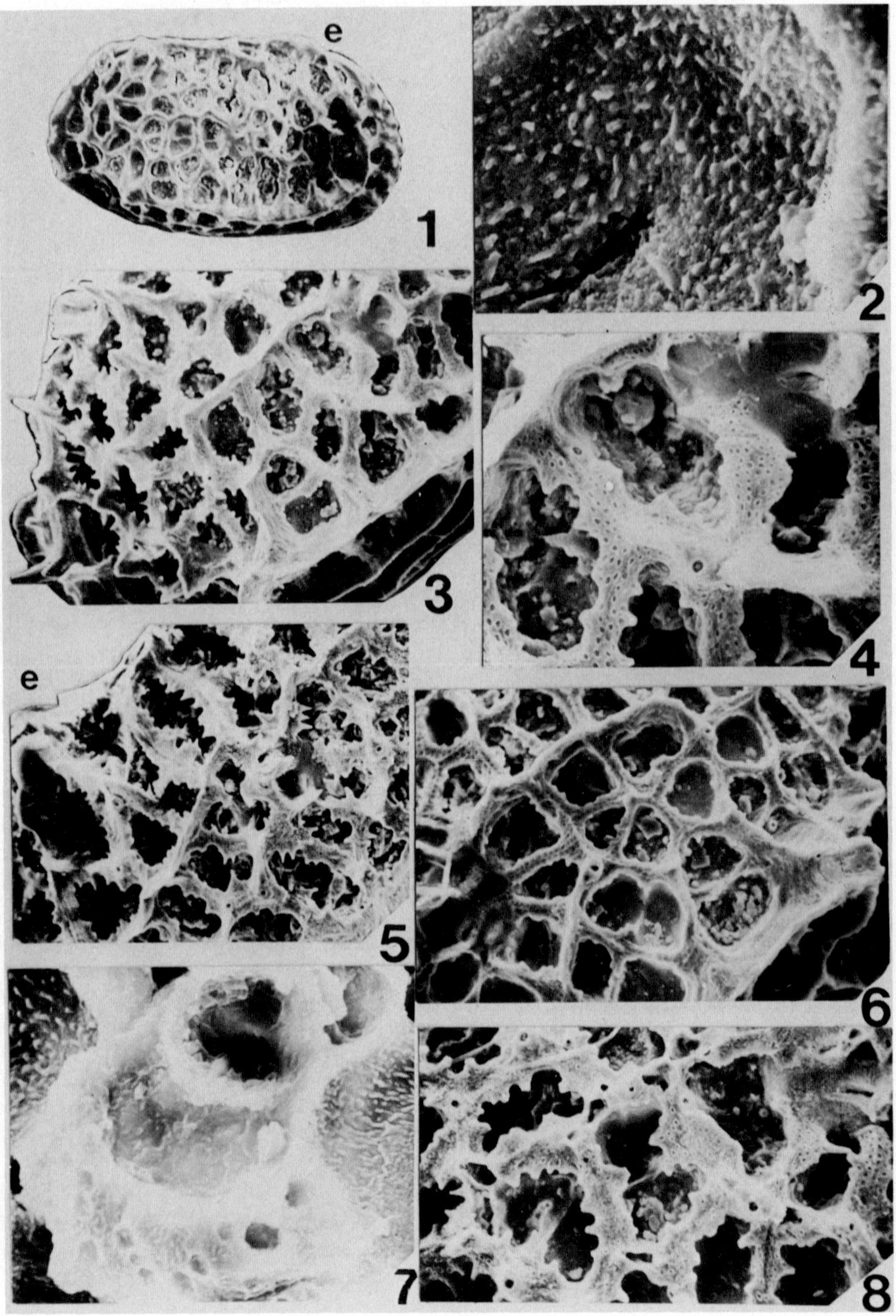

Thus, according to Peypouquet (1979), *Krithe* adapting to seawater deficient in dissolved O_2 need extra blood cells to respire effectively and these can occupy the vestibular regions of the inner lamellae — the lower the external dissolved O_2 concentration the larger the vestibules. Conversely, *Krithe* adapted to well-oxygenated water do not need this extra space and the vestibules are accordingly reduced.

A number of other environmental factors need to be considered in evaluating the hypothesis. These include turbidity, temperature, CO_2, salinity, pH, food supply, the carbonate equilibrium, the molting cycle (especially the molt into adulthood) and sediment type.

Turbidity does not seem to be an important variable in the deepwater zones where *Krithe* live, but temperature certainly varies considerably. Off SE Australia in the Miocene, for example, shelf surface water temperatures were about 23°C (Gill 1968), and the temperature of wind-driven, seasonal upwelling waters along the shelf edge would have been much less, about 10°C. But in crustaceans, temperature has no influence at isosmoticity, i.e. it does not affect osmoconformers (Dorgelo 1981), and *Krithe* is a holomarine osmoconformer, with a hemolymph slightly hypertonic to the external medium as in barnacles (Barnes & Klepal 1974). Similarly, crustacean respiratory systems have low sensitivity to CO_2 (Lockwood 1968). The salinity of shelf edge and epibathyal (including upwelling) waters differs hardly at all in physiological terms, i.e. by less than one part per thousand. Changes in pH do occur, especially at the sediment/water interface (*Krithe* is a benthic detrital feeder) and a lowering of pH from the alkaline side — owing to organic matter breakdown, for example — would tend to release O_2 from the blood (Wolvekamp & Waterman 1960). This effect agrees with the hypothesized mechanism. The carbonate equilibrium is a physico-chemical phenomenon which affects the shell but does not seem to have much direct relevance for an animal's respiration. At the adult molt, hemoglobin is probably transferred from the blood to the ovaries (and testes?) so that after egg-laying it might be at two-thirds its normal concentration (Lockwood 1968). But this factor affects animals in well-oxygenated and poorly oxygenated waters equally. Hemoglobin synthesis is dependent also upon food supply (Lockwood 1968). Thus, in regions which have little food supply except during upwelling pulses there would be an effect due to this factor which, again, is in the hypothesized direction, since upwelling waters are deficient in dissolved O_2. Finally, when the sediment substrate is rich in ferrous iron this tends to increase the production of hemoglobin by animals in oxygen deficient water (Lockwood 1968).

Summing up, the four significant factors — dissolved O_2, pH, food supply and substrate ferrous iron (commonly present in offshore shelf and epibathyal slope sediments derived from the continents) — all affect HB synthesis in the same, additive, direction. This result

---

Figure 2. *Bradleya praemackenziei* Whatley and Downing, 1983: (1) Adult male (phenotypic, aggraded), right valve, lateral view, × 100; (2) Same specimen as (1), detail of reticule aggradation, × 3500; (3) Adult male (genotypic), right valve, detail of reticule spines, × 200; (4) Same specimen as (3), detail of valve ornament, × 500; (5) Adult female (genotypic), left valve, detail of reticule spines, × 200; (6) Juvenile female, left valve, detail of reticular pattern, × 250; (7) Adult female (phenotypic, degraded), detail of degradation on reticulum, × 2000; (8) Adult female (phenotypic, degraded), detail of valve ornament, × 350; 'e' indicates the eye tubercle.

provides substantial support for the hypothesis proposed by Peypouquet (1977) in order to account for the observed variations in the vestibules of different individuals of the same *Krithe* and *Parakrithe* species.

The major application for Peypouquet's ideas is to the interpretation of paleobathymetry and upwelling phenomena. Peypouquet et al. (1982) indicated *Krithe* vestibula provide data for paleobathymetric interpretation of the Antares site (Aquitaine) during the Eocene/Oligocene crisis, and Donze et al. (1982) used the technique to determine paleobathymetry in the NE Tunisian Kef section (Maastrichtian to Early Eocene). McKenzie & Peypouquet (1984) use *Krithe* and *Parakrithe* to hypothesize upwelling in the Fossil Beach Miocene of Victoria (same locality as noted earlier).

The significance of this work for biogeographers is that it adds detail to the interpretation of oceanic paleohydrology. Several authors (notably McKenzie 1974, Neale 1976, Feldmann, this volume) have used oceanic paleocurrents to account for distributions of assemblages they have studied in the Australian Tertiary and Late Cretaceous respectively.

2.3 *Available light*

The physiological responses by ostracodes (deepwater marine, interstitial and cave-dwelling taxa) to diminution in available light is loss of eyes. The relationship has been particularly well-demonstrated by Benson (1975a: text-figures 13, 23) with respect to the deepwater, psychrospheric fauna. His data indicate an upper depth threshold for blindness of around 600 m and a lower threshold for sightedness at around 900 m. The interpretation of blindness is easy, because blind species have no eye tubercle in the anterodorsal region of each valve, whereas forms which can see (shelf shallow water and slope epibathyal taxa) carry more or less well-developed eye tubercles (Figs.1 and 2).

The combination of characteristics referred to in these two last sections along with other factors associated with deepwater ostracodes (greater size, thinner shell, increasing spinosity) allow relatively precise determinations of paleobathymetry which are supplemented by reference to the data from Recent studies such as that by van Morkhoven (1972) for the northwest Gulf of Mexico. The distinctiveness of this deepwater fauna since the advent of the world oceanic psychrosphere during the Eocene has been demonstrated numerically by Benson (1975b); and for the Recent of Sahul Shelf, NW Australia by McKenzie (1976).

2.4 *Osmoregulation*

Ostracoda, like other crustaceans can be divided into osmoconformers and osmoregulators (McKenzie 1981: Fig.28.3). Generally, the marine species, mostly cytheracean Podocopida, are regarded as conformers, but osmoregulators include both marine and continental taxa. For example, the holoeuryhaline genus *Cyprideis* is an osmoregulator with an adaptive capacity for salinities ranging from less than 3‰ to nearly 100‰.

The phenotypic response by *Cyprideis* and several related genera to reduced salinities is the development of carapace nodes, and the noding sites seem to be genetically controlled (Sandberg 1964). Any study of the distribution of this taxon needs to take these factors into account, otherwise one risks falling into the error of earlier workers who used the different specific ephithets torosa and littoralis for the same common European *Cyprideis* species — the name *C.torosa* (Jones 1950) has priority. A major relevant study is that

Table 2. Paratethyan relict assemblages, becoming more dominated eastwards by freshwater species (data from McKenzie 1981).

Sea/lake	Total species	Marine-derived	Freshwater-derived
Black-Azov	113	105	8
Caspian	33	27	6
Aral	11	4	7
Issuk-Kul	10	3	7

by van den Bold (1976) for Neogene Cyprideini of the Caribbean. The biogeographic pathways which he figures (van den Bold 1976 Figs.3-5) had to be elucidated for about ten species, each containing nodose morphotypes and varying in its temporal-spatial distribution; and the *Cyprideis* complex as a whole had to be separated from some closely similar taxa in the genus *Hemicyprideis.* The end result of his analysis was the first satisfactory correlation of brackish-water beds in the Caribbean Neogene with their coeval marine facies.

Another group of osmoregulatory ostracodes are those belonging in the podocopid Cypridacea which inhabit saline lake (athalassic) environments. These have been well studied by De Deckker (1981).

McKenzie (1981) used both cytheracean and cypridacean saline lake Ostracoda to develop a genetic classification of salt lakes (Table 2) en route to which he discussed the gradual breakup of Paratethys into isolated basins differing in their salinities and faunas. This biogeogrphi phenomenon dominated the Neogene history of eastern Europe and south-central Asia.

But the latest data on osmoregulation emphasize physiological considerations. Dorgelo (1981) refers to the inverse relationship between blood osmoconcentration and temperature as possibly due to movement of water from the hemolymph to the cells, following metabolite accumulation in the cells. This leads to higher intracellular water content and lower circulating blood volume at low temperature (Dorgelo 1981).

Aladin (1983) used modern microcrioscopical techniques to identify two different types of hemolymph osmoregulation in Ostracoda (hyperosmotic and amphiosmotic). *Cyprideis torosa*, because of its wide salinity tolerance, is a typical amphiosmotic ostracode; whereas species such as *Limnocythere inopinata, Candona marchica, Cyclocypris laevis* and *Plesiocypridopsis newtoni* are hyperosmotic. Hemolymph regulation is shown biochemically to occur via salt sequestration in the non-calcified membraneous part of the inner lamella. Aladin (1983) notes that upon salinity increase, 'ostracodes with well-developed ability for amphiosmotic regulation of hemolymph will increase in numbers but those with the ability for hyperosmotic regulation only will decrease in numbers until their complete disappearance'. The opposite case (salinity decrease) is exemplified by the classic Neogene shrinkages of Paratethys referred to earlier (Table 2).

2.5 *Food supply*

It is obvious that food supply is a critical factor in determining the survival and distributions of taxa. The recent emphasis has been genetic, linking food supply to particular environmental situations which can, secondarily, influence dispersal/vicariance. Landsat imagery has made it obvious that the mouths of great rivers are regions of nutrient enrichment and abundant food supply. Similarly, upwelling waters indicate high-nutrient zones which attract phyto- and zooplankton, then fish. As this rich organic matter reaches the bottom it triggers a bloom in the benthic detrital scavenging fauna of which ostracodes are an important component. In Peypouquet et al. (1982), nutrients derived both from major rivers (ancestral to the modern Dordogne and Garonne) and from upwelling pulses are interpreted to explain particular ostracode assemblages of the Eocene/Oligocene crisis in Tertiary Aquitaine.

2.6 *Trace elements*

Several researchers have looked at the trace elements composition of ostracode carapaces in the expectation that this would lead to development of some environmentally sensitive criteria. Cadot et al. (1975) noted that variations in $MgCO_3$ concentration in the carapace calcite were correlated with temperature — there is a tendency to secrete calcite with less magnesium in cold water than in warm water. Currently, Chivas et al. (1983) have embarked on an experimental program using inductively-coupled Argon Plasma Spectrometry to study concentrations of Ca, Mg, Sr and O^{18} from waters and the shells of ostracodes which inhabit them in order to find relationships between water chemistry/ temperature/shell chemistry.

2.7 *Solute composition*

Forester (1983) demonstrated that solute composition rather than salinity was the most critical factor in controlling the occurrence of two mutually exclusive *Limnocythere* species in the numerous North American lakes which he sampled. One species was shown to be confined to Na carbonate-bicarbonate waters depleted in Ca, the other to Na, Mg, Ca sulphate-chloride waters depleted in bicarbonate.

Application of the methodology depends upon more complete understanding (i.e. sampling) of the distributions of modern species. This done, the detailed analysis of Neogene and Holocene assemblages can be made by direct analogy while older distributions/ assemblages can be interpreted more generally by generic-level comparisons.

As with the previous subsection, such experimentation seems most applicable to distributions of freshwater and athalassic taxa.

3 EVOLUTIONARY CONSIDERATIONS

One result of the more detailed approach to environmental analysis in assessing the influence of environment upon biogeography has already been documented. This is the realization that Ostracoda are environmentally plastic organisms which produce a variety of eco-phenotypes in response to environmental variations. When this spatial variation is considered with the addition of the time variable, any unit within the study group has to be evaluated both as phenon (ecotypic variant) and taxon (genotypic 'variant'). Ostracoda are virtually unique among crustaceans because they have a huge fossil data base (over 50 000 taxa). Thus the elucidation of ecotypic and genotypic responses in Ostracoda is not the occasional effort to understand a fortunately well-preserved assemblage which is substantially isolated in time from other assemblages but the everyday preoccupation of any biogeographer working with these very abundant fossil as well as living animals.

Colin & Danielopol (1980) faced this kind of challenge when they reviewed the biogeography and evolution of the Timiriaseviinae, a group of limnocytherid cytheracean Podocopida with a worldwide distribution and a geological record ranging from the Triassic to the Recent. Their method was first to establish several 'evolutionary tendencies' and next to show that some of them developed irregularly and independently in different lineages. This satisfies the conditions for mosaic evolution provided plesiomorphic features and apomorphic features occur simultaneously. They also show that at least one of these 'tendencies', the development of a broodchamber, must have existed as a potential in the gene complex of different stocks before its realization at different times and on different continents. It will soon become apparent that mosaic evolution is not rare in the rich geological record of Ostracoda.

4 GENETIC STUDIES

Gradually, the genetics of Ostracoda are becoming better known although it is regrettable that all the effort has concentrated upon continental species (Hartmann 1968, Bianchi Bullini & Bullini 1972, Tetart 1978). One specialist has initiated research to correct this imabalance by commencing work on genetics of the marine genus *Leptocythere* (Guillaume pers. comm.). The major problems have been technical, owing to the extremely small size of ostracode chromosomes. Preparation slides have to be scanned at time consuming high magnifications (around 1000) in order to identify the mitoses. Furthermore, the ostracode body is completely enclosed within two calcitic valves which adversely affects fixation. Geneticists therefore shied away from tackling the group and it was left to ostracodologists themselves to develop the necessary expertise. The results, however, have been worth the wait. Tetart (1978) studied three superfamilies — Darwinulacea, Cytheracea and Cypridacea.

The Darwinulacea have only acrocentric chromosomes — so small (about 1) that they lie on the fringe of existing techniques. This chromosome pattern establishes the primitiveness of darwinulaceans but does little to elucidate the complex record and world-wide zoogeography of the group which seems to have been parthenogenetic for most of its history (McKenzie 1971).

The karyotype of *Limnocythere* (Cytheracea) differs only numerically (N = 7) from that of cypridacean species (Tetart 1978) and is of little direct help in developing an understanding of the distributions in space-time of such genera as *Cyprideis*, with over 30 Paratethyan Neogene species (Krstić 1968a,b), and *Chinocythere* with over 80 Chinese Paleogene species (Bojie 1978).

In contrast, karyotype details are now known for over 30 cypridacean species (Hartmann 1968, Tetart 1978). The patterns are complex but fall into two classes, symmetric and asymmetric. The latter are shown to derive from the former, the usual modifications being via loss of a partner chromosome or by centric fusions. Chromosome loss is typical of males but is also known from some parthenogenetic female karyotypes. Centric fusion type modifications occur in great diversity and lead to the speculation that they represent evolutionary processes which are geologically young and even contemporary, as for example when several different karyotypes are determined within the same deme (Tetart 1978).

It is rational to recognize an evolutionary series consisting first of intrapopulational (deme) variation, leading to variability in different demes and thence to fixation of the variation through isolation of the different environments in space, water chemistry and time. The final stage is the acquisition of a new symmetry by the isolated variants. These conclusions provide a key to understanding the development of mosaic evolution such as that studied by Colin & Danielopol (1980) and cited in the previous section.

The genetic data can also be used to explain parthenogenetic variations if it is assumed that the chromosome remaining in the parthenogenetic karyotype behaves as a sex chromosome. Thus, the genus *Isocypris* occurs as several parthenogenetic species in Europe which presumably derive from the bisexual southern African species but have evolved parthenogenetically following their spatial isolation from the ancestral stocks. The earliest fossil record of a parthenogenetic *Isocypris* is from the Pleistocene of India (Singh 1969) and it was referred to a living southern African species. On this evidence, all the six parthenogenetic European species have evolved parthenogenetically only during the Quaternary and include at least one very young taxon (*Isocypris foxi* Moroni from the Italian ricefields); and there is another parthenogenetic Quaternary species in North America [*Isocypris longicomosa* (Furtos)].

5 BIOGEOGRAPHY

5.1 Marine distributions

McKenzie (1967) and Neale (1976) associated the biogeography of many marine Cretaceous-Cenozoic Ostracoda with the evolution of Tethys. Subsequently, McKenzie (1973, 1974) stressed the importance also of the West Wind Drift in southern (40°) latitudes and of peripacific migrations in contributing to Australasian faunas, an approach similar to that of Fleming (1962). McKenzie (1978) noted that migrations via the Tethyan corridor occurred in phase with the schedules of rifting and suturing required by the adherents of Continental Drift and Plate Tectonics. Later, McKenzie (1980) pointed out that particular loci, including the Panamanian filter and the suture zone between Australasia and the Indonesian Arc favoured species interactions (radiations, extinctions, migrations) which added complexity to distributional patterns. Recently, McKenzie (1983) considered distributions of Paleozoic-Cenozoic Tethyan Ostracoda with reference also to Expanding Earth theory and concluded that a vast wedge-shaped Permo-Triassic Tethys (required by Plate Tectonicists) was incompatible with known distributions of such Paleozoic groups as Phosphatocopida, Leperditicopida, Entomozoacea and Eridostraca. Instead, Tethys from the Cambrian onwards may well have been a thermospheric, sublatitudinal, elongate, relatively narrow and shallow intercontinental sea which did not develop oceanic characteristics until the Jurassic (McKenzie 1983) and did not become psychrospheric until the Late Eocene (Benson 1975b). After the terminal Miocene (Messinian) salinity crisis, its Mediterranean sector became thermospheric in the Pliocene and has remained relatively warm at depth, with the modern Mediterranean fauna being mainly derived from the Atlantic (Spain/Portugal and Morocco) via the Gibraltar sill (Benson 1979). The Tethyan corridor was thus of fundamental importance in determining the biogeographical relationships of ostracode faunas from the Paleozoic to Early Cenozoic as can be illustrated by Cretaceous-Paleogene distributions.

In the North Atlantic sector during the Early Cretaceous there were good correspondences between the American (North and South) and West African/European faunas. Generic commonality was over 60% initially, diminished slightly to 50% during the Aptian/Albian, regressed to 20% during the Cenomanian/Turonian but improved again, to 49%, in the latter Late Cretaceous (Tambareau 1982b: Table 1).

On the other hand, the opening of the South Atlantic (in the Early Cretaceous) did not lead to important migrations there vis à vis Tethys until the end of the Cretaceous, because of blocking by the emerged Rio Grande Rise and Walvis Ridge (Tambareau 1982a). In the Southern Ocean, Dingle (1981) has recorded the relationships of Late Cretaceous (Campanian and Maastrichtian) ostracodes of South Africa to those of Argentina, West Africa, Tanzania and Western Australia showing that the South African assemblages most nearly resemble those of Western Australia (27% similarity in the Campanian). Meanwhile, Guha (1976) drew attention to close links between the Late Mesozoic faunas of Madagascar, South Africa and India, and lately (pers. comm.) extended this observation to include the Australian Early Cretaceous (Aptian-Albian) faunas described by Krommelbein (1975). Thus, the relationships of the Austral Province (South America, South Africa/Madagascar, India, Australia) were closer with the Indo-Pacific margins of Tethys, during the early Cretaceous, than with the Atlantic margin. The most characteristic widely distributed Cretaceous ostracodes of this southern province belong in the Majungaellini.

Along the northern margins of Tethys faunal similarities exist throughout the Late Cretaceous-Palaeogene. Wide ranging taxa include *Cytheretta*, Buntoniinae, *Alatacythere, Uroleberis.* The trachyleberidid genera described by Deroo (1966) can be recognised all the way to India (Jain 1981) and even to China (Yang pers. comm.). Some lineages of *Cytheretta*, its importance in stratigraphy and biogeography underlined by Keen (1972), go all the way from southeastern England to Yunnan. *Schizocythere*, a useful Paleogene fossil in Europe, is equally useful in the Middle East from Saudi Arabia to India (Siddiqui 1981). *Saida*, stratigraphically important in northwestern Germany during the Cretaceous-Paleogene (Herrig 1968), is also a useful marker in the Late Cretaceous-Paleogene of Australia (McKenzie 1974). Major papers which underline these general (wide-ranging) faunal similarities refer to Britain (Neale 1978, Keen 1978), northwestern Europe (Herrig 1966, Pietrzeniuk 1969, Uffenorde 1979), Poland (Szczechura 1965, 1977), the Ukraine and Caucasus (Sheremata 1969, Lubimova 1965), and western China (Huang 1979, Yang pers. comm.).

Provincialism in Tethys first became important during the Late Cretaceous (Bate & Bayliss 1969, Bate 1972, Neale 1976, Tambareau 1982a) with accompanying endemism. Among the indices of the increasing provincialism can be included *Loculicytheretta*, which is restricted to middle (Mesogean) Tethys in the Paleogene, and Buntoniinae which characterise the Cretaceous of northern Tethys but are absent or virtually so from the Indopacific sector (Bate & Bayliss 1969). Nevertheless, many faunal elements remain common throughout Paleogene Tethys ranging from the Gulf Coast to Australasia (McKenzie 1967).

5.2 *Transitional facies assemblages*

The best known large faunas which exemplify transitional facies are the *Cypridea* (Purbeckian-Wealden) assemblages (e.g. Krommelbein 1966), faunas of the Paratethys (mainly Neogene) referred to earlier and the Paleogene Bohai fauna (Bojie 1978). This

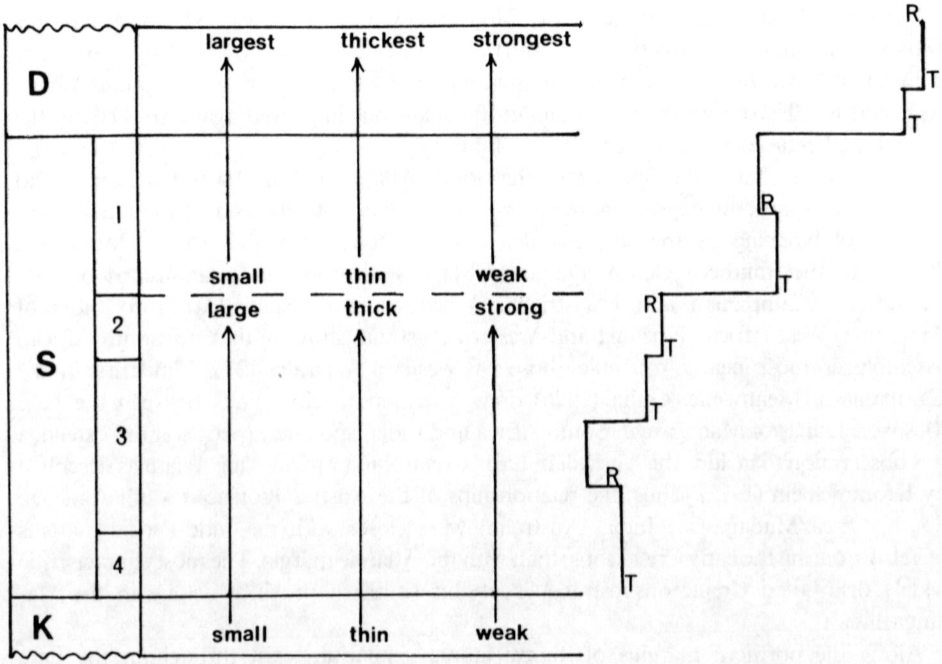

Figure 3. Variation in the Paleogene ostracode fauna of Bohai (Bojie 1978, Hou 1982). K – Kongdiang Formation, separated by an unconformity from the Cretaceous; S – Shahejie Formation (1, 2, 3, 4 are Members in this Formation); D – Dongying Formation, separated by an unconformity from the Neogene; T – transgression; R – regression. The pattern on the right represents the increase/decrease in numbers of new species (from Table 4, Bojie 1978). Consistent general trends in size, shell thickness and ornament are indicated. Total fauna of 425 spp. is more than 90% endemic. Several genera are so large that they must include numerous spatial as well as temporal genotypes and phenotypes.

latter includes 425 species and 23 new genera and is further remarkable in that most of the fossils have come from wells and bores, there being virtually no Paleogene outcrops in the region. The Paleogene section is separated by unconformities from the underlying Cretaceous and the overlying Neogene. It measures some 4000-5000 m. As indicated by the many new genera, the fauna is largely endemic; and two-thirds of the species belong in only nine genera.

Chinese workers have succeeded in defining several consistent, biostratigraphic trends which can be linked to both ecophenotypic and evolutionary development. The trends are expressed by size, shell thickness, ornament and the increase in new species (Bojie 1978, Hou 1982). They are shown diagrammatically in Figure 3. The Bohai fauna initiated in widespread freshwater ostracodes of small size, with thin shells and a weak ornament. The facies then became transitional (including oil shales, gypseous and salt beds, and biolimestones) with a relatively steady increase in new species matched to increasing transgression (except for a minor regressive interlude at the close of Kongdiang Formation deposition) until a major regressive crisis at the end of the period represented by Shahejie Formation unit 2. In the subsequent period, transgression dominated and the fauna, which had relapsed to morphotypes similar to those of the Kongdiang Formation, reached its apical development, being characterised by the largest, thickest shelled and most heavily ornamented taxa in the entire sequence, before a final regressive crisis.

In terms of the environmental parameters and evolutionary modes discussed in preceding sections, the Bohai Ostracoda can be interpreted as illustrating phenotypic response to increased calcite saturation, higher temperatures and greater food supply all in shallow well oxygenated facies in some of which the phytal substrate was so dense that they became diagenetically altered into oil shale beds (major source rocks for the associated Shengli Oilfield). There is also a definite indication that increasing transgression correlates positively with increase in numbers of new species, possibly by providing a greater variety of niches in which ecophenotypes could develop and then evolve as distinct taxa. The same interpretation could be given to increases in numbers of species relative to increasing transgression elsewhere in the world, e.g. Aquitaine. But the Bohai example is especially clear because of the high percentage of endemism (over 90%), so that the relationship of speciation to transgression is not diluted by migrations from nearby areas of similar facies. There is no evidence that such areas existed in northeastern China. Indeed, the only assemblages with possible relationships to the Bohai fauna are those described by Swain (1947) from the Salt Lake Basin, Utah, but it seems unlikely that significant migration from Utah to Bohai occurred.

5.3 *Continental faunas*

In recent years, there has been a large increase in the number of papers dealing particularly with saline lake and basin faunas. In Europe, this effort has been mainly in connection with the Paratethyan basins in which Ostracoda are the most important subsurface biostratigraphic markers. Elsewhere, however, the emphasis has been placed on the study of existing saline lakes with the promise of uniformitarian application of the results at least into the Neogene as in the work of Forester (1983) and Chivas et al. (1983) recorded earlier.

But most of the existing information regarding the biogeography of continental Ostracoda relates to freshwater faunas. In particular, the wide distribution of many parthenogenetic taxa has always been difficult to explain. McKenzie (1971) hypothesized that many ostracodes reproduce bisexually in their native environments but switch to parthenogenetic reproduction when dispersed beyond these.

The hypothesis can be readily tested either by studying the distributions of individual genera in continental areas or by looking at entire assemblages of isolated habitats (such as islands) into which the ostracodes must have been passively dispersed, mainly via dessication resistant eggs and such intermediate vectors as winds, birds, fish and man. Other environments of this type include farm dams, water troughs, glaciated terrains of Europe and North America, and ricefields (McKenzie 1971).

Thus, Neale (1983) considered the distribution of the very large genus *Candona* which has well over 100 species described mainly from the northern hemisphere. He pointed out that the bisexual populations were confined to southern, non-glaciated Europe whereas the parthenogenetic populations occurred in former glaciated terrains (Hartmann 1968). The major controlling factor appeared to be temperature, but Hartmann would agree (discussion in McKenzie 1971) that the parthenogenetic taxa form the dispersal fringe of *Candona* in its bid to reoccupy former habitats. The genus evolved during the Cretaceous in the northern hemisphere from which it had been driven out by the advancing Pleistocene glacial fronts. An excellent illustration of this idea is provided by Diebel & Pietrzeniuk (1975) who show that bisexual populations of *Limnocythere inopinata* occurred widely during a Pleistocene interglacial epoch in the same region of Germany where today only parthenogenetic populations of this species can be found.

A burgeoning literature appears to have considered every possible passive dispersal agent (Fox 1965, Löffler & Leibetseder 1966, McKenzie & Hussainy 1968, Delorme & Donald 1969, Kornicker & Sohn 1971, De Deckker 1977, Sohn & Kornicker 1979, Victor et al. 1981). For this reviewer, the agent most responsible for the passive dispersal of freshwater ostracodes intercontinentally is man. A corollary of this view is that many observed world distribution patterns (McKenzie 1971) are very young indeed; in many cases they do not extend back more than several hundred years. This new field of ostracode research is strictly historical biogeography.

The Italian ricefields of Lombardy and Padana are the best available example to test the idea of man's importance as a major dispersal agent for freshwater Ostracoda. This was first suggested by Fox (1965), subsequently by Moroni (1967) and McKenzie (1971); and Neale (1977) ascribed the affinities between the ricefields Ostracoda of Sri Lanka and those of Indonesia to exchanges of rice seeds and rice plants in the past.

Northern Italy is preferred as a study region for several cogent reasons. First, like the dispersed Ostracoda under consideration, rice is an introduced crop in Italy. Second, the Italian historical record is very good especially during the period when Renaissance dukes vied with each other in prestige and social conscience. Third, it is highly likely that some ostracodes of the Italian ricefields were introduced originally via other crops and then found their way into the ricefields ecosystem; most of these other introductions are also securely dated.

The dispersal of Ostracoda into ricefields is really a facet of island biogeography. Although islands in the general sense are thought of as bodies of land surrounded by water, ricefields are aquatic islands surrounded by land. It is true that they are accessible via irrigation canals but few ostracode species are common to both habitats (they include *Cypridopsis* spp.) so that the island character of ricefields, certainly as regards taxa introduced from other countries is hardly diminished by this factor.

Moreover, the Italian ricefields are not merely spatial aquatic islands, which appear and disappear seasonally in an annual cycle, they are also aquatic islands in the historical sense because rice is not native to Italy.

Chiappelli (1930) in his monograph on rice cultivation in Italy, listed some 39 varieties of *Oryza sativa* (rice) together with the years in which they were introduced and the countries of origin when known. As might be expected, most of the varieties were asiatic in origin — China, Indonesia and Japan — with others coming from the USA (six varieties since 1920), Spain (four varieties, three of them introduced near 1930 the other 'recently') and Malagasy (one variety, introduced near 1930). The earliest definite date of introduction given by Chiappelli (1930) was 1828, for the variety Bertone introduced from China, but a few were cited simply as 'old'. They included the varieties Ostiglia and Chinese Ostiglia — the varietal name coming from an important river port on the Po, between Mantova and Ferrara in the extensive Lombardy plain (Fig.4).

Rice is known to have been cultivated on the Lombardy plain since 1475. But risiculture was not practised near Ferrara (Fig.4) until after 1287 because rice is not listed as a crop in the Statuto Ferrariae of that date. Probably its introduction was associated with the bonifactions — clearing of forests and drainage of swamps into the Po and Adige

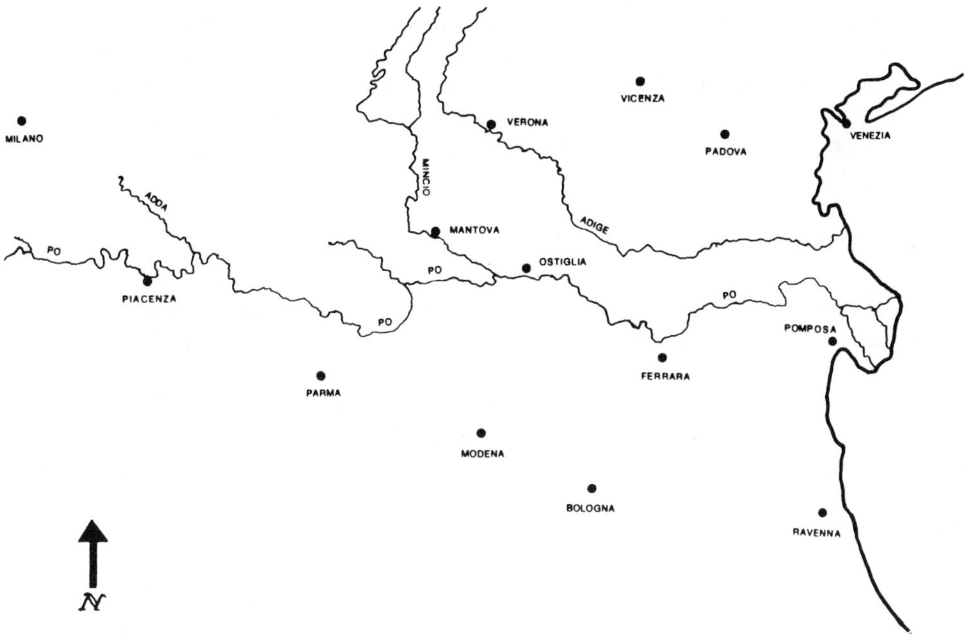

Figure 4. Sketch map of the Lombardy plain, northern Italy, from Milano to Venezia. Scale: 1 cm = 15 km.

by a network of canals — carried out by the dukes of Ferrara in the early 15th century (Zucchini 1967).

The nearest ricefields to Italy at the beginning of the 15th century were those of Macedonia, where rice has probably been cultivated since the time of Alexander; and from Macedonia the links run via Asia Minor, Persia and India to southeast Asia and China. There rice cultivation is extremely ancient going back over 7000 years (Barraclough 1978, Bai 1982).

Rice, of course, is not the only crop to have been introduced into northern Italy; indeed, the introductions began even before Roman times since it is known that the Etruscan settlements harvested wheat (introduced from Asia Minor). Hemp, mulberry, and peach were all introduced before the fall of Rome (Avellini et al 1980, Bignardi 1978, Coggi 1939). Plants used for color dyeing in textile productin, cotton and sugar cane were introduced following the Crusades (Borlandi 1950, Gleijeses 1977). The discovery of the Americas led to introductions of tobacco, tomato, potato and sunflower in the 16th and 17th centuries (Zucchini 1967:177-178). Finally, the development of botanical gardens (beginning with one at Genoa) from the 17th century onwards was another avenue by which exotic plants and animals, including Ostracoda, could have been introduced into Italy.

With such a plethora of crops introduced from every quarter of the globe it is scarcely remarkable that foreign species are often found in Italian ricefields. The dessication-resistant eggs of cypridacean ostracodes are so small (<50 μm) that they might easily be imported into Italy with seed of exotic cereals and plants, in soil packed around cuttings, in dust brought in with trade goods, and by travellers or migrating peoples. While most of these taxa are tropical and subtropical species, their survival in Italy is not surprising because even at the northern Italian latitudes ambient temperatures of around 25-35°C are typical in the Lombardy ricefields, especially during summer (June-August).

Table 3 lists some undoubted foreign ostracode species, their regions of origin and plants through which they might have been introduced. They represent about 30% of the total known Italian ricefields ostracode fauna. Once introduced, they could easily move into the aquatic islands which constitute the Italian ricefields ecosystem by one or other of the numerous dispersal mechanisms cited earlier.

6 CONCLUSIONS

This review has endeavoured to place ostracode biogeography into a contemporary context by emphasising research areas which are current and, occasionally, controversial.

It is evident that the relationship between shell ornament/thickness and chemistry of the ambient water is an area of major effort and importance (Peypouquet 1979, Forester 1983). It has significance for dispersing phenotypes and genotypes not only in relatively open marine systems such as Tethys (Ducasse & Rousselle 1978, 1979) but also in transitional facies that are spatially isolated and have a large endemic fauna — including several genera with numerous ecophenotypes — such as the Paleogene of Bohai, China (Bojie 1978).

In a scientific climate where there is general acceptance of Continental Drift and Sea Floor Spreading, some biogeographers are using distributions of fossil Ostracoda to evaluate opposing configurations of continents and oceans proposed by Plate Tectonicists and Earth Expansionists; others are associating speciation with sea level changes. Thus, McKenzie (1983) supposes that for most of the Paleozoic and until the Jurassic Tethys may have been a shallow, latitudinal, intercontinental sea rather than the vast, wedge-shaped, Permo-Triassic ocean favored by Plate Tectonics. Similarly, marine incursions (transgressions) are now linked confidently with increase in speciation, and regressions with faunal impoverishment (Bojie 1978), the ostracode data being consistent with those for assemblages of other taxa (Hou 1982). The small size, ubiquity and abundance of Ostracoda throughout the Phanerozoic make them excellent organisms for testing such ideas. If the biogeographic data are inconsistent with theoretical expectations then it is the latter which must change to match the observational evidence.

The distributions of athalassic and freshwater taxa continue to attract specialists. Much of this research is based on solid economic incentives. Thus, elucidation of the history and biogeography of Paratethyan ostracodes (Krstic 1968a,b) is applicable in petroleum exploration because Ostracoda are reliable paleoecologic and biostratigraphic markers (see Fig.1). Further, historical research on man as a dispersal agent for ostracodes is useful for improving understanding of the ricefields ecosystem in which they are a principal component of the heleoplankton (Table 3). It seems clear that the post-human distributions are significantly different from the pre-human ones.

Finally, the technically difficult field of genetic research has begun to develop (Tetart 1978) and promises to offer a satisfying rationale for such problems as parthenogenetic speciation and mosaic evolution.

ACKNOWLEDGEMENTS

Research on the historical biogeography of Italian ricefields Ostracoda was carried out

Table 3. Foreign Ostracoda introduced into Italian ricefields, their areas of origin and some possible plant vectors.

Genera/species	Area(s) of origin	Possible plant vectors
Ilyocypris australiensis	Australia	Eucalyptus, rice
Notodromas persica	Central Asia	Citrus, rice, mulberry, peach, wheat
Chlamydotheca incisa	South America	Maize, potato, tomato, tobacco, sunflower
Dolerocypris sinensis	China, SE Asia	Hemp, peach, mulberry, rice, spices, sugar cane
Tanycypris pellucida	SE Asia	Hemp, rice, spices, sugar cane
Strandesia (3 spp.)	Africa, Asia	Citrus, hemp, rice, spices, saffron, sugar cane
Stenocypris major	India, Ceylon	Citrus, cotton, hemp, rice, spices, sugar cane
Ilyodromus viridulus	Australia	Eucalyptus, rice
Isocypris (2 spp.)	Africa	Cotton, saffron, wheat
Cypretta (3 spp. at least)	Mainly Asia	Citrus, cotton, hemp, rice, spices, sugar cane

as Visiting Professor at the Istituto di Ecologia, Parma, Italy, during study leave from Riverina College of Advanced Education in 1982. Research with J.P.Peypouquet on Australian Tertiary Ostracoda was sponsored by ARGS Grant No.E80 15287. Miss Michele Fromholtz typed the manuscript, part of which was discussed with colleagues at the Smithsonian Institution, Washington, DC, during a short term visit in 1982. Dr G.McKenzie, Riverina College of Advanced Education, checked through part of section 2.2.

ADDENDUM

Recently, several papers have reinforced the paleoecologic significance of Ostracoda and documented the rapport between living taxa and their environment. In particular, the review by Carbonel et al. (in press) details progress on many aspects of the paleoecology of limnic ostracodes, citing numerous samples from Europe and the USSR, the Americas and Africa for virtually the entire geologic record of these taxa.

Meanwhile, Aladin (1983, 1984) has continued his microcrioscopical research on ostracode osmoregulation in more or less brackish-marine environments; and Tones (1983) has established that even freshwater ostracodes tend to remain slightly hypertonic to the ambient medium as concentration increases.

Research on ostracode shells as records of ecosystem fluctuations has been cited with respect to continental taxa. Recently, Bodergat (1983) has provided much experimental data on euryhaline (*Cyprideis*) and marine taxa as well, including a study of some psychrospheric species. A general summary of factors affecting the biogeography of the latter was recently provided by Benson et al. (1984). These authors concluded that the well-documented cosmopolitanism of modern deep-sea ostracode/faunas initiated with development of global oceanic thermal stratification in the Eocene, prior to which such fauna were more provincial.

REFERENCES

Aladin, N.V. 1983. Salinity adaptations and osmoregulation abilities of the Ostracoda from the Caspian and Aral Seas. *Zool. J.* 62: 51-57 (in Russian).

Al-Furaih, A.A.F. 1980. *Upper Cretaceous and Lower Tertiary Ostracoda (Superfamily Cytheracea) from Saudi Arabia.* Riyadh: University Libraries.

Avellini, L., R.Finzi & M.Palazzi 1980. *L'Emilia-Romagna una Regione.* Bologna: Zanichelli.

Bai, S.Y. 1982. *An Outline History of China.* Beijing: Foreign Languages Press.

Barnes, H. & W.Klepal 1974. The general biology of *Verruca stroemia* (O.F.Müller). IV. Effect of salinity and temperature on survival behaviour, and osmotic relations. *J. Exp. Mar. Biol. Ecol.* 14: 37-46.

Barraclough, G. 1978. *The Times Atlas of World History.* Sydney: Collins.

Bate, R.H. 1972. Upper Cretaceous Ostracoda from the Carnarvon Basin, Western Australia. *Spec. Paps. Palaeont.* 10: 1-85.

Bate, R.H. & D.D.Bayliss 1969. An outline account of the Cretaceous and Tertiary Foraminifera and of the Cretaceous ostracods of Tanzania. *Proceedings of the Third African Micropaleontological Colloquium.*

Benson, R.H. 1975a. Morphologic stability in Ostracoda. *Bull. Amer. Paleont.* 65: 13-46.

Benson, R.H. 1975b. The origin of the psychrosphere as recorded in changes of deepsea ostracode assemblages. Lethaia 8: 69-83.

Benson, R.H. 1979. In search of lost oceans: a paradox in discovery. In J.Gray & A.J.Boucot (eds.), Historical Biogeography, Plate Tectonics and the Changing Environment. Corvallis: Oregon State University Press.

Bianchi Bullini, A.P. & L.Bullini 1972. Osservazioni sul corredo chromosomico di *Eucypris virens* e di *Cypris bispinosa* (Crustacea Ostracoda). *Lincei Rend. Sci. Fis. Math. Nat.* 52: 432-426.

Bignardi, A. 1978. *Le Campagne Emiliane nel Rinascimento e nell'Eta Barocca.* Bologna: Forni.

Bojie 1978. *Early Tertiary Ostracode Fauna from the Coastal Region of Bohai.* Beijing: Science Press (in Chinese).

Bold, W.A.van den 1976. Ostracode correlation of brackish-water beds in the Caribbean Neogene. *Proceedings of the VII Caribbean Geological Conference.*

Borlandi, F. 1950. Nota per la storia della produzione e del commercio di una materia prima. Il guado nel Medio Evo. In A.Giuffrè (ed.), *Studii in Onore di Gino Luzzatto.* Vol.I. Milan: Giuffrè.

Cadot, H.M., R.L.Kaesler & W.R.van Schmus 1975. Application of the electron microprobe analyser to the study of the ostracode carapace. *Bull. Amer. Paleont.* 65: 577-585.

Chiappelli, R. 1930. Le varietá di riso coltivate in Italia: e cenni sulla botanica, sulla genetica, sulla selezione. *Quad. Staz. Speriment. Risicolt. Vercelli* 4: 1-133.

Chivas, A.R., P.de Deckker & J.M.G.Shelly 1983. Magnesium, Strontium, and Barium partitioning in nonmarine ostracode shells and their use in paleoenvironmental reconstructions — a preliminary study. In R.F.Maddocks (ed.), *Application of Ostracoda.* Houston: Univ. Houston Geosci.

Coggi, S. 1939. Juta e febre autrachiche. *Pubbl. Fed. Nazl. Fasc. Indust. Tess. var. Cap. Milano* 1939: 1-559.

Colin, J.P. & D.L.Danielopol 1980. Sur la morphologie, la systématique, la biogéographie et l'évolution des ostracodes Timiriaseviinae (Limnocytheridae). *Paleobiol. Continent.* 11: 1-52.

De Deckker, P. 1977. The distribution of the 'giant' ostracods (Family: Cyprididae Baird, 1845) endemic to Australia. In H.Loffler & D.L.Danielopol (eds.), *Aspects of Ecology and Zoogeography of Recent and Fossil Ostracoda.* Hague: Junk.

De Deckker, P. 1981. Ostracods of athalassic salt lakes. A review. In W.D.Williams (ed.), *Salt Lakes.* Hague: Junk.

Deevey, G.B. 1983. Planktonic ostracods (Myodocopa, Halocyprididae) from six *Eltanin* cruises in South Pacific and Antarctic waters. *J. Crust. Biol.* 3: 409-416.

Delorme, L.D. & D.Donald 1969. Torpidity of freshwater ostracodes. *Canad. J. Zool.* 47: 997-999.

Deroo, G. 1966. Cytheracea (Ostracodes) du Maastrichtien de Maastricht (Pays Bas) et des régions voisines; résultats stratigraphiques et paléontologiques de leur étude. *Med. geol. Stich.* Ser.C2(2): 1-197.

Diebel, K. & E.Piertrzeniuk 1975. Mittel- und jungpleistozäne Ostracodenfaunen des Raums Potsdam-Brandenburg-Rathenow im stratigraphischer und ökologischer Sicht. *Z. geol. Wiss. Berlin* 9: 1197-1233.

Dingle, R.V. 1981. The Campanian and Maastrichtian Ostracoda of south-east Africa. *Ann. S.Afr. Mus.* 85: 1-181.

Donze, P., J.P.Colin, R.Damotte, H.J.Oertli, J.P.Peypouquet & R.Said 1982. Les Ostracodes du Campanien terminal à l'Eocène Inférieur de la Coupe du Kef, Tunisie Nord-Occidentale. *Bull. Cent. Rech. Explor.-Prod. Elf-Aquitaine* 6: 273-335.

Dorgelo, J. 1981. Blood osmoregulation and temperature in crustaceans. In W.D.Williams (ed.), *Salt Lakes.* Hague: Junk.

Ducasse, O. & L.Rousselle 1978. *Hammatocythere oertlii* (Ducasse) (Ostracodes). Espèce polymorphe de l'Eocène du Blayais. *Bull. Inst. Geol. Bassin d'Aquitaine* 24: 3-35.

Ducasse, O. & L.Rousselle 1979. Les *Hammatocythere* (Ostracodes) de l'Oligocène aquitain. *Bull. Inst. Geol. Bassin d'Aquitaine* 25: 221-225.

Fleming, C.A. 1962. New Zealand biogeography, a palaeontologist's approach. *Tuatara* 10: 53-108.

Forester, R. 1983. Relationship of two lacustrine ostracode species to solute composition and salinity: implications for paleohydrochemistry. *Geology* 11: 435-438.

Fox, H.M. 1965. Ostracod Crustacea from ricefields in Italy. *Mem. Ist. ital. Idriobiol. Pallanza* 18: 205-214.

Gill, E.D. 1968. Oxygen isotope palaeotemperature determinations from Victoria, Australia. *Tuatara* 16: 56-61.

Gleijeses, V. 1977. *La Storia di Napoli.* Napoli: Soc. Editrice Napoletana.

Gou, S.Y., S.Y.Zheng & B.R.Huang 1983. *Pliocene Ostracode Fauna of Leizhou Peninsula and Northern Hainan Island, Guangdong Province.* Beijing: Science Press (in Chinese).

Guha, D.K. 1976. On some Mesozoic Ostracoda from subcrops of Banni, Rann of Kutch, India. *Proceedings VI Indian Colloquium of Micropalaeontology and Stratigraphy.*

Hartmann, G. 1968. Ostracoda (3. Lieferung). In *Bronn's Klassen und Ordnungen des Tierreichs, 5 (Arthropoda), 1. Abt. Crustacea 2 Buch IV*: 409-568.

Hazel, J.E. 1971. Paleoclimatology of the Yorktown Formation (Upper Miocene and Lower Pleistocene) of Virginia and North Carolina. In H.J.Oertli (ed.), *Paléoécologie des ostracodes, Bull. Cent. Rech. Pau-SNPA* 5 suppl.: 361-375.

Herrig, E. 1966. Ostracoden aus der Weissen Schreibkreide (Unter-Maastricht) der Insel Rugen. *Paläont. Abh., Abt.A, Paläozool.* 2: 695-1024.

Herrig, E. 1968. Zur Gattung *Saida*Hornibrook (Ostracoda, Crustacea) in der Oberkreide. *Geologie* 17: 964-981.

Hou, Y.T. 1982. Cretaceous-Tertiary biostratigraphy and petroleum prospecting of eastern China. *Proceedings of the Symposium on Petroleum Geoscience, Academica Sinica.* Beijing: Science Press (in Chinese).

Huang, B.R. 1979. *Report of the Scientific Expedition in the Mount Jolmo Lungma Region.* Nanjing: Institute of Geology and Palaeontology (in Chinese).

Jain, S.P. 1981. Checklist of Ostracoda from India — 1. Cenozoic Ostracoda (Marine). *J. Palaeont. Soc. India* 25: 85-105.

Jones, P.J. & K.G.McKenzie 1980. Queensland Middle Cambrian Bradoriida (Crustacea): new taxa, palaeobiogeography and biological affinities. *Alcheringa* 4: 203-225.

Kaesler, R.L. & K.C.Lohmann 1976. Phenotypic variations of populations of *Krithe producta* with environment. *Abh. Verh. naturwiss. Ver. Hamburg* 18/19 (Suppl.): 279-286.

Keen, M.C. 1972. Mid-Tertiary Cytherettinae of North-west Europe. *Bull. Brit. Mus. (Nat. Hist.) Geol.* 21: 261-349.

Keen, M.C. 1976. An evolutionary study of two homeomorphic Tertiary cytherid ostracod genera. *Abh. Verh. naturwiss. Ver. Hamburg* 18/19 (Suppl.): 319-323.

Keen, M.C. 1978. The Tertiary-Paleogene. In R.H.Bate & E.Robinson (eds.), *A Stratigraphical Index of British Ostracoda.* Liverpool: Seel House.

Kornicker, L.S. 1981. Revision, distribution, ecology and ontogeny of the ostracode Superfamily Cyclasteropinae (Myodocopida: Cylindroleberididae). *Smithson. Cont. Zool.* 319: 1-548.

Kornicker, L.S. & I.G.Sohn 1971. Viability of ostracode eggs egested by fish and effects of digestive fluids on ostracode shells: ecologic and paleoecologic implications. In H.J.Oertli (ed.), *Paléoécologie des Ostracodes, Bull. Cent. Rech. Pau-SNPA* 5 suppl.: 125-135.

Krommelbein, K. 1966. On 'Gondwana Wealden' ostracodes from NE Brazil and West Africa. *Proceedings of the Second West African Micropaleontological Colloquim.* Leiden: Brill.

Krommelbein, K. 1975. Ostracoden aus der Kreide des Great Artesian Basin, Queensland, Australia. *Senck. leth.* 55: 455-483.

Krstić, N. 1968a. Ostracodes des couches Congeriennes: 1. *Cyprideis* I. *Bull. Mus. Hist. Nat. Belgrade* Ser.A 23: 107-151.

Krstić, N. 1968b. Ostracodes des couches Congeriennes: 3. *Cyprideis* II. *Bull. Mus. Hist. Nat. Belgrade* Ser.A 23: 153-183.

Lockwood, A.P.M. 1968. *Aspects of the Physiology of Crustacea.* Edinburgh: Oliver & Boyd.

Löffler, H. & J.Leibetseder 1966. Daten zur Dauer des Darmdurchganges bei Vogeln. *Zool. Anz.* 177: 334-340.

Lubimova, P.S. 1965. Ostracodes of the Lower Cretaceous deposits of the Precaspian Depression. *VNIGRI Trans.* 244: 1-199 (in Russian).

Martin, J.L. 1977. Le métabolisme du magnésium chez *Cancer irroratus* (Say, 1817) (crustacé décapode). *Bull. Inst. Oceanogr. Paris* 53.

McKenzie, K.G. 1967. The distribution of Caenozoic marine Ostracoda from the Gulf of Mexico to Australasia. In G.A.Adams & D.V.Ager (eds.), *Aspects of Tethyan Biogeography, Syst. Assoc. Publ.* 7: 219-238.

McKenzie, K.G. 1971. The palaeozoogeography of freshwater Ostracoda. In H.J.Oertli (ed.), *Paleoécologie des Ostracodes, Bull. Cent. Rech. Pau-SNPA* 5 suppl.: 207-238.

McKenzie, K.G. 1973. Cenozoic Ostracoda. In A.Hallam (ed.), *Atlas of Palaeobiogeography.* Amsterdam: Elsevier.

McKenzie, K.G. 1974. Cenozoic Ostracoda of southeastern Australia with the description of *Hanaiceratina*new genus. *Geoscience and Man* 6: 153-182.

McKenzie, K.G. 1976. Sahul Shelf assemblages and the evolution of Post-Palaeozoic Ostracoda. *Abh. Verh. naturwiss. Ver. Hamburg* 18/19 (Suppl.): 215-228.

McKenzie, K.G. 1978. Biogeographic patterns in Australian Cainozoic Ostracoda, with the description of *Orlovibairdia*new genus. *J. Palaeont. Soc. India* 20: 279-288.

McKenzie, K.G. 1980. A new subfamily from Gorong, Seram, Molluccas, with the description of *Renaudcypris* new genus (Crustacea, Ostracoda). *Bull. Mus. Natn. Hist. Nat. Paris* 4° sér.2, sect.A 2: 507-515.

McKenzie, K.G. 1981. Palaeobiogeography of some salt lake faunas. In W.D.Williams (ed.), *Salt Lakes.* Hague: Junk.

McKenzie, K.G. 1983. Palaeozoic-Cenozoic Ostracoda of Tethys. *Boll. Soc. paleont. ital.* 21: 311-326.

McKenzie, K.G. & S.U.Hussainy 1968. Relevance of a freshwater cytherid (Crustacea, Ostracoda) to the Continental Drift Hypothesis. *Nature* 220: 806-808.

McKenzie, K.G. & J.P.Peypouquet 1984. Oceanic palaeo-environment of the Miocene Fyansford Formation from Fossil Beach, near Mornington, Victoria, interpreted on the basis of Ostracoda. *Alcheringa* 8: 291-303.

Morkhoven, F.P.C.M.van 1972. Bathymetry of Recent marine Ostracoda in the northwest Gulf of Mexico. *Trans. Gulf Coast Assoc. Geol. Socs.* 22: 241-252.

Moroni, A. 1967. Ostracodi delle risaie italiane. Sistematica, Ecologia, Distribuzione geografica. *Studium Parmense*: 1-79.

Neale, J.W. 1976. Cosmopolitanism and endemism — an Australian Upper Cretaceous paradox. *Abh. Verh. naturwiss. Ver. Hamburg* 18/19 (Suppl.): 265-274.

Neale, J.W. 1977. Ostracods from the ricefields of Sri Lanka (Ceylon). In H.Löffler & D.L.Danielopol (eds.), *Aspects of Ecology and Zoogeography of Recent and Fossil Ostracoda.* Hague: Junk.

Neale, J.W. 1978. The Cretaceous. In R.H.Bate & E.Robinson (eds.), *A Stratigraphical Index of British Ostracoda.* Liverpool: Seel House.

Neale, J.W. 1983. The Ostracoda and Uniformitarianism. I. The later record: Recent, Pleistocene and Tertiary. *Proc. Yorks. Geol. Soc.* 44: 305-326.

Peypouquet, J.P. 1977. Les Ostracodes et la connaissance des paléomilieux profonds. Application au Cénozoique et l'Atlantique nord-oriental. Thèse Doctorat d'Etat des Sciences, l'Université de Bordeaux I.

Peypouquet, J.P. 1979. Ostracodes et paléoenvironnements. Méthodologie et application aux domaines profonds du Cénozoique. *Bull. B.R.G.M.* Sect.IV, No.1 1979: 3-79.

Peypouquet, J.P., O.Ducasse & J.Gayet 1982. Les Ostracodes et la paléogéographie, la paléohydrologie et la paléoclimatologie lors de la 'crise Eocène/Oligocène' dans l'Atlantique nord-oriental. In E.M.Gallitelli (ed.), *Paleontology, Essential of Historical Geology.* Modena: Mucchi.

Peypouquet, J.P., O.Ducasse & L.Rousselle 1981. Morphogenesis and Environment. Theoretical and practical aspects from *Hammatocythere*: Paleogene Ostracoda of the Aquitaine Basin. *International Symposium on Concepts an Methodology in Paleontology, Barcelona.*

Pietrzeniuk, E. 1969. Taxonomische und biostratigraphische Untersuchungen am Ostracoda des Eozan 5 im Norden der Deutschen Demokratischen Republik. *Paläont. Abh. Abt.4 Paläozool.* 4: 1-162.

Pora, E. 1969. L'importance du facteur rhopique (équilibre ionique) pour la vie aquatique. *Verh. internatl. Ver. Limnol. Stuttgart* 17: 970-986.

Rosenfeld, A. 1982. The secretion process of the ostracode carapace. In R.H.Bate, E.Robinson & L.M.Sheppard (eds.), *Fossil and Recent Ostracods.* Chichester: Horwood.

Rousselle, L., O.Ducasse, P.Genot, M.Ringeade & A.M.Saubade 1980. Les populations locales fossiles 'marquers géobiologiques des paléoenvironnements et leurs transformations'. *Bull. Soc. Geol. France* (7)22: 213-218.

Sandberg, P.A. 1964. The ostracod genus *Cyprideis*in the Americas. *Stockholm Contribs. Geol.* 12: 1-178.

Sewell, R.B.S. 1940. The extent to which the distribution of marine organisms can be explained by and is dependent on the hydrographic conditions present in the great oceans, with special reference to the plankton. *Proc. Linn. Soc. London* 152: 256-286.

Sheremata, V. 1969. *Paleogene Ostracoda of the Ukraine.* Lvov: University Press (in Russian).

Siddiqui, Q.A. 1981. Some species of the genus *Schizocythere* from the Early Tertiary shelf sea of Pakistan. In J.W.Neale & M.D.Brasier (eds.), *Microfossils from Recent and Fossil Shelf Seas.* Chichester: Horwood.

Simpson, G.G. 1940. Mammals and Land Bridges. *J. Wash. Acad. Sci.* 30: 137-163.

Singh, D. 1969. Some additional ostracodes from the Upper and Lower Karewas, Kashmir. *Bull. Indian Geol. Assoc.* 2: 1.

Sohn, I.G. & L.S.Kornicker 1979. Viability of freeze-dried eggs of the freshwater *Heterocypris incongruens.* In N.Krstić (ed.), *Proceedings of the VII International Symposium on Ostracodes, Belgrad* 1.

Swain, F.M. 1947. Tertiary non-marine Ostracoda from the Salt Lake Formation, northern Utah. *J. Paleont.* 21: 518-528.

Szczechura, J. 1965. Cytheracea (Ostracoda) from the uppermost Cretaceous and lowermost Tertiary of Poland. *Act. Paleont. Polon.* 10: 451-564.

Szczechura, J. 1977. Ostracods from the Upper Eocene of east Poland. *Act. Paleont. Polon.* 22: 55-92.

Tambareau, Y. 1982a. Les ostracodes et l'histoire géologique de l'Atlantique Sud au Crétacé. *Bull. Cent. Rech. Explor.-Prod. Elf-Aquitaine* 6: 1-37.

Tambareau, Y. 1982b. Les ostracodes et l'évolution de l'Atlantique Nord au Crétacé. *Bull. Soc. Geol. France* (7)24: 1077-1085.

Tetart, J. 1978. Les garnitures chromosomiques des Ostracodes d'eau douce. *Trav. Lab. Hydrobiol. Grenoble* 69/70: 113-140.

Turpen, J.B. & R.W.Angell 1971. Aspects of molting and calcification in the ostracod *Heterocypris. Biol. Bull.* 140: 331-338.

Uffenorde, H. 1979. On the bio- and ecostratigraphical distribution of Ostracoda in the Oligo-/Miocene of the southern North Sea Basin. A preliminary report. In R.Vinken (ed.), *Project 124: The Northwest European Tertiary Basin* 5: 19-38.

Victor, R., K.W.Dance & H.B.N.Hynes 1981. Drift of ostracod crustaceans in adjacent intermittent and permanent streams. *Hydrob.* 80: 219-223.

Wolvekamp, H.P. & T.H.Waterman 1960. 2. Respiration. In T.H.Waterman (ed.), *The Physiology of Crustacea. Vol.I Metabolism and Growth.* New York: Academic Press.

Zucchini, M. 1967. *L'agricoltura Ferrarese Attraverso i Secoli.* Rome: Volpe.

Addendum

Aladin, N.V. 1983. On displacement of the critical salinity barrier in the Caspian and Aral Seas, the Branchiopods and Ostracoda taken as examples. *Zool. J.* 62: 689-694 (in Russian).

Aladin, N.V. 1984. Salinity adaptations and osmoregulation abilities of Ostracoda from Black and Azov Seas. *Zool. J.* 63: 185-190 (in Russian).

Benson, R.H., R.E.Chapman & L.T.Deck 1984. Paleoceanographic events and deep-sea ostracodes. *Science* 224: 1334-1336.

Bodergat, A.M. 1983. Les ostracodes, temoins de leur environnement approche chimique at écologique en milieu lagunaire et oceanique. *Docs Labs Geol. Lyon* 88: 1-246.

Carbonel, P., J-P.Colin, D.L.Danielopol, H.Loffler & I.Neustrueva (in press). The paleoecology of limmic ostracodes, a review of some major topics. *Palaeogeogr., Palaeoclimatol., Palaeoecol. spec. Vol.*

Tones, P.L. 1983. *Megalocypris* ingens Delorme (Ostracoda) in Saskatchewan saline lakes: osmoregulation and abundance. *Hydrobiologia* 105: 133-136.

Subject index

Taxonomic index